Optical Mineralogy

Second Edition

Optical Mineralogy

Second Edition

David Shelley
Department of Geology, University of Canterbury
Christchurch, New Zealand

Elsevier
New York • Amsterdam • Oxford

Elsevier Science Publishing Co., Inc.
52 Vanderbilt Avenue, New York, New York 10017

Sole distributors outside the United States and Canada:

Elsevier Science Publishers B.V.
P.O. Box 211, 1000 AE Amsterdam, The Netherlands

First Edition published in 1975 by Elsevier Science Publishers B.V., Amsterdam

Library of Congress Cataloging in Publication Data

Shelley, David.
Optical mineralogy.
Rev. ed. of: Manual of optical mineralogy. 1975.

Bibliography: p.
Includes index.
1. Optical mineralogy—Handbooks, manuals, etc.
I. Shelley, David. Manual of optical mineralogy. II. Title.
QE369.06S5 1985 549'.125 84-21214
ISBN 0-444-00838-1

Current printing (last digit):
10 9 8 7 6 5 4 3 2 1

Manufactured in the United States of America

To Iola

Contents

Preface

The accumulation of mineral data proceeds at an ever-increasing rate, as evidenced by the greatly expanded, 10-volume (?) edition of Deer, Howie, and Zussman presently being produced. Much new information has been gained with sophisticated tools such as the microprobe, the use of which is now almost obligatory for any detailed analysis of complex minerals such as pyroxene and garnet. Despite such developments, optical mineralogy retains its role as the primary method of mineral identification in rocks and for observing petrographic detail, and of necessity it forms a major part of university courses in geology. The subject embraces various aspects of crystallography, optical theory, microscopy, and mineral systematics, and there are numerous texts on all of these. However, it is easy to be diverted by such texts from the main purpose of the subject which is to identify minerals successfully, and I believe the student is best served by a one-volume reference containing an integrated treatment of all the information normally required. It is with this in mind that I have written the following text, and I hope it will prove useful for all stages of undergraduate and later work.

Readers who are familiar with *Manual of Optical Mineralogy* (1975) will recognize that it forms the basis of this new book. The flavor is much the same; indeed, quite large sections of the previous work have been used wholesale. Nevertheless, substantial changes and additions have been made, mainly in response to the suggestion that I had been rather too thorough in my previous efforts at keeping theory down to the bare essentials. Thus the chapter on principles has been enlarged to include, among other things, a more thorough treatment of retardation, interference figures, and optic-axis dispersion. In the *Manual* I separated principles from techniques as far as was possible, and this I continue to do. But here the sections on techniques are expanded so that they now fill three chapters instead of one. Thin-section and grain-mount procedures receive separate treatments, and the description of the universal stage is joined by a completely new section on the uses of the spindle stage. There is more emphasis on the accurate measurement of refractive indices.

To assist students in the early stages of training, a number of step-by-step exercises have been interspersed in appropriate places in the text. Naturally, these cannot substitute for a properly supervised course of study, but they should provide a second approach for those wanting to help themselves, and perhaps they will serve as a means of revising the subject for those too long away from the microscope.

Another addition is the explanation of stereographic projection, appended to the chapter on crystallography. This is built on further in the chapter on universal and spindle stages.

When the publishers asked me to consider a new edition, my first thought was to include color photographs of minerals as seen down the microscope. Elsevier kindly agreed to this, and the result is the separate color section of 40 plates. They usefully illustrate many points of theory and technique and, I hope, provide an instant guide to the appearance of the common rock-forming minerals.

I have taken the opportunity to update the mineral systematics. Such updating ranges from changing the refractive indices of some minerals in the third decimal place to the more important revisions of terminology for minerals such as the amphiboles. Five minerals have been added to the descriptions: eudialyte/eucolite, aenigmatite, palygorskite/sepiolite, mordenite, and scheelite. As in the *Manual,* minerals are assigned a number in order of description to facilitate rapid cross-referencing between the Determinative Tables of Chapter 8 and the descriptions of Chapter 9.

Because this book is as much text as laboratory manual, the word "manual" has been dropped from the title.

Finally, one cannot present a book such as this without acknowledging the considerable debt to all those mineralogists and petrographers whose data are compiled here, those fellow geologists who critically commented on the *Manual,* and my immediate colleagues, particularly Dr. S. D. Weaver and Mr. D. Smale, for patiently answering all the questions given them during the writing of this new book. My thanks too to Albert Downing, responsible for the photographic work herein, and Lee Leonard, who again drafted all the figures.

Preface to the First Edition

Optical mineralogy forms a major part of most university courses in geology, and is a prerequisite for petrography. In order to cover all those aspects of crystallography, theory, technique, procedure, and systematics used in the practice of the subject, most teachers find it necessary to recommend several texts. These not only prove expensive to the student, but are often rather too detailed and abstruse for practical work. The prime intention of this book, therefore, is to provide a handy one-volume reference to all the information normally required in the laboratory; it is hoped that the book will be useful as such for all stages of undergraduate and later work.

Many students find difficulty with their microscope work, and are not helped by the many texts in which principles and techniques are almost irretrievably mixed. In this book, the underlying principles are separated from the laboratory techniques; also separately described are the laboratory procedures that should enable the student to develop an efficient but rigorous routine for identifying minerals.

In attempting to explain optical theory it is easy to lose sight of the prime purpose of the exercise, that is, to identify minerals successfully. All texts written for mineralogists have simplified theory to some extent, and here it is kept to the bare essentials necessary for understanding the interrelationships of optics and crystallography. Hence, anisotropic minerals are discussed solely in terms of the uniaxial and biaxial indicatrices, no reference being made to wave-front theory.

Chapter 4 explains the standard techniques used to identify minerals in general petrographic work; included is a section on universal-stage methods. The techniques for thin-section work and grain-mount studies are treated separately where they differ significantly.

Each mineral described in Chapter 7 is assigned a number in order of description, and this facilitates rapid cross-reference between the descriptions and Determinative Tables of Chapter 6. The mineralogical data have been brought up to date as far as possible by reference to the literature through *Mineralogical Abstracts*. However, data appertaining to mineralogical oddities which distort the normal ranges of properties have been omitted. In general, the depth of treatment is that considered suitable for undergraduate and routine petrographic work. Information on dispersion in minerals has been omitted since in my experience, very few workers use this property for identification. More orientation diagrams than usual are provided. These are designed to help and encourage the student to check the properties of minerals in several orientations. Except for a few minerals such as the feldspars, information on paragenesis is brief, this being a subject more appropriately dealt with in petrological or theoretical mineralogy texts.

Three more theoretical texts that are thoroughly recommended to students as supplements to this book are *An Introduction to Crystallography* by F.C. Phillips (1971), *Optical Crystallography* by E.E. Wahlstrom (1969), and *An Introduction to the Rock-Forming Minerals* by W.A. Deer, R.A. Howie, and J. Zussman (1966).

Inevitably, a book such as this owes a considerable debt to the many mineralogists and petrographers whose data have been compiled here. Acknowledgment to all is impossible, and is made only where diagrams have been taken directly from their work. I should like to thank Dr. G.J. van der Lingen and Dr. J. Bradshaw for commenting on some parts of the text, and especial thanks go to Lee Leonard for draughting all the figures.

David Shelley

October 1974

Copyright Acknowledgments

I am greatly indebted to the authors and the following journals and publishers for permission to reproduce their diagrams. Credit to the authors is given in the figure captions.

The *American Mineralogist* published by the Mineralogical Society of America.

The *Mineralogical Magazine* published by the Mineralogical Society of Great Britain.

Nature published by MacMillan Journals Ltd.

The *American Journal of Science*.

John Wiley and Sons, Inc.

E. Schweizerbart'sche Verlagsbuchhandlung.

Schweizerische Mineralogische und Petrographische Mitteilungen published by Verlag Leeman.

Longman Group Limited.

Birkhäuser Verlag.

I am also grateful to Harvard University Press for permission to use the quote at the beginning of Chapter 3.

1

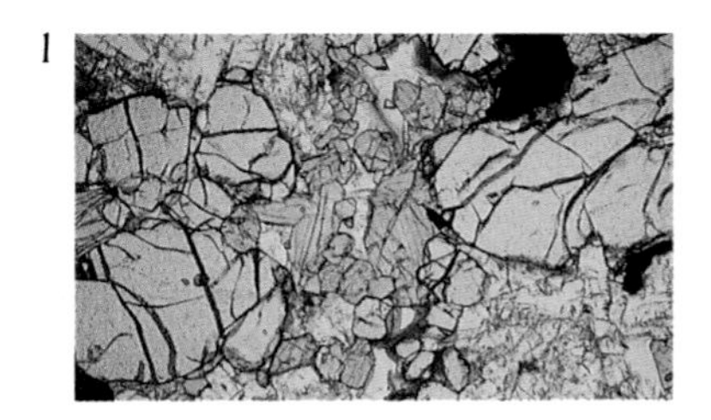

2

3

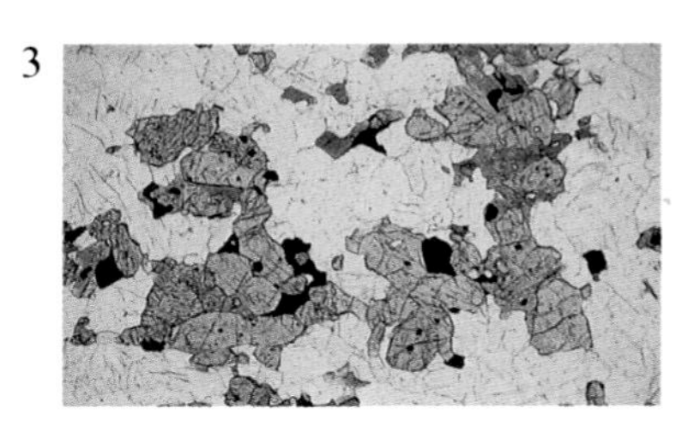

4

PLATES 1 (plane-polarized light) and **2** (crossed-polarized light) *Gabbro, Rakaia River, New Zealand. View measures 3.3 × 2.2 mm.*

This rock illustrates the three common igneous minerals olivine, augite, and plagioclase.

Olivine is seen as large, colorless, high relief, irregularly cracked crystals; no cleavage is evident. In Plate 2 the olivine demonstrates a range of retardation from very low (dark-grey grain left center, which would provide an optic-axis interference figure) through second-order yellow (largest grains) and blue (top left) to third-order green (bottom left center). The latter grain demonstrates the birefringence to be at least 0.040.

The pink, moderate-high relief, well-cleaved crystals in Plate 1 are *augite*. Note the octagonal pyroxene section towards the top of the view. In Plate 2, the augite shows a moderate birefringence of at least 0.020 (second-order blues). Note the twin planes in several grains. The grain just below center showing first-order grey is suitable for obtaining an optic-axis interference figure.

Plagioclase is present as the colorless, low-moderate relief material with first-order greys and whites in Plate 2. Note the typical igneous feldspar grain shapes dominated by (010) parallel to the repeated twin lamellae.

PLATES 3 (plane-polarized light) and **4** (crossed-polarized light) *Norite, Bluff, New Zealand. View measures 3.3 × 2.2 mm.*

Orthopyroxene and plagioclase are the dominant minerals in this igneous rock.

The *orthopyroxene* forms subhedral crystals which are pleochroic in pinks and greens, have a high relief and good cleavages. Darker green crystals of *hornblende* are present (top center and bordering the orthopyroxene to the left and right). The low birefringence is demonstrated in Plate 4 by first-order interference colors up to orange-yellow. Note that the reds and blues are interference colors of the more highly birefringent hornblende. Two low-interference color grains of orthopyroxene (bottom left and bottom right) display very fine exsolution lamellae of clinopyroxene.

The low-moderate relief, colorless material of Plate 3 is entirely *calcic-plagioclase*, and its characteristic first-order grey/white interference colors are displayed in Plate 4; note the repeated twin lamellae.

Opaque minerals are also present.

PLATES 5 (plane-polarized light) and **6** (crossed-polarized light) *Hornblende Gabbro, Bluff, New Zealand. View measures 0.9 × 0.6 mm.*

This rock is composed mainly of hornblende and plagioclase. Plate 5 illustrates *hornblende* as subhedral, green-brown pleochroic, moderate-high relief, well-cleaved crystals. Note that both the amphibole cleavages (angle of 125°) are displayed in two crystals, but only one cleavage is generally visible, and none is in the crystal at top right. The pleochroism is evident from the variety of colors which depend on crystal orientation. Note that this hornblende is almost colorless when it displays a single cleavage perpendicular to the E–W polarizer vibration direction (grain at center). The interference colors for the hornblende range up to the sensitive tint red, and demonstrate at least a moderate birefringence. A useful optic-axis figure

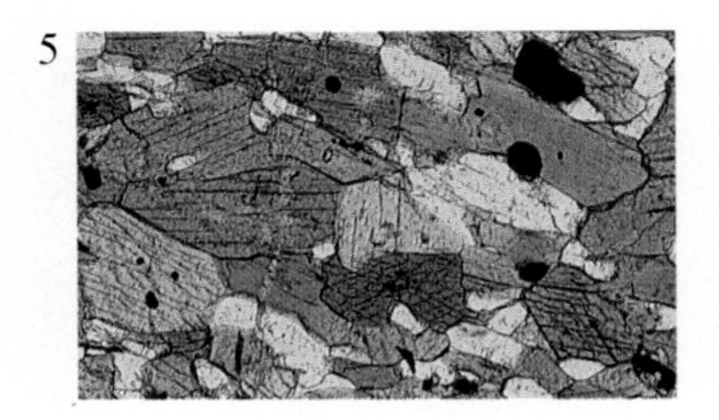

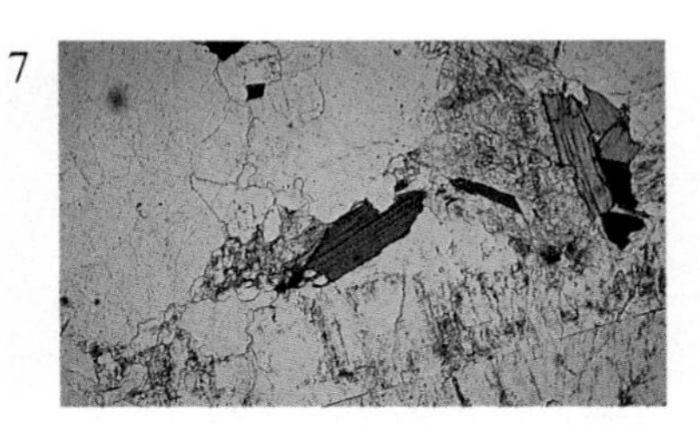

would be obtained from the low birefringent grain (right center).

Colorless, low-moderate relief grains of *calcic-plagioclase* are present, and in Plate 6, the low birefringence (first-order grey/white) and twinning are evident.

Opaque mineral grains are also present.

PLATES 7 (plane-polarized light) and 8 (crossed-polarized light)
Gneissic Granite, NE of Westport, New Zealand. View measures 3.3 × 2.2 mm.

Microcline, plagioclase, quartz, biotite, and muscovite make up this rock.

The *biotite* occurs as subhedral, brown, pleochroic, well-cleaved grains, with a moderate relief. Pleochroism is demonstrated by the fact that the two darker grains near the center have cleavages closer to the E–W polarizer vibration direction than the lighter grains on the right. Note the mottled extinction effect and red/green interference colors of the biotite in Plate 8. It is not immediately possible here to decide on the order of color because of the absence of sharp color rings.

Colorless, low-relief *microcline* at bottom of view displays a simple Carlsbad twin together with the characteristic cross-hatch twin pattern in the one Carlsbad-twin part (Plate 8). The simple twin plane, as well as fine-grained alteration, is visible in Plate 7.

The colorless, low-relief and featureless *quartz* (top left) displays its normal range of first-order interference colors up to creamy white (top center).

Plagioclase (top right), small amounts of *muscovite* (colorless in Plate 7, bright interference colors in Plate 8), and *opaques* are also visible.

PLATES 9 (plane-polarized light) and 10 (crossed-polarized light)
Separation Point Granite, Nelson, New Zealand. View measures 3.3 × 2.2 mm.

As well as microcline, plagioclase, quartz, and biotite, this rock contains abundant *titanite* (sphene) shown in Plate 9 as two wedge-shaped, very-high relief, pink-brown crystals. Both display a very high order interference color in Plate 10 (distinguished from first-order colors by inserting a sensitive-tint plate).

Biotite occurs as greenish-brown, pleochroic, moderate relief, subhedral crystals with a cleavage parallel to the crystal length. Note that biotite is darker the nearer the cleavage direction is to the polarizer vibration direction (E–W here). Note also that the crystals are warped and display mottled extinction. Interference colors are pink and green, but the order of color cannot immediately be discerned here.

The low-relief, colorless, and slightly altered background material is dominated by *microcline* and *sodic-plagioclase*.

PLATES 11 (plane-polarized light) and **12** (crossed-polarized light) *Separation Point Granite, Nelson, New Zealand. View measures 3.3 × 2.2 mm.*

The common accessory mineral *apatite* is well illustrated in this granite as relatively small colorless, moderate-relief crystals. It occurs in the feldspar (lower left center) and among the brown biotite (upper right center). The crystals are subhedral and display no cleavage. Note the very low birefringence shown in Plate 12, the crystals displaying first-order greys only.

Biotite (altered in part to *chlorite*) in Plate 11 displays its pleochroism from light brown (cleavage at high angle to E–W polarizer) to dark brown (cleavage at low angle to polarizer). The moderate relief crystals are altered to green chlorite in layers parallel to the cleavage. The crystal nearest the center is mainly chlorite.

Most of the low-relief, colorless, slightly altered background material is *microcline* which displays the characteristic cross-hatch twinning in crossed-polarized light (Plate 12). Some *plagioclase* (bottom left and center) and *opaque minerals* are also present.

PLATES 13 (plane-polarized light) and **14** (crossed-polarized light) *Hauynophyre, Kaiserstuhl, W. Germany. View measures 3.3 × 2.2 mm.*

This undersaturated igneous rock is dominated by *haüyne* (of the *sodalite* group), as seen in Plate 13 as low-moderate relief, colorless, six-sided euhedral crystal sections. Inclusions of opaque material are arranged in zones parallel to crystal faces. Note the isotropic nature of the haüyne—all crystals are black in Plate 14.

Melanite garnet is also present as very dark brown (almost opaque) six-sided crystal sections which have a very high relief almost obscured by their intense color. Like the haüyne, the garnet is isotropic, all crystals being black in crossed-polarized light.

The high-relief, small, yellowish-green crystals in the matrix are *clinopyroxene,* and they display a low to moderate birefringence in crossed-polarized light (Plate 14). Most of the low-relief, colorless matrix material is haüyne.

PLATE 15 (crossed-polarized light) *Basaltic Tuff, Samoa. View measures 3.3 × 2.2 mm.*

A variety of *zeolites* are illustrated filling vesicles in this volcanic rock. The zeolite in the large vesicle has a low birefringence, and consists of radiating clusters which have grown inwards from the vesicle wall.

PLATE 16 (plane-polarized light) *Incipiently metamorphosed sandstone, Upper Kyeburn, New Zealand. View measures 0.9 × 0.6 mm.*

This rock illustrates typical matrix recrystallization resulting from prehnite/pumpellyite facies metamorphism. Shown are a number of colorless detrital *quartz* grains, set in a matrix which comprises fine-grained, turbid, high-relief, pleochroic, blue/green *pumpellyite,* together with *sheet silicates* grown with a preferred orientation (E–W) at the edges of the quartz grains. This mode of occurrence is typical of metamorphic pumpellyite; the mineral is usually too fine grained to allow a complete optical characterization in routine work.

PLATES 17 (plane-polarized light) and **18** (crossed-polarized light) *Prehnite-quartz veins, S. Auckland, New Zealand. View measures 0.9 × 0.6 mm.*

These veins cut Manaia Hill Group volcanogenic sediments, and typify prehnite/pumpellyite facies metamorphism. The euhedral, tabular, colorless, moderate-relief *prehnite* crystals (relief has been enhanced by closing the lower diaphragm) are set amongst low-relief, colorless and featureless *quartz*. Note the cleavage parallel to the length of the central prehnite crystal, and the large relatively shapeless section (top right) which is cut more or less parallel to the cleavage. In crossed-polarized light we observe one crystal in extinction and a range of interference colors up to second-order blue. The large section subparallel to tablet surfaces (top right) displays a relatively low interference color, and it would probably provide a useful figure. Note the typical grey to creamy white interference colors of the quartz.

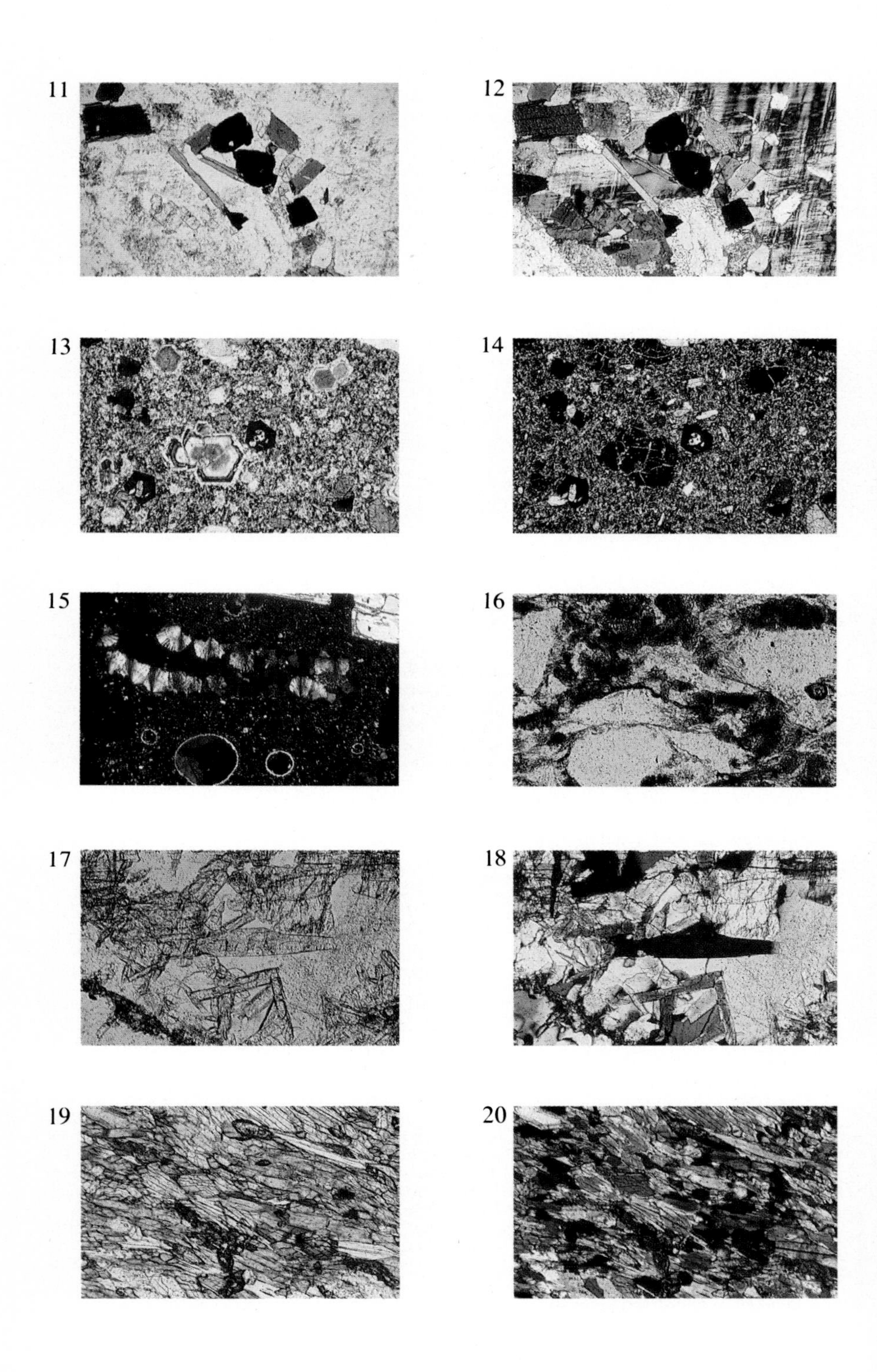
11
12
13
14
15
16
17
18
19
20

21

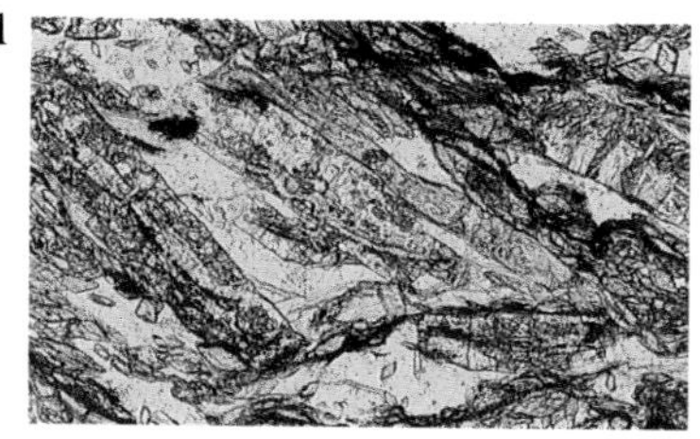

22

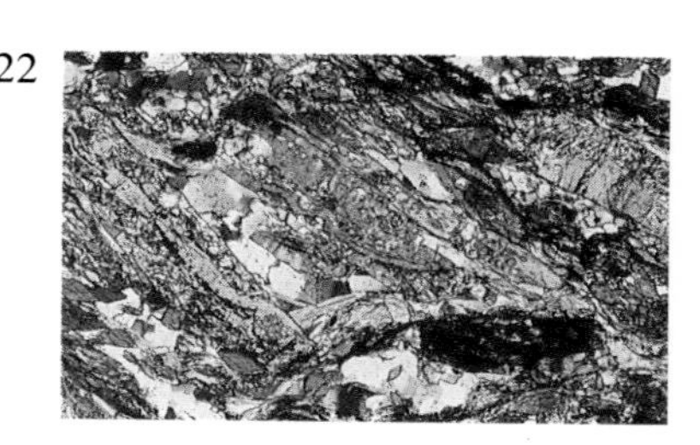

23

24

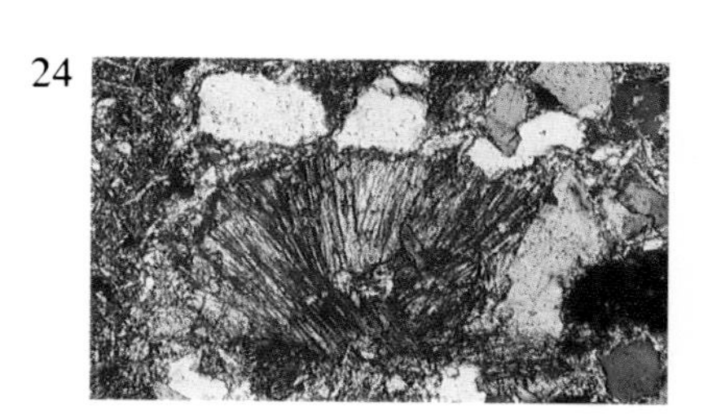

PLATES 19 (plane-polarized light) and **20** (crossed-polarized light) *Glaucophane Schist, Franciscan Group, California. View measures 0.9 × 0.6 mm.*

The distinctive pleochroism of *glaucoplane* from colorless to pale blue to violet is illustrated in Plate 19. Grains that display two typical cleavages at 125°, with the acute angle more or less E–W, consistently display the violet color. Prismatic sections with their lengths oriented E–W are consistently blue. The polarizer vibration direction is E–W. Note the general preferred orientation of grains which defines the schistosity of this rock. In Plate 20, the moderate birefringence of the glaucophane is evident from the second-order blues.

Also present are a few colorless grains of *muscovite,* discernible in Plate 20 from the mottled extinction (right lower center).

PLATES 21 (plane-polarized light) and **22** (crossed-polarized light) *Lawsonite Schist, Franciscan Group, California. View measures 0.9 × 0.6 mm.*

Four colorless, subhedral, (010)-dominated plates of *lawsonite* trend NW–SE and display a moderate-high relief. The moderate birefringence is shown in Plate 22 by the second-order blue.

Also visible is a background of low-relief, low-birefringence *quartz,* euhedral 125° cross sections of almost colorless *glaucophanic amphibole* (bottom left and top right), and low-moderate relief, colorless *muscovite* showing pink interference colors in Plate 22 (lower center and top left).

PLATES 23 (plane-polarized light) and **24** (crossed-polarized light) *Incipiently recrystallized sandstone, Franciscan Group, California. View measures 0.9 × 0.6 mm.*

Because of low temperatures, metamorphic effects in low geothermal gradients are often incipient, even when pressures have been extremely high. Illustrated here is a radiating aggregate of prismatic *jadeite,* a clinopyroxene diagnostic of ultrahigh-pressure/low-temperature metamorphism, set amongst detrital *quartz* grains barely affected by metamorphism. Note that the jadeite is colorless, has a moderate-high relief, a low birefringence, and highly inclined extinction (overlapping groups of prisms oriented NW–SE and NE–SW extinguish incompletely and display anomalous blue colors). The jadeite has developed from some relatively unstable material, probably feldspar. Compare the effects in this rock with the incipient recrystallization of the pumpellyite-bearing sandstone (Plate 16).

The quartz of Plate 24 shows the usual range of interference colors from dark grey to creamy white.

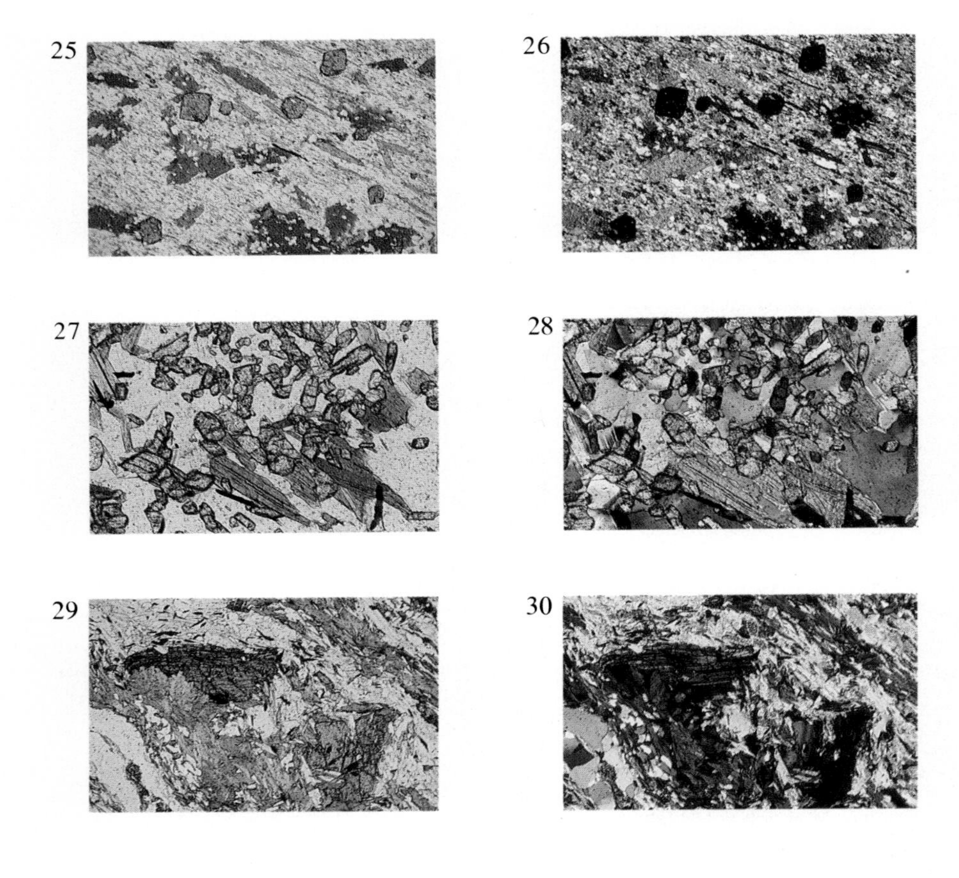

PLATES 25 (plane-polarized light) and **26** (crossed-polarized light)
Garnet Schist, Southern Alps, New Zealand. View measures 3.3 × 2.2 mm.

The schistosity in this rock, defined by quartz and muscovite, is overgrown by posttectonic garnet and biotite. The *garnet* occurs as euhedral, high-relief crystals in Plate 25; their isotropic nature is seen in Plate 26. Note the very fine inclusion trains parallel to the schistosity of the fine matrix. Brown *biotite*, with a moderate relief, displays a random orientation so that we see basal sections (bottom center), and sections with sharply defined cleavage traces (center left). In crossed-polarized light (Plate 26) the basal section is pseudo-isotropic, but sections with sharply defined cleavages display bright second- or third-order interference colors. Note the abundant inclusions.

Several pale-green cross sections of *chlorite* (with a moderate relief) are shown from near the center towards the right of the view. In crossed-polarized light, these green crystals have very low interference colors, some of which are anomalously brown, typical of positive chlorite. The fine-grained matrix is dominated by low-relief *quartz* and moderate-relief *muscovite*.

PLATES 27 (plane-polarized light) and **28** (crossed-polarized light)
Gneiss, Fiordland, New Zealand. View measures 0.9 × 0.6 mm.

Plate 27 illustrates numerous high-relief subhedral grains of *epidote* set in low-relief, colorless *plagioclase*. Well cleaved, brown, pleochroic *biotite* is also shown. Plate 28 illustrates interference colors for the epidote ranging up into the second order. The distinctive lack of first-order white is well shown by a number of grains displaying yellow interference colors grading straight into grey at their thin edges. The biotite in Plate 28 displays its characteristic mottled extinction, and thin edges to the grains allow interference-color rings to be counted (up to third-order green at right center). Plagioclase has first-order grey interference colors and repeated twins (left center) as visible in Plate 28.

PLATES 29 (plane-polarized light) and **30** (crossed-polarized light)
Chloritoid Schist, Connemarra, Eire. View measures 2.7 × 1.8 mm.

The two superficially similar minerals *chloritoid* and *chlorite* often occur together as illustrated here. Two large crystals of pale blue-green, high-relief chloritoid are shown. One towards the top left displays a cleavage or parting E–W; the other (bottom right) is relatively featureless. In crossed-polarized light, both crystals display very low interference colors, and the inclined extinction and repeated twinning is visible in the top left crystal. In contrast, the chlorite has a lower relief, a more obvious cleavage, and stronger pleochroism. In crossed-polarized light the chlorite displays anomalous blue-purple interference colors typical of low-birefringent, negative varieties.

Other minerals illustrated include *quartz* (lower left) and *muscovite* which has a low-moderate relief, bright, mainly pink interference colors, and mottled extinction.

PLATES 31 (plane-polarized light) and **32** (crossed-polarized light)
Stilpnomelane Schist, New Zealand. View measures 0.9 × 0.6 mm.

Stilpnomelane is a sheet silicate found in low-grade metamorphic rocks; it is superficially similar to biotite. Sections through a group of *stilpnomelane* plates in Plate 31 display the pleochroism so that the brown color is darkest (almost opaque) when the crystal length is E–W (parallel to the polarizer vibration direction). Stilpnomelane differs from biotite in its more robust appearance, its poorer cleavage, lack of mottled extinction, and usually more intense color.

The crystals are set in a metamorphic mosaic of colorless, low-relief, low-birefringence *quartz;* some small pale crystals of high birefringence *biotite* and low birefringence *chlorite* can also be seen.

PLATES 33 (plane-polarized light) and **34** (crossed-polarized light)
Tremolite, Southern Alps, New Zealand. View measures 0.9 × 0.6 mm.

Tremolite is a strongly prismatic amphibole characteristic of greenschist facies metamorphism. Illustrated is a pure tremolite rock displaying the prismatic shapes, one good example of a cross section with 125° cleavages, and the general moderate relief and lack of color. A moderate birefringence is evident from the second-order interference colors of Plate 34.

PLATES 35 (plane-polarized light) and **36** (crossed-polarized light)
Staurolite Schist, Connemarra, Eire. View measures 3.3 × 2.2 mm.

Staurolite is a metamorphic mineral characterizing the amphibolite facies. It has a distinctive yellow color and high relief, as illustrated in Plate 35. Note the pleochroism shown by slightly different colors in various grains. No cleavage is evident. The low birefringence (first-order grey-yellow colors) are displayed in Plate 36.

Other minerals present include pleochroic brown *biotite* (note the section more or less parallel to cleavage which shows a low-interference color; this would provide an off-centered optic-axis figure), *plagioclase* (low-relief, colorless, and multiple twinning, as seen at lower right center of Plate 36), and *opaques*.

PLATES 37 (plane-polarized light) and **38** (crossed-polarized light)
Sillimanite Gneiss, E of Greymouth, New Zealand. View measures 0.9 × 0.6 mm.

This rock occurs in the marginal zone of a granite. Typical of high temperature metamorphism, *sillimanite* (variety *fibrolite*) is seen as a swarm of fine, needle-shaped crystals cutting brown *biotite* and colorless *quartz*. Note the moderate relief and lack of color of the sillimanite which shows interference colors up into the second order in Plate 38. The biotite is pleochroic (grains with cleavage E–W parallel to the polarizer vibration direction are darkest), and the quartz has its characteristic featureless appearance. Plate 38 shows the mottled extinction and third-order colors of biotite, and the first-order grey to creamy white of quartz.

PLATES 39 (plane-polarized light) and **40** (crossed-polarized light)
Dolomite, Coverham, South Island, New Zealand. View measures 3.3 × 2.2 mm.

Dolomite often grows in sediments as euhedral rhombs, and as such serves well to illustrate the principal properties of the trigonal carbonates. Like calcite, dolomite twinkles (changes relief on rotation of the stage) and this effect is shown by the euhedral crystals of Plate 39. Some rhombs display a moderate relief, others a low relief, and all of the latter would change to a moderate relief on rotation of the stage. In Plate 40, the very high birefringence is evidenced by the very high order interference colors. Unlike first-order white, these will not change on insertion of an accessory sensitive-tint plate. Some grains of lower partial birefringence display moderate order pink and green interference colors.

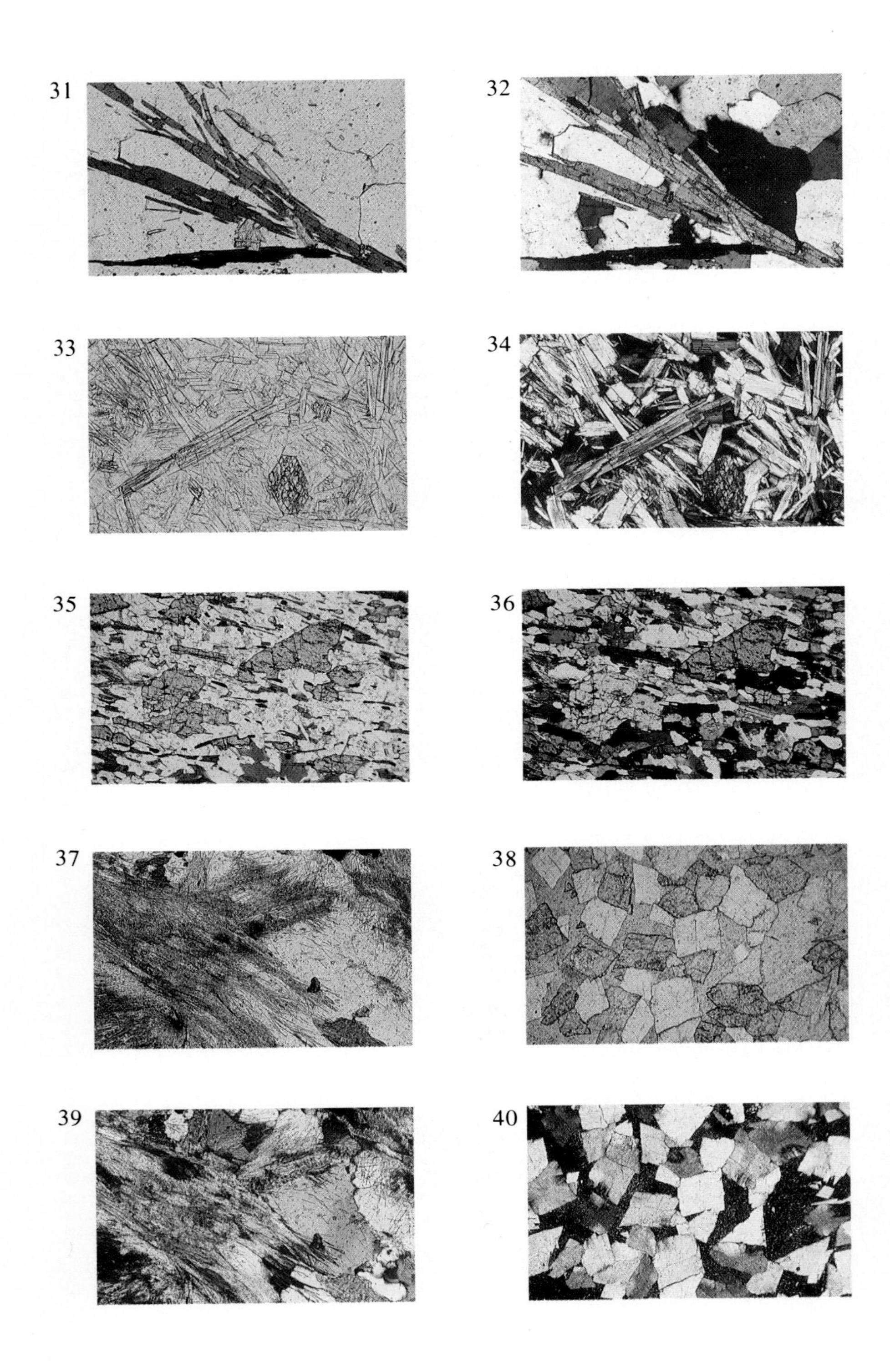
31
32
33
34
35
36
37
38
39
40

Abbreviations and Symbols Used in the Text

a, b, c	Crystal axes and cell edges
α, β, γ	Angles between the positive ends of the b and c, a and c, and a and b crystal axes, respectively
α, β, γ	Least, intermediate, and greatest refractive indices in biaxial crystals *Note:* the particular meaning of α, β, and γ in various parts of the text is always obvious or made clear by a note
α', γ'	Refractive indices in a general section of a biaxial crystal where $\alpha < \alpha' < \beta < \gamma' < \gamma$
ω	The ordinary ray or refractive index in uniaxial crystals
ε	The extraordinary ray or refractive index in uniaxial crystals
ε'	A ray or refractive index in uniaxial crystals intermediate between ω and ε
n	Refractive index of isotropic material
RI	Refractive index
δ, δ'	Birefringence, partial birefringence
R	Retardation
X, Y, Z	Vibration directions in biaxial crystals of the fastest, intermediate, and slowest rays, respectively; equivalent to the refractive indices α, β, and γ
OAP	Optic axial plane
+ve, −ve	Positive, negative
$2V_x$, $2V_z$	The optic axial angle about X or Z
$2E$	The apparent optic axial angle in air
$r < v$, $r > v$	Measures of the optic-axis dispersion
NA	Numerical aperture

t	Thickness of section
CB	Canada balsam
λ	Wavelength
ca.	Approximately
∥	Parallel to
⊥	Perpendicular to
∧	Angle between two faces or directions
$<$, $>$	Less than, greater than
(), {}, []	Face, form, and zone symbols enclosing Miller indices (*see* Chapter 1)

The abbreviations and symbols used in this book reflect a common, but by no means universal, usage in North America. Students will encounter other symbols in the literature, and Table 0.1 summarizes these. To Table 0.1 I add some comments:

1. The former use of Np, Nm, and Ng (Winchell, 1937) has disappeared almost completely.
2. The use of zone axes [100], [010], and [001] is always an acceptable alternative to *a, b,* and *c* or *x, y,* and *z*, though their use on crystal drawings already marked with (100), (010), and so on for faces, is liable to be confusing.
3. The recommendation of the International Mineralogical Association that *X, Y,* and *Z* be used for crystal axes seems not to have been adopted by mineralogists.
4. The authoritative Deer, Howie, and Zussman usage of different symbols for crystal axes and cell dimensions is designed with X-ray crystallography in mind. Many mineralogists wish to maintain an equivalent distinction between vibration directions and refractive indices, especially for biaxial crystals.
5. One sometimes encounters the use of α and γ for the fast and slow directions in uniaxial minerals (e.g., Williams, Turner, and Gilbert, 1982).

Table 0.1 Examples of the Use of Differing Symbols in the Literature[a]

	Crystal axes	Cell dimensions	Refractive indices	Optical directions
This book *American Mineralogist* *Canadian Mineralogist*	a, b, c	a, b, c	n, ω, ε α, β, γ	ω, ε (or O, E) X, Y, Z
Deer, Howie, and Zussman *Mineralogical Magazine*	x, y, z	a, b, c	n, ω, ε α, β, γ	ω, ε α, β, γ
Mineralogical Abstracts	a, b, c	a, b, c	n, ω, ε α, β, γ	ω, ε α, β, γ
International Mineralogical Association (*Can. Min.* 1978; 16:113–117)	X or [100] Y or [010] Z or [001]	a, b, c	n, ω, ε ? α, β, γ	ω, ε ? α, β, γ
Phillips and Griffen (1981)	a, b, c	a, b, c	$n, n_\omega, n_\varepsilon$ $n_\alpha, n_\beta, n_\gamma$	O, E X, Y, Z
Winchell (1965)	a, b, c	a, b, c	n, n_o, n_e n_x, n_y, n_z	O, E X, Y, Z
Tröger (1979)	a, b, c	a, b, c	n, n_O, n_E n_X, n_Y, n_Z	O, E X, Y, Z
Winchell (1937)	a, b, c	a, b, c	N, ω, ε N_p, N_m, N_g	O, E X, Y, Z

[a] The usage of symbols in journals, as reported above, is not without variation. The *Canadian Mineralogist,* in particular, displays considerable variety so that x, y, and z (1982; 20:239), as well as X, Y, and Z crystal axes (1981; 19:643), may be found.

Optical Mineralogy

Second Edition

Chapter 1

An Introduction to Crystallography

Crystals may be investigated in a number of ways, the most obvious being to analyze their form and symmetry in hand specimen. More fundamentally perhaps, the atomic structure of crystals can be studied with X-rays. This book is concerned primarily with another branch of crystallography in which the optical properties of crystals are studied. With the aid of the polarizing microscope, these properties enable us to identify the common rock-forming minerals quickly and accurately.

Useful as optical properties are, to apply them successfully, it is necessary to understand their relationship to the more general aspects of crystallography. The following account introduces the general crystallographic concepts and terminology required for optical work.

Crystals

A crystal is a solid substance characterized by a regular internal arrangement of its constituent atoms. This internal arrangement can be determined using X-rays. The flat surfaces (*crystal faces*) that bound well-shaped crystals are disposed in such a way that they reflect the internal arrangement of the atoms. Crystals may be distinguished as *euhedral,* if they are bounded by crystal faces, *subhedral,* if they are partially so bounded, or *anhedral,* if no crystal faces are present.

The Interfacial Angle

Two crystals are seldom exactly alike in shape. For various reasons that need not concern us here, certain crystal faces may be better developed in one crystal than in others. Nevertheless, it is always found that the angle

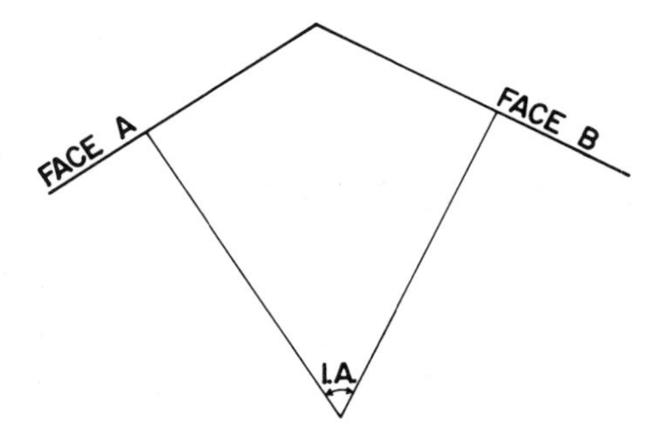

Figure 1.1 The interfacial angle (IA) to two crystal faces *A* and *B*.

between two particular faces for a specific mineral is constant, regardless of how well developed the faces are. This directly reflects the regularity of the internal arrangement of atoms. The angle is expressed as the *interfacial angle,* which is the angle between the normals to the two faces (Figure 1.1).

Crystal Symmetry

Almost all crystals possess a degree of symmetry, which may be expressed in terms of either axes, planes, or a center of symmetry. When considering the symmetry elements of a crystal, it is important to disre-

Figure 1.2 Examples of symmetry axes in crystals.

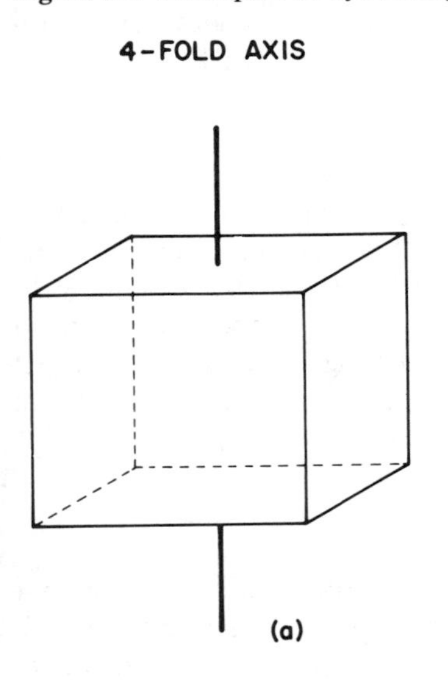

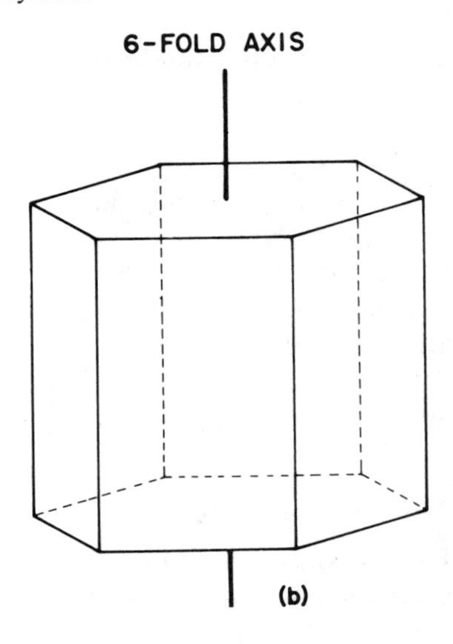

gard the imperfections of growth, and the unequal development of faces. Symmetry is judged on the angular relationships between crystal faces.

Axes of Symmetry

If the (perfect and regular) crystal has an identical disposition of faces in two or more positions when rotated about a line, that line is termed an axis of symmetry. There are twofold (diad), threefold (triad), fourfold (tetrad), and sixfold (hexad) axes of symmetry in various minerals. As examples, a fourfold axis is shown in a simple cube (Figure 1.2a), and a sixfold axis in a simple hexagonal prism (Figure 1.2b).

Planes of Symmetry

If the (perfect and regular) crystal can be divided into two mirror–image halves, the plane that so divides it is termed a plane of symmetry. A simple cube has nine such planes of symmetry, and one of these is shown in Figure 1.3.

Center of Symmetry

If the (perfect and regular) crystal has a central point through which a line in any direction will emerge at an identical point on either side of the crystal, it is said to have a center of symmetry. A simple cube, for exam-

Figure 1.3 **(a)** One of the nine planes of symmetry of a cube and **(b)** the fourfold inversion axis of a tetragonal sphenoid that relates faces *A* and *B*.

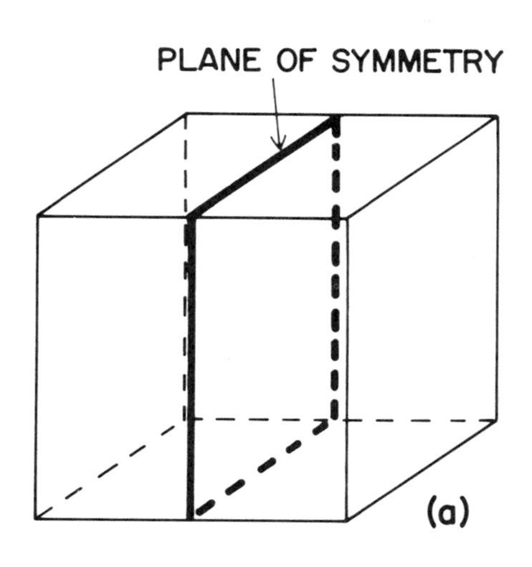

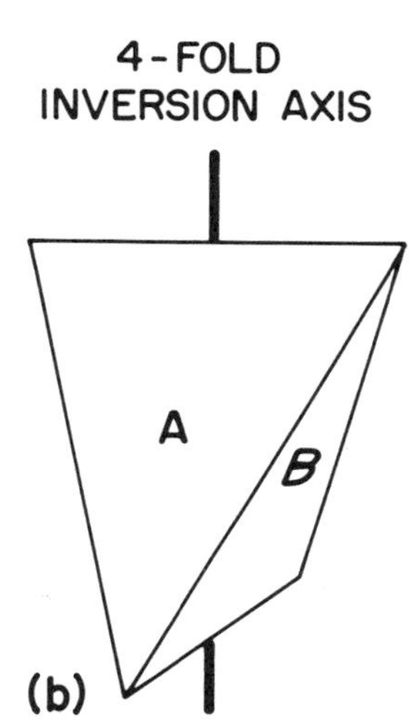

ple, clearly has such a center. A center of symmetry may also be visualized as a onefold axis of rotary inversion (see below).

Axes of Rotary–Inversion Symmetry

Two-, three-, four-, and sixfold axes of rotary–inversion symmetry refer to a combination of rotation about a line by 180°, 120°, 90°, or 60°, respectively, combined with inversion through the center. Thus, in Figure 1.3b, faces *A* and *B* of a tetragonal sphenoid are related by a 90° rotation combined with inversion through the center. A onefold axis of rotary inversion (360° rotation plus inversion) is the same as a center of symmetry.

Crystal Systems

Depending on the particular symmetry elements of a crystal, it may be classified as belonging to one of seven crystal systems. Each crystal system can be described in terms of three or four *crystal axes* known as *a, b,* or *c,* the choice and disposition of which are a consequence of symmetry. The following crystal systems are normally distinguished (Figure 1.4). Note that the symmetry axes referred to may be either simple rotary or rotary–inversion axes, except in the orthorhombic system for which no rotary–inversion axes exist.

Cubic system. Three crystal axes: a_1, a_2, a_3, all equal and at right angles to each other. All cubic minerals have four triad axes of symmetry.

Tetragonal system. Three crystal axes: two equal, a_1, a_2, and a third, generally unequal, *c*, all at right angles to each other. The *c*-axis is always a tetrad axis of symmetry.

Hexagonal and trigonal systems. Four crystal axes: three equal, a_1, a_2, a_3, which lie in a plane at 120° from each other, and a fourth, generally unequal, *c*, which is at right angles to the plane of the *a*-axes. The *c*-axis is a hexad axis of symmetry in the hexagonal system, and a triad in the trigonal system.

Orthorhombic system. Three crystal axes: *a, b, c,* in general all unequal, and all at right angles to each other. The *c*-axis (and often the *a*- and *b*-axes) is a diad axis of symmetry.

Monoclinic system. Three crystal axes: *a, b, c,* in general all unequal. *b* is at right angles to the *ac*-plane. The angle between *a* and *c* is not a right angle, and the obtuse angle between them is termed β (not to be confused with the refractive index β). The *b*-axis is always a diad axis of symmetry.

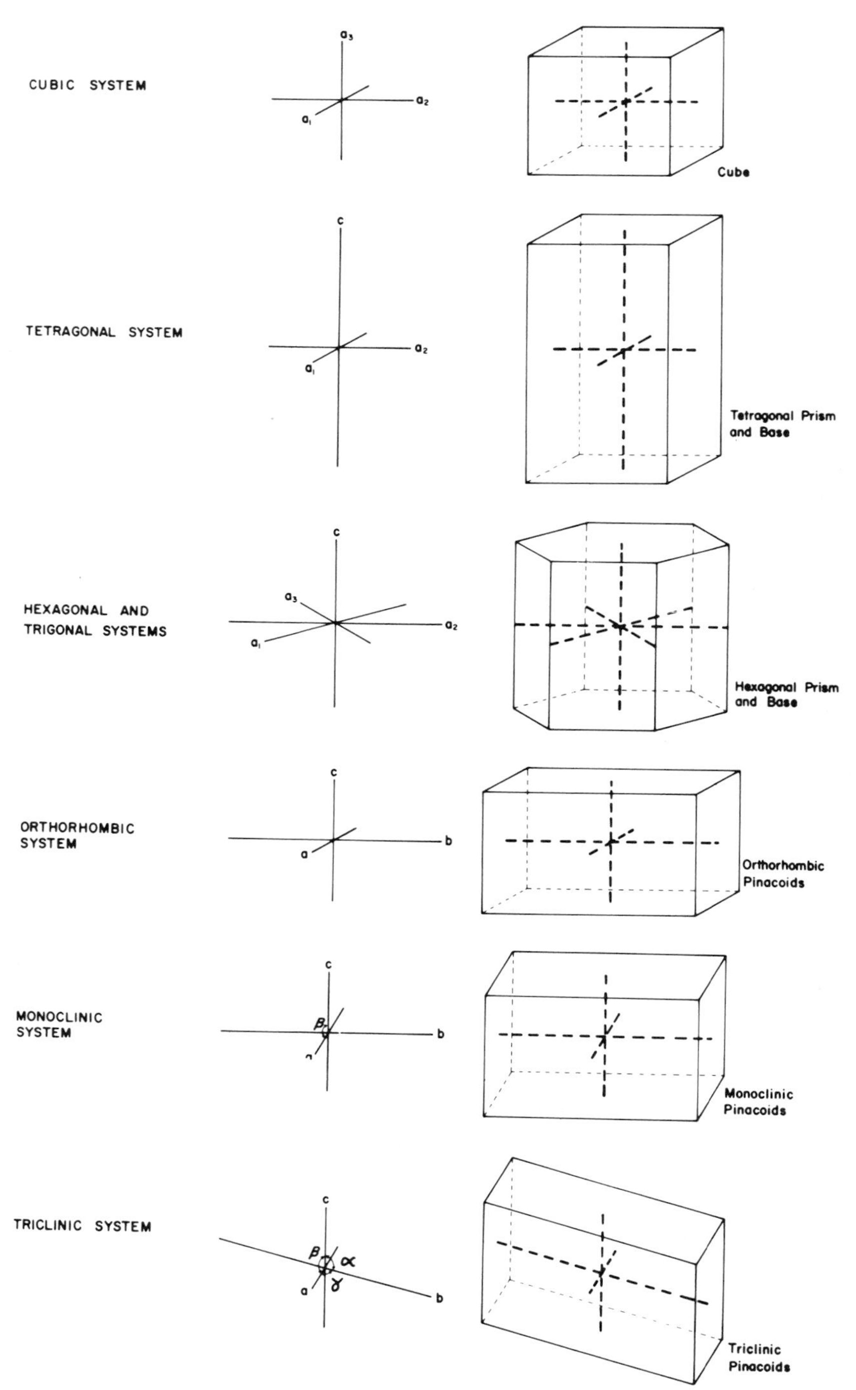

Figure 1.4 Crystal axes and some simple forms of the crystal systems.

Triclinic system. Three crystal axes: *a, b, c,* in general all unequal, and none of them at right angles to each other. The angles between the positive ends (Figure 1.4) of *b* and *c, a* and *c,* and *a* and *b,* are termed α, β, and γ, respectively (not to be confused with the refractive indices α, β, and γ). Only a center of symmetry is possible.

More detailed information on the combinations of symmetry elements possible in each system (leading to a total of 32 *crystal classes*) can be found in standard crystallography texts (Phillips, 1971; Whittaker, 1981).

Unit Cells and Axial Ratios

The basic atomic structure that is repeated regularly in a crystal is termed the *unit cell.* X-ray studies enable the size of the unit cell to be determined. Its dimensions are expressed in terms of lengths along the direction of the crystal axes *a, b,* and *c*. The *axial ratio* simplifies these absolute lengths into relative values. For tetragonal, hexagonal, or trigonal minerals, the axial ratio is expressed in terms of the ratio $c:a$, *a* being taken as 1. For orthorhombic, monoclinic, or triclinic minerals, the axial ratio is expressed in terms of the ratio $a:b:c$, *b* always being taken as 1.

Miller Indices

There are several ways in which the angular relationship of a crystal face to the crystal axes can be expressed. The most common method is by means of Miller indices. These denote the reciprocals of the distances by which a crystal face intercepts the crystal axes, the distances being measured in units proportional to the axial ratio. Three indices (or four for the hexagonal and trigonal systems) are given, one for each of the crystal axes, and they are always expressed as whole numbers or zero. If a face is parallel to a crystal axis, it intercepts it at infinity, the reciprocal of which is zero. If a face intercepts the negative end of a crystal axis, a bar is placed over the number. A few examples will serve to clarify their use.

Cube (Figure 1.5a)

Face *A* cuts a_1, but is parallel to a_2 and a_3. The reciprocals of the intercept distances are therefore a_1 (1), a_2 ($\frac{1}{\infty} = 0$), a_3 ($\frac{1}{\infty} = 0$). The Miller indices are (100). Similarly, face *B* is (010) and face *C* (001).

Pyritohedron (Figure 1.5b)

Face *A* is parallel to a_3, and cuts a_1 at half the distance from the origin as a_2. The relative intercept distances are a_1 (1), a_2 (2), a_3 (∞), and the reciprocals are $\frac{1}{1}$, $\frac{1}{2}$, $\frac{1}{\infty}$, which expressed as whole numbers (Miller indices)

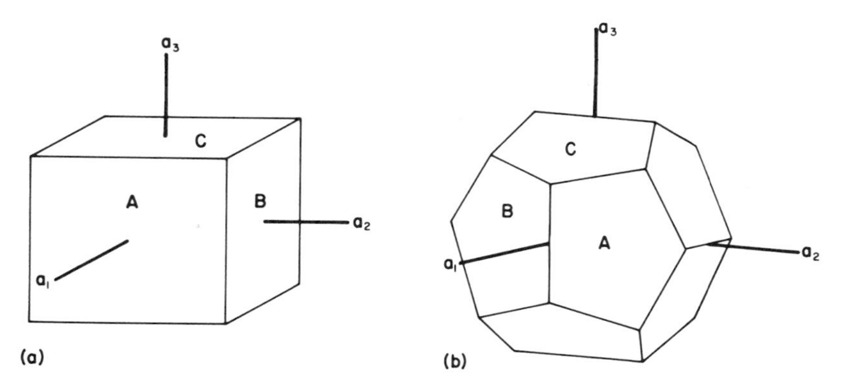

Figure 1.5 The text explains how to derive Miller indices for the crystal faces *A*, *B*, and *C* of a cube **(a)** and pyritohedron **(b)**.

is (210). Face *B* intercepts at the relative distances of 1, $\bar{2}$, and ∞, and the Miller indices are (2$\bar{1}$0). Face *C* intercepts at the relative distances 2, ∞ and 1, and the Miller indices are (102).

Tetragonal Pyramid (Figure 1.6)

Face *A* is parallel to a_2 and intercepts the c and a_1 axes at unequal distances. However, these distances are in the same relative proportion as the axial lengths as expressed in the ratio $c : a = 2.01 : 1$ (Figure 1.6b). The

Figure 1.6 The text explains how to derive Miller indices for the faces *A* and *B* of a tetragonal pyramid.

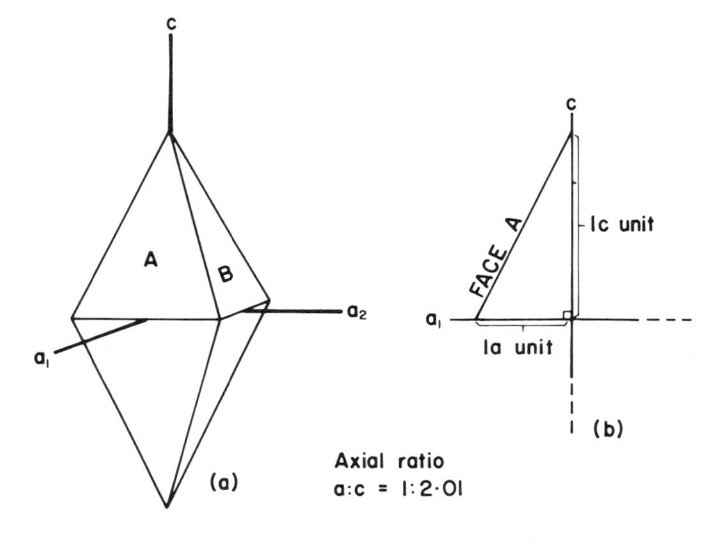

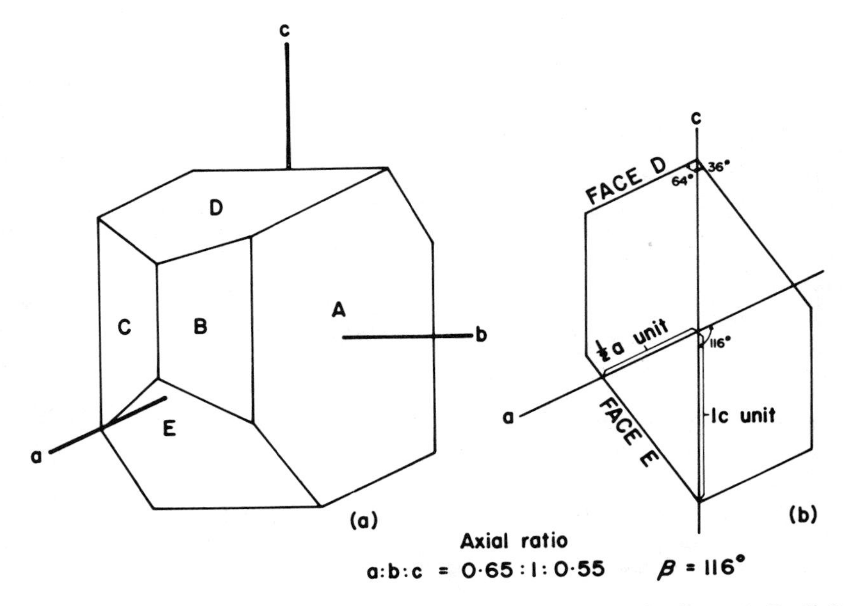

Figure 1.7 The text explains how to derive Miller indices for the faces *A, B, C, D*, and *E* of a monoclinic feldspar crystal.

Miller indices are therefore (101). Face *B* is parallel to a_1 but intercepts *c* and a_2 at distances in the same proportion as the axial lengths, and is therefore (011).

Monoclinic Feldspar Crystal (Figure 1.7)

Face *A* intercepts *a* and *c* at ∞, and the Miller indices are (010). Face *B* intercepts *a* and *b* at distances directly proportional to the unit lengths, and is parallel to *c*. The Miller indices are therefore (110). Face *C* is identical to *B* except that it cuts the negative end of *b*, and is therefore $(1\bar{1}0)$. Face *D* is parallel to both *b* and *a* (Figure 1.7b), and is therefore (001). Face *E* is parallel to *b*, but cuts *a* and *c*. The intercept distance along *a* is half the unit length compared to its intercept along *c* (Figure 1.7b). Therefore the relative intercept lengths can be expressed as 1, ∞ and $\bar{2}$, which as the reciprocal Miller indices is $(20\bar{1})$.

Note: The Miller indices of individual crystal faces are always placed in parentheses, e.g. (101). Miller indices may be more generally expressed as (hkl) or (hkil), where h + k + i always equals zero.

Crystal Forms

A crystal form is a face or a group of faces that have identical relationships to the crystal axes by virtue of the crystal symmetry. The form can be expressed in terms of the Miller indices of one representative face, the Miller indices being placed in braces, e.g. {100}.

A cube form, for example, is denoted by {100}, which refers not only to (100), but also to (010), (001), ($\bar{1}$00), (0$\bar{1}$0) and (00$\bar{1}$), because all these faces are identical with respect to symmetry. A form is *closed* if it can completely enclose space and exist by itself. *Open* forms do not enclose space, and must exist in combination with other forms.

Some form names in common use are described below.

Cube {100}. The closed form comprising the six sides of the simple cube, all of which are identical in terms of symmetry.

Other cubic-system forms. There are 15 forms in the cubic system, each of which has several faces, all identical in terms of symmetry. These include the octahedron {111}, the rhombdodecahedron {110}, the tetrahexahedron {hk0}, and the trapezohedron {hll}, all of which are closed forms.

For crystals other than cubic, the following are common forms.

Pyramid. A form comprising several nonparallel faces that meet at a point, e.g. {111} in the tetragonal system.

Rhombohedron. A closed form comprising six faces whose intersection edges are not at right angles, e.g. {10$\bar{1}$1}.

Prism. An open form of several faces all of which are parallel to the same axis; this axis is most often *c*. Common form indices are {110} and {10$\bar{1}$0}.

Pinacoid. An open form comprising two parallel faces, e.g. {010} in the monoclinic system.

Crystal Zones

A zone is a group of crystal faces that intersect in a set of parallel edges. The direction of the parallel edges is known as the zone axis, and is expressed as coordinates derived from Miller indices, and placed in square brackets [uvw]. If the Miller indices of two faces ($h_1k_1l_1$) and ($h_2k_2l_2$) in a zone are known, then the zone axis is derived as follows:

$$u = (k_1 \times l_2) - (l_1 \times k_2)$$

$$v = (l_1 \times h_2) - (h_1 \times l_2)$$

$$w = (h_1 \times k_2) - (k_1 \times h_2)$$

For example, the faces (210) and (110) belong to a zone defined by an axis [001]. Zone axes are commonly cited instead of crystal axes. Thus [100] is the same as the *a*-axis, [010] the *b*-axis, and [001] the *c*-axis.

Crystal Habit

Crystals, even of one species, vary considerably in shape, depending on rates of growths, impurities present during growth, and a host of other factors. Nevertheless, particular species are characterized by certain shapes called crystal habits. Well-known terms that are self-explanatory are *fibrous, acicular* (needlelike), *columnar, tabular, scaly, micaceous.* In addition, form names are used if a particular form is well developed, hence *cubic, prismatic, pyramidal,* etc.

Cleavage, Fracture, and Parting

Many crystals break easily along smooth planes, which are parallel to possible crystal faces, usually simple index ones, and across which the interatomic bonding is relatively weak. Such planes are termed cleavage planes. Cleavages are repeated by the symmetry of a crystal in exactly the same way as faces. It will be found, in the systematic description of minerals, that the indices of a cleavage are often placed in braces like crystal forms. Hence in a cubic mineral, {100} cleavage refers not only to (100), but also to the (010) and (001) cleavages, all of which are identical in terms of symmetry. On the other hand, in a triclinic mineral with little or no symmetry, {100} refers to a single plane (100) because there are no other planes identical in terms of symmetry.

A cleavage may be described as *perfect, good, distinct, imperfect,* or *poor,* depending on its ease of development.

Fracture refers to the shape of surfaces formed by breaking a crystal along directions other than cleavages. If a crystal has a number of perfect cleavages, fracture may be difficult to observe. Fracture may be *conchoidal* (shelllike surfaces), *even* (subplanar), *uneven,* or *hackly* (jagged).

Parting refers to planes of separation which bound twin planes or exsolution lamellae in a crystal. It may be confused with cleavage.

Twinned Crystals

A twinned crystal is formed of two or more individuals of the same species, joined together according to a definite law. They may be joined together as *contact twins,* simply united by a common plane, or as *penetration twins,* where they appear to cross each other in a complex, but symmetrical way. *Simple twins* consist of just two individuals. Repeated or *polysynthetic twins* consist of several individuals, often in lamellar form.

The geometric relationship between the individuals of a *twinned crystal* can be described in terms of *twin axes* or *twin planes.* A twin axis is a line of rotation about which one twin can be brought into the orientation of the other (usually by a rotation of 180°). A twin axis is the most convenient way of describing most twins. A twin plane is a plane of reflection across which the twins are mirror images. Obviously, a twin plane cannot be parallel to a plane of symmetry of the crystal because the mirror images across such a plane are identical. The twin plane must not be confused with the *composition plane,* which is the plane that actually unites the two individuals, though in some cases the two planes coincide.

Normal twins have a twin axis normal to the composition plane. *Parallel twins* have a twin axis that lies in the composition plane parallel to a crystal edge. *Complex twins* can be visualized as a combination of normal and parallel twinning, the twin axis lying in the composition plane normal to a crystal edge (for examples refer to the plagioclase-feldspars).

Twin crystals may form in several ways. *Growth* or *primary twins* form during the growth of the crystal, often at the beginning of growth. *Secondary twins* form subsequent to crystal growth. For example, *deformation twins* result from the rotation of part of the crystal into a twin orientation during deformation. Calcite develops deformation twins very easily, if squeezed in a vice. *Transformation twins* result from the change of symmetry of certain crystals on changes of temperature and pressure. For example, monoclinic alkali-feldspar may invert to a triclinic form on cooling, and in so doing will nucleate numerous twins in a cross-hatched arrangement (see the section on feldspars). Leucite twins in a similar way on changing from cubic to a lower symmetry on cooling.

Stereographic Projection

Crystal symmetry is judged on angular relationships between crystal faces, not on the particular dimensions of those faces. Such angular relationships are not difficult to visualize for simple and regular crystal forms such as the cube and hexagonal prism, but for less regular, less symmetric, and multifaceted crystals, a method of representing three-dimensional angular relationships is desirable. Likewise, structural geology requires a method of representing the angular relationships between bedding planes, schistosities, and other surfaces, and optical mineralogy a means of representing various optical directions. For all these purposes, stereographic projection using a stereographic net (Figure 1.8) is employed.

Much of the routine in optical mineralogy can be done without stereographic projection, but for universal- and spindle-stage work (Chapter 6), it is absolutely essential, and the following account provides an introduction to the subject. For examples of its use in describing the 32 crystal classes, the reader is referred to Phillips (1971) and Whittaker (1981).

The principle of stereographic projection can be understood from Figure 1.9. This illustrates in (a) a line which in space is oriented 30° from vertical and passes through the center of a sphere. Its point of intersection with the outer surface of the sphere (lower hemisphere) is projected to the opposite pole of the sphere intersecting the equatorial section at point *A*. A stereographic net represents this equatorial section and is graduated in

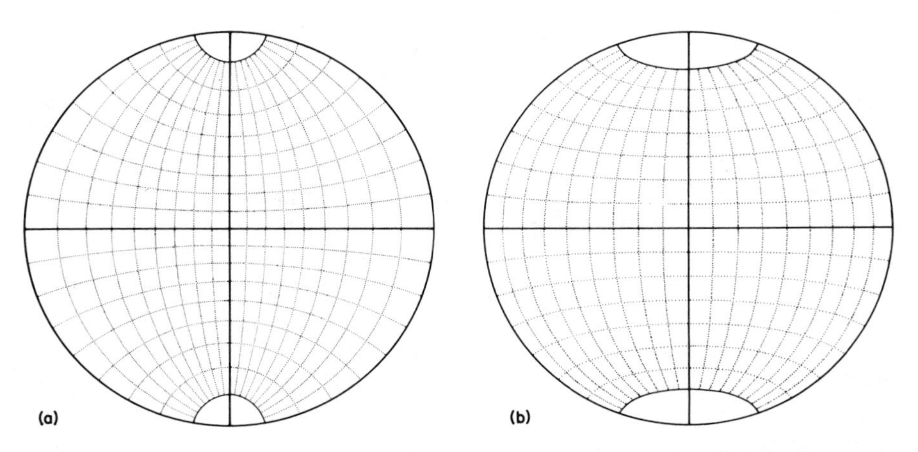

Figure 1.8 **(a)** The true stereographic or Wulff net and **(b)** the slightly distorted equal area or Schmidt stereographic net.

degrees along both the "north to south" ("N–S") and "east to west" ("E–W") diameters so that the projected points for lines plunging at any angle along these directions can be plotted. Thus point A in Figure 1.9c represents the projection of the line plunging west and 30° from vertical shown in Figure 1.9a. Point B of Figure 1.9c represents a similar projection point for a line plunging west 60° from vertical.

Also represented on a stereographic net are the projections for a family of planes that pass through the center of the sphere and intersect in the N–S diameter of the equatorial section. Thus, in Figure 1.9b, a plane that dips 60° to the east can be projected from the outer surface of the lower hemisphere to the opposite pole so that it intersects the equatorial section in a circular arc $A'B'C'D'$. Such a circular arc is called a *great circle,* and the stereographic net (Figure 1.8) shows a set of them representing planes that dip at various angles between 0° and 90° both east and west. Obviously a vertical plane projects onto the net as a straight diameter, and a horizontal plane as the circumference of the net. The plane dipping 60° eastwards, shown in Figure 1.9b, is plotted on the stereographic net for the lower hemisphere projection in Figure 1.9c.

The other feature of a stereographic net is the family of *small circles* centered about the north and south points of the equatorial section. These represent the projections onto the net of a set of conical surfaces. Hence the small circle highlighted in Figure 1.9c represents two cones of semi-angle 40°, each with its axis horizontal and N–S, and with their apices passing through the center of the sphere.

Regardless of whether one is plotting the positions of crystal axes, crystal faces, or structural data in geology, the most important point to remember about stereographic projection is that all planes and lines are visualized as passing through the center of the sphere; they are com-

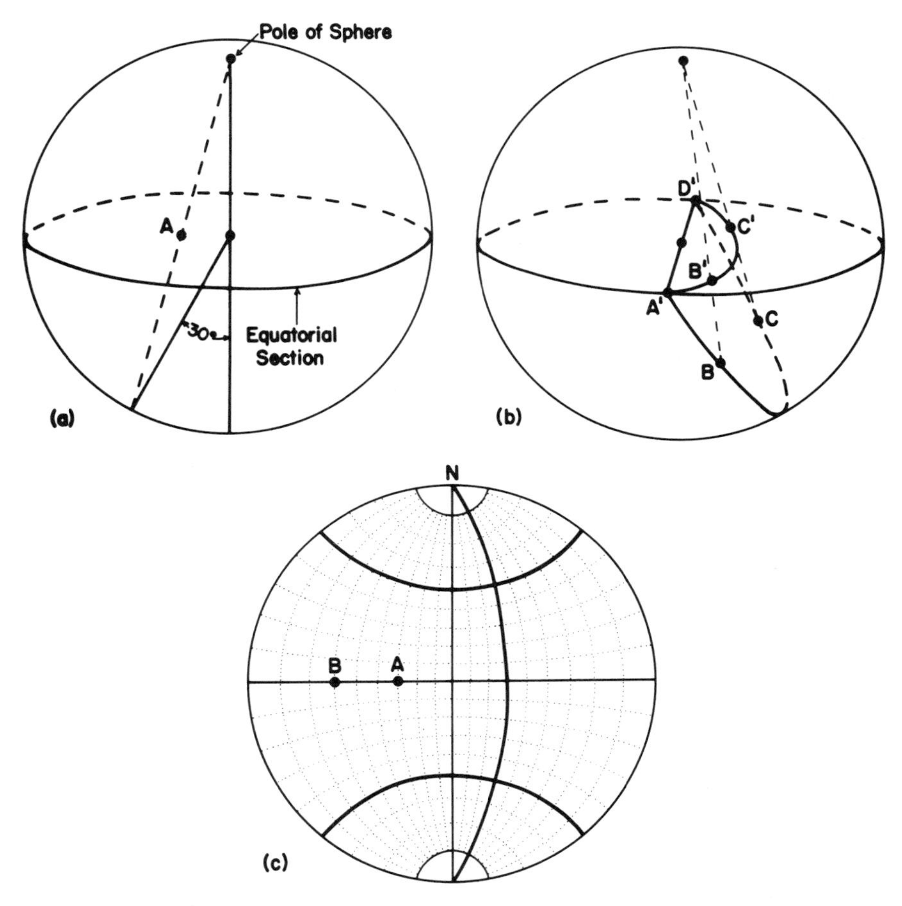

Figure 1.9 Principles of lower-hemisphere stereographic projection (see text for details).

pletely divorced from other spatial considerations. In this way, angular relationships and only angular relationships are represented.

A *lower hemisphere* stereographic projection is used in Figure 1.9. In fact, the same lines, planes, and conical surfaces can be projected onto the equatorial surface of a sphere from the *upper hemisphere,* and this is illustrated in Figure 1.10. In practice, crystallographers generally use upper hemisphere projections because one naturally looks at a crystal from above. Geologists naturally think of reference surfaces and lines projecting downwards into the earth and consequently use lower hemisphere projections. The universal stage (Chapter 6) is used in structural petrology to gather crystallographic data, which is then plotted with a lower hemisphere projection. A slightly distorted version of a stereographic net called the *equal area* or *Schmidt net* (Figure 1.8b) is com-

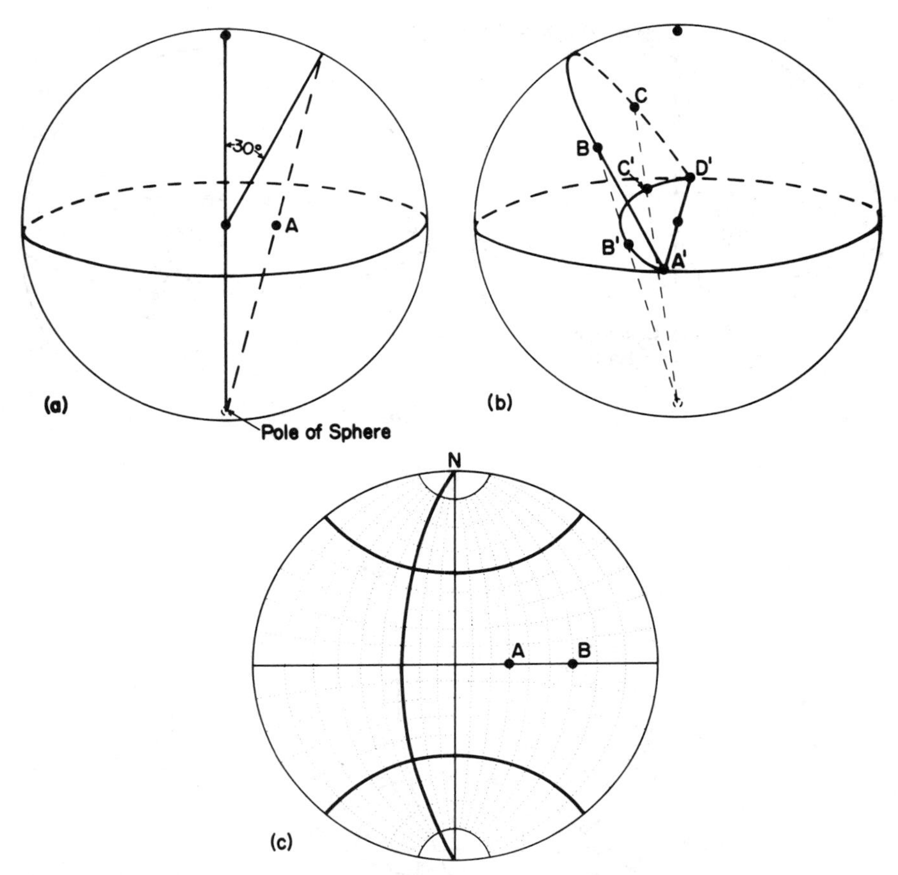

Figure 1.10 Principles of upper-hemisphere stereographic projection (see text for details).

monly used for geological applications. On this, equal areas projected from the reference sphere remain roughly equal (Turner and Weiss, 1963), though in consequence the great and small circles are no longer true circular arcs. The type of projection used must always be specified.

Examples of the Use of Stereographic Projection

I show here in principle how to plot lines, planes, and poles to planes, and how to rotate data into a different orientation. For particular conventions in plotting crystallographic and geological data, the reader is referred to standard texts of crystallography and structural geology. Examples of the application of stereographic projection to universal and spindle stage work are given in Chapter 6.

In the following account all projections are from the lower hemisphere because this is most frequently used in petrological applications. The top and bottom of the stereographic net will be labeled "north" and "south" and the left and right "west" and "east."

(a) *Lines*. Lines plunging at any angle towards N, E, S, or W can be plotted directly on the net using the graduations provided. For example, in Figure 1.11a, a horizontal E–W line plots at *A* on either side of the net, a line plunging 30° towards the W plots at *B*, and a line plunging 50°N plots at *C*.

The operation is more involved for lines of other orientation, and it is necessary to use a sheet of tracing paper placed over the net, pinned to the center (Figure 1.11b). A reference mark is placed on the tracing paper over the perimeter of the net at N. For a line plunging 40°SW, trace a line

Figure 1.11 How to plot lines in stereographic projection. See text for details.

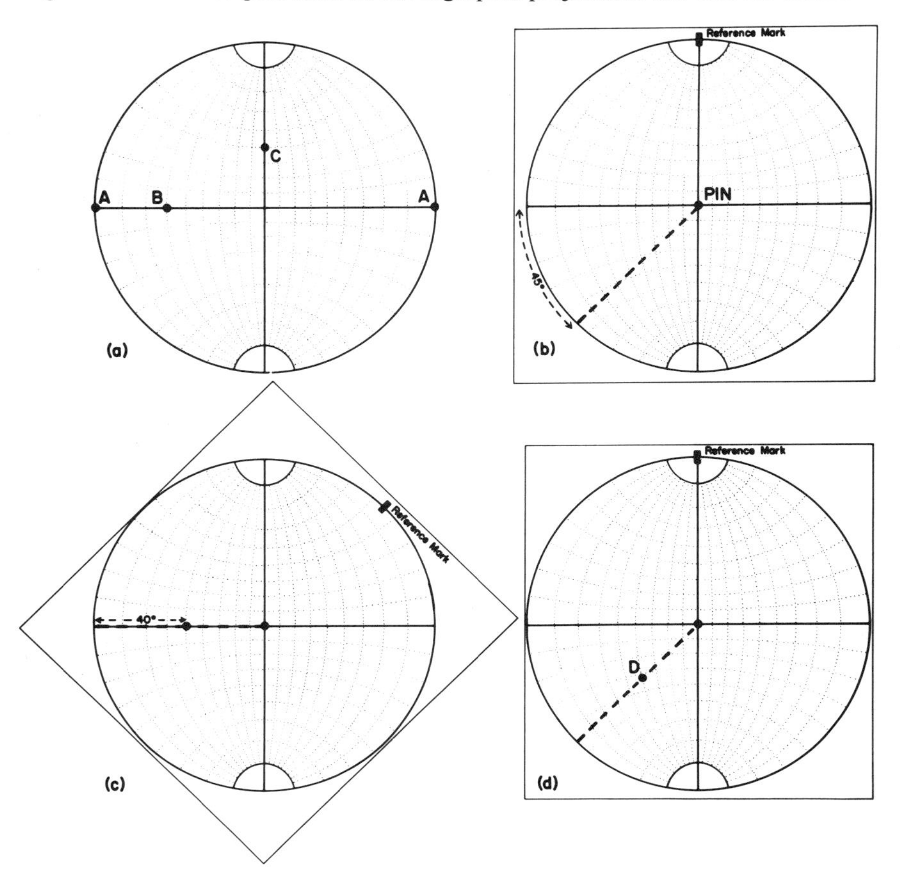

from the center of the net exactly towards the SW using the graduations at the perimeter of the net as a guide (Figure 1.11b). Next, rotate the tracing paper so that this line is in the E–W position, and measure the 40° angle of plunge in from the perimeter (Figure 1.11c). Finally, return the tracing paper to its original position, which now gives the desired projection of the plunging line as point *D* (Figure 1.11d).

(b) *Planes*. Planes dipping W or E can be plotted directly using the great circles. Thus in Figure 1.12a, a plane dipping 60°W projects as great circle *A*, a plane dipping 10°E as great circle *B*. A horizontal plane plots as the perimeter of the net.

Tracing paper placed over the net is required for other orientations. For a plane dipping 80°SW, first trace a line from the center towards the SW

Figure 1.12 How to plot planes and poles to planes in stereographic projection. See text for details.

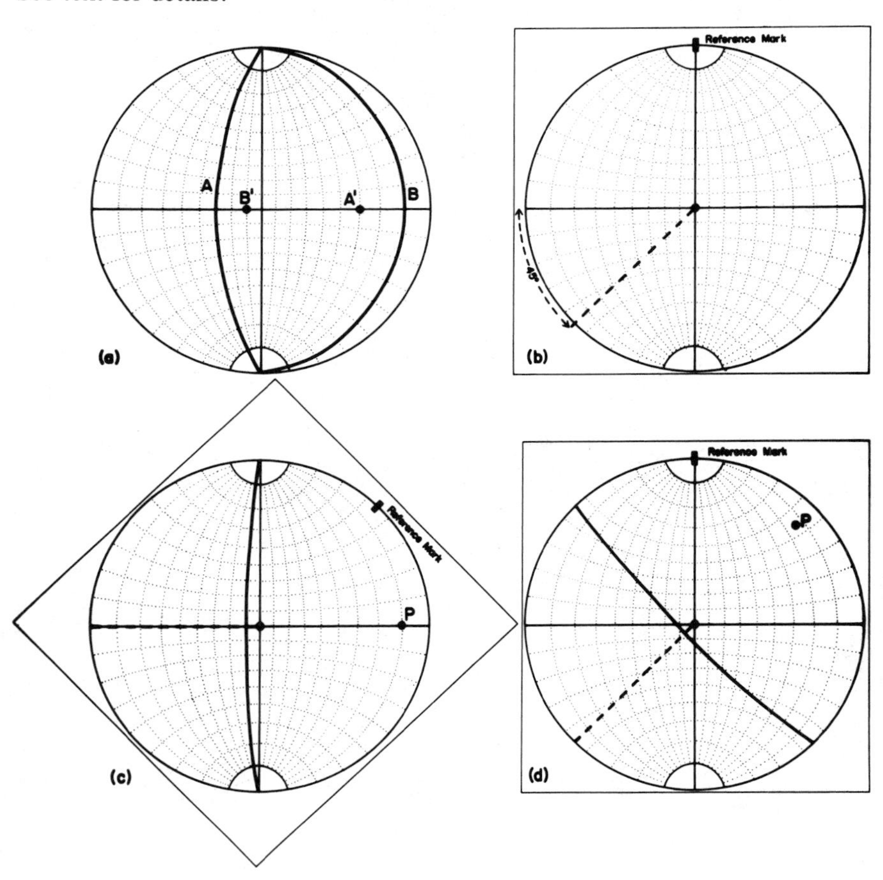

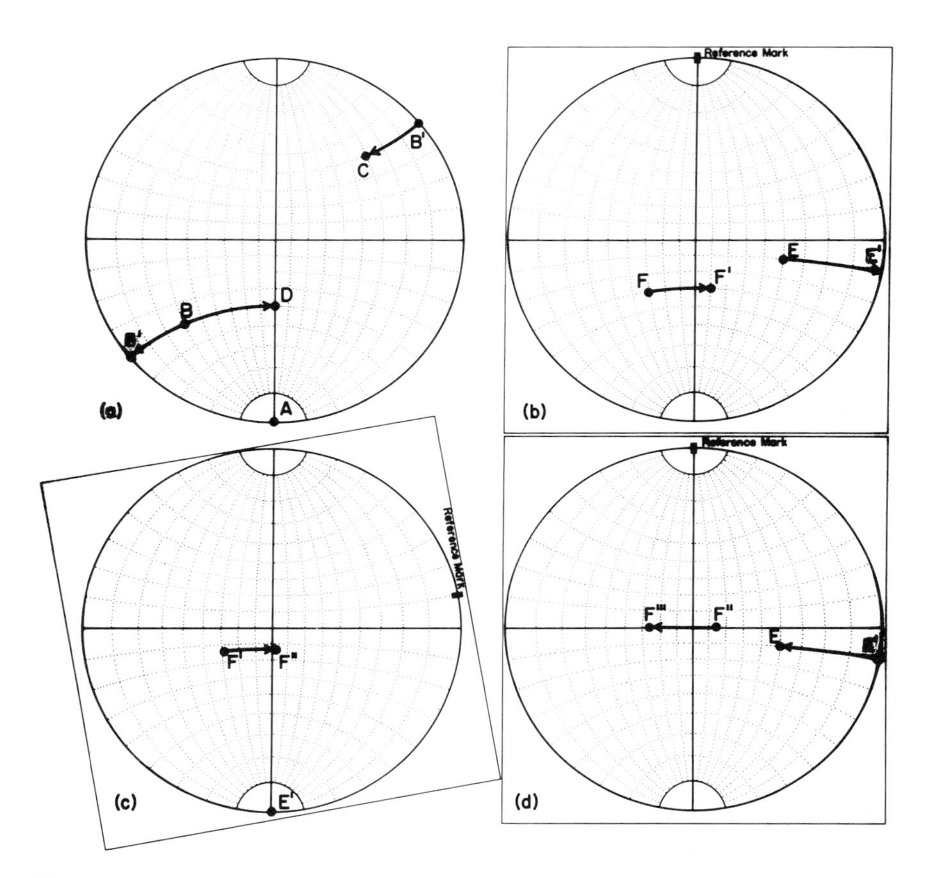

Figure 1.13 How to rotate data in stereographic projection into a new orientation. See text for details.

using the graduations at the perimeter as a guide (Figure 1.12b). Rotate the tracing paper until this line is E–W, and trace on the 80° great circle (Figure 1.12c). Return the tracing paper to its original position, which now gives the desired projection of the plane (Figure 1.12d).

(c) *Poles to planes.* When plotting crystal faces or geological data such as bedding, it is often most convenient to plot the unique pole (direction perpendicular) to the plane rather than the plane itself. This is easily achieved by counting 90° along the E–W diameter of the net. For example, Figure 1.12a shows the positions of the poles A' and B' to the great circle projections of planes A and B, respectively. The positions of poles to planes, which do not dip directly E or W, are found using tracing paper as shown in Figure 1.12c and d. Thus the position of the pole P to the plane dipping 80°SW can be found with the tracing paper as in Figure 1.12c by counting 90° along the E–W diameter from the great circle.

(d) *Rotation of stereographic plots*. Often it is necessary to rotate crystallographic or geological data into a new orientation, and this is done using the small circles of the net. For example, to rotate *B* in Figure 1.13a about the axis *A* by 60°, simply move the pole *B* 60° along the small circle using the graduations along the circle as a guide; rotation in the one direction takes *B* to *D*, but rotation in the other takes it through 30° to *B'* whence it continues on the opposite side of the net a further 30° to *C*.

Any other rotation axis, which plots on the perimeter of the net but not in a N or S position, must first be put into one of those positions (together with the poles to be rotated) using an overlay of tracing paper; the procedure described for Figure 1.13a is then followed after which the tracing paper is returned to its original position.

In the general situation the rotation axis will not lie along the perimeter, and it is then necessary to perform a subsidiary rotation first. In Figure 1.13b we wish to rotate pole *F* 30° about axis *E*. First rotate both *E* and *F* along small circles by the same amount until *E* has reached the perimeter at *E'*. Rotate the tracing paper so that *E'* is at the S of the net, and rotate *F'* 30° in the desired direction along a small circle to *F''*. Return the tracing paper to its original position, and reverse the initial rotation so that *E'* returns to *E*. *F''* is rotated by the same number of degrees to *F'''*, which represents the solution to the problem.

The easiest way to rotate a great circle projection of a plane is to represent that plane by its pole, which after rotation is transferred once more into great circle projection.

Chapter 2

The Polarizing Microscope

The polarizing microscope (also known as the petrological or petrographic microscope) is the principal piece of equipment used by the geologist to observe the optical properties of minerals. There are numerous microscopes available on the market today and they vary considerably in their construction details. This applies particularly to the illumination and substage condenser systems, some of which are designed for routine, uncritical work, whereas others require or allow detailed adjustments to be made. The student will have to learn how to adjust these systems correctly for the particular microscope he is provided with, and he should read the appropriate microscope manual.

The purpose of this chapter is to familiarize the student with the terminology and function of the various parts of the microscope. A Zeiss microscope (1970—RP 48 model) and a Swift microscope (1961 model) are used for illustration (Figures 2.1 and 2.2). The detailed manipulation of the microscope for observing particular optical properties is described in the appropriate parts of the succeeding chapters.

The Rotating Stage

All petrological microscopes are fitted with a rotating stage which is graduated in degrees. A vernier scale is usually fitted adjacent to the stage so that tenths of a degree of rotation can be measured. A clamp enables the stage to be fixed in any position.

The Polarizer and Analyzer

In modern microscopes, sheets of polaroid are used to produce polarized light (q.v.). One sheet, called the polarizer, is placed below the stage, and on the Zeiss and Swift microscopes allows light to pass through it vibrat-

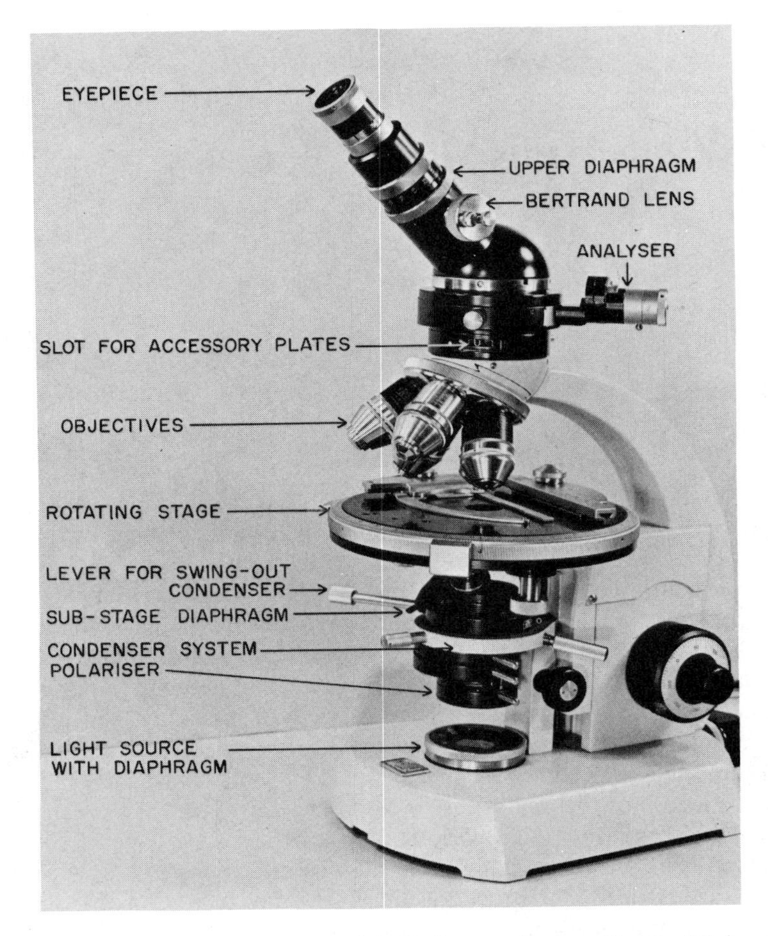

Figure 2.1 Polarizing microscope model RP 48, manufactured by Carl Zeiss, Federal Republic of Germany (1970).

ing E–W (parallel to the E–W cross-hair of the eyepiece). Some other microscopes (e.g., older Leitz models) have the polarizer oriented with a N–S vibration direction, although the E–W orientation is now becoming standard. The polarizer is normally left fixed in position, but on most microscopes it can be removed or rotated if necessary.

The analyzer is a second sheet of polaroid placed between the objective and the eyepiece. It allows light to vibrate in a direction (N–S on the Zeiss and Swift) at right angles to the polarizer vibration direction. It is designed to be either swung in or out of the beam of light, depending on the particular optical properties to be observed. On some more sophisticated microscopes (e.g., the Zeiss) the analyzer can also be rotated.

The student should periodically check: (1) that the polarizer and ana-

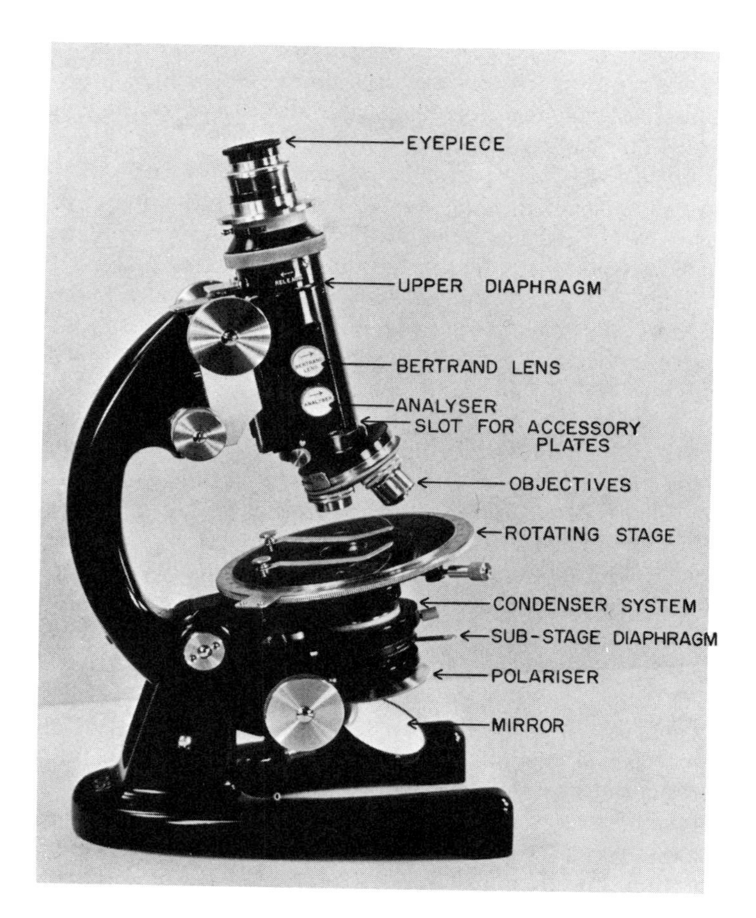

Figure 2.2 Polarizing microscope manufactured by James Swift and Sons Ltd., England (1961).

lyzer vibration directions are at right angles to each other, and (2) that their vibration directions are oriented parallel to the cross-hairs of the eyepiece. This is easily done as follows: (1) no light should reach the eye when both polarizer and analyzer are placed in the beam of light (with no mineral specimen on the stage); and (2) prismatic grains of a mineral with straight extinction (q.v.), such as sillimanite, should extinguish when the prismatic edges are parallel to the cross-hairs.

The Objectives

These are the lenses used for magnifying the specimen on the stage. Three, four, or five are normally supplied as standard equipment and include objectives of low power (×2–×4) for viewing large fields of view, of medium power (×10) for general optical work, and of high power (×40–×50) for observing interference figures and very small grains. On most modern microscopes, the objectives are fitted to a rotating nosepiece which allows quick and easy change of magnification.

The *depth of focus* of a particular objective refers to the vertical distance that can be seen simultaneously in focus with the objective. It may be increased by closing the substage diaphragm. The *working distance* is the distance between the objective lens and the specimen, when the specimen is in focus. The *numerical aperture* (NA) is a measure of the cone of light that enters the objective lens when in focus. NA is equal to $n \sin \mu$, where n is the refractive index of the medium between the specimen and the objective ($n = 1$ for air), and μ is the half-angle of the cone of light.

There are several types of objective designed to overcome various optical aberrations. Chromatic aberration is caused by the separate resolution of the colors of the spectrum resulting in color fringes. Spherical aberration results in the image being focused in different planes for light passing through the margins and center of the lens. *Achromatic* objectives correct for the chromatic aberration. *Planachromats* also correct for the spherical aberration, and are preferable, especially for low magnifications.

The Eyepiece (or Ocular)

The eyepiece is the lens fitted to the top of the microscope tube, and it magnifies and focuses the image produced by the objective lens. The eyepiece assembly contains two cross-hairs, and is slotted into the microscope tube so that the cross-hairs are oriented E–W and N–S, i.e. parallel to the vibration directions of the polarizer and analyzer. The lens of the eyepiece can be raised or lowered in its mount so that the cross-hairs are focused.

Most eyepieces have a magnification of ×8 or ×10. The magnification produced by both eyepiece and objective is obtained by multiplying the two separate magnifications. The student should note the sizes of the fields of view for various combinations of eyepiece and objective. These can be measured using a stage micrometer.

Condenser System

The condenser consists of a number of lenses positioned below the stage, which can be adjusted to vary the beam of light impinging on the specimen. The condenser systems incorporated into different microscopes vary considerably, and the student should refer to the appropriate microscope manual for details.

For ordinary viewing, the focal plane of the condenser system should coincide with the focal plane of the objective. This is achieved by raising or lowering the condenser until the maximum resolution is obtained.

On some microscopes (including the Zeiss model of Figure 2.1), there is an additional swing-out condenser, used only when observing interference figures. It should be used with the ordinary condenser system fully raised.

Substage Diaphragm

An iris diaphragm is incorporated in most condenser systems and should be adjusted for ordinary viewing as detailed in the following section on illumination. For observing relief and Becke lines, it is usually necessary to close this diaphragm partially.

Illumination System and Mirror

Illumination may be by daylight, but normally some form of electric lamp is used. If the light source is an ordinary tungsten-filament bulb, a blue filter should be used to reduce the yellow color of the light. A ground-glass plate may be necessary to diffuse the light from high-intensity bulbs and prevent an image of the light filament being seen. Some microscopes such as the Zeiss have built-in illuminators, but others such as the Swift use a mirror to direct light along the microscope tube.

Most microscopes require some adjustments to be made to the illumination system depending on the type of observation to be made. If a mirror is used, the first adjustment should be to ensure that it is tilted so as to reflect light directly along the microscope tube (many students forget to do this!). The mirror has on one side a plane surface which is normally used; the other concave surface is sometimes used to improve the illumination. For ordinary viewing the important requirement is to adjust the illuminator and substage diaphragm until an evenly illuminated field of view is obtained.

With more expensive microscopes, *critical illumination* (or the best compromise of contrast and resolution) can be achieved in the following way: partially close the diaphragms of the light source and substage system; move the substage system up or down until the diaphragm of the light source is seen in focus; open that diaphragm until its image just fills the field of view; insert the Bertrand lens and view the image of the substage diaphragm; when closed, its image should be in the center of the field of view—center the substage system if not; then open the diaphragm until its image fills the field of view. These steps are repeated for each objective, but it is not usually possible to fulfill the last step with the higher magnification objectives. With many student microscopes, some of the adjustments mentioned above are not possible. In any case, less than critical illumination is often preferred in routine work because the substage system has to be fully raised to obtain interference figures, and many workers leave it in this position for convenience. Further comment on regulation of the conditions of illumination may be found in Hartshorne and Stuart (1970).

The Bertrand Lens

The Bertrand lens magnifies and focuses interference figures and is swung into position in the microscope tube when so required. An alternative means of viewing interference figures is to remove the eyepiece and look down the microscope tube at the high-power objective lens, preferably with the aid of a pinhole stop inserted at the top of the tube.

Upper Diaphragm

Both the Swift and Zeiss microscopes have an iris diaphragm in the upper part of the microscope tube. It is used in conjunction with the Bertrand lens, and should be partially closed if interference figures from very small grains are to be observed. Many other microscopes incorporate a fixed diaphragm in the Bertrand lens, and adjustment is not possible.

Accessory Plates

Accessory plates consist of mineral sections of a thickness such that they produce a known amount of retardation (q.v.). They are used for studying interference figures and the retardation produced by mineral specimens. When required, they are inserted into the microscope tube in a slot between the objectives and the analyzer.

Centering the Microscope

When the stage is rotated, the axis of rotation should coincide with the center of the field of view. This is achieved on the Zeiss microscope by adjusting a collar on the barrel of each objective. The Swift is adjusted by means of two screws set into the microscope tube just above the objectives. Some other microscopes have centering screws fitted to the stage. Unless each objective is centered independently, as with the Zeiss, it is important to ensure that the centering is most precise for the highest-power objective normally used. To center a microscope, the point about which an object is seen to rotate when the stage is rotated must be brought to the center of the cross-hairs by adjusting the centering screws. If the microscope is badly out of center, center first for low power, then more precisely for high power.

The substage system can be centered using a low-power objective and by closing the lower diaphragm. The small area of light visible should be exactly centered at the intersection of the cross-hairs. If not, adjust the centering screws of the substage system.

The Bertrand lens can also be adjusted on more expensive microscopes. To center it, use a high-power objective in plane-polarized light and focus at first on a dark object in a thin section. Then insert the Bertrand lens and increase the distance between the stage and objective until a telescopic view of the dark object is clearly seen. Move the object until it is at the intersection of the cross-hairs, then rotate the stage. If the object moves away from the center of the field of view, make the appropriate adjustment to the centering screws of the Bertrand lens.

Care of the Microscope

The microscope should be protected from dust as much as possible. The lenses should be cleaned with special lens tissue, or with a soft brush, preferably one attached to a small bellows for blowing away coarse material. If a liquid cleaner is necessary, use xylol or benzene but not alcohol which dissolves the cements in some objectives. Liquid cleaners should be used sparingly and in a well-ventilated room to reduce any possible health risk.

The microscope should always be carried by holding the main arm support, not by holding the microscope tube.

When focusing on specimens using high power, the working distance is so small that special care must be exercised. The objective should be lowered until it almost touches the specimen (observed from the side), and then slowly raised until brought into focus. When viewing with the high-power objective use only the fine-focusing adjustment.

Chapter 3

Principles of Optical Mineralogy

> To ask whether a given representation (of polarised light) is "true" is futile. It must suffice that the representation assists ready recollection of the behavior and permits any solution of the various problems encountered.*

Most texts integrate the discussion of optical theory with a description of laboratory techniques. For easier laboratory reference, this book separates as far as possible the background theory from its practical application. Thus, the present chapter discusses in principle the nature of the four most important optical properties of minerals: viz., refraction and RI, birefringence, the optical indicatrix, and color. The succeeding chapters describe how these properties are measured in practice, and how they may be interrelated with such properties as cleavage and crystal shape.

All books on optical mineralogy simplify optical theory to some extent, and this book is no exception. This chapter provides what, in my experience, is just enough theory for successful laboratory work in identifying minerals.

Polarized Light

The optical properties of minerals are most simply explained in terms of the electromagnetic theory of light. According to this theory, light may be discussed in terms of wavelengths (Figure 3.1a) and vibrations perpendicular to the direction of propagation. White light comprises a gradational series of wavelengths from 390 nm (violet) to 770 nm (red), and may be separated into its component colors by the well-known triangular glass prism. The colors of the visible spectrum are familiar as the colors of the

---

* From W. A. Shurcliff, *Polarised Light* (Cambridge, Mass., 1962), page 1, by permission of Harvard University Press.

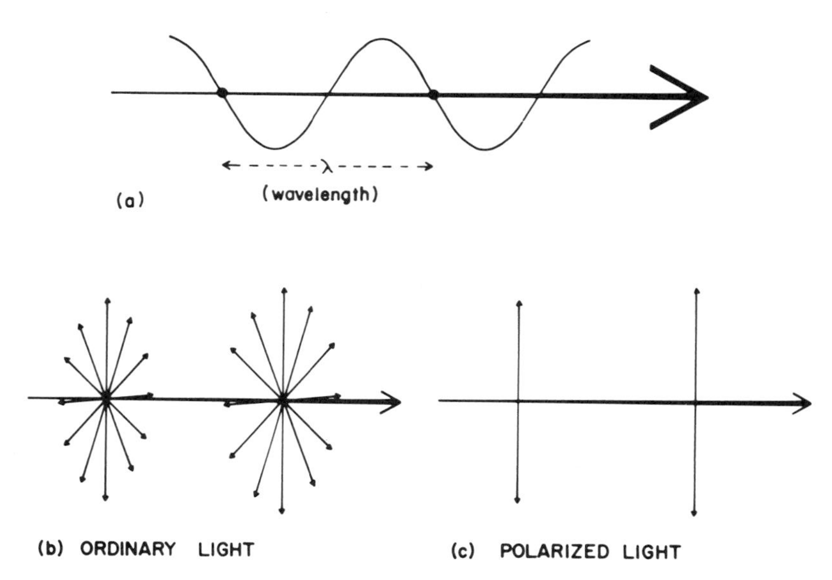

Figure 3.1 Concepts of: **(a)** wavelengths of light, **(b)** vibration directions of ordinary light, and **(c)** the single vibration direction of polarized light.

rainbow, and they are listed in Figure 3.2. If light has a very restricted range of wavelength, it is termed monochromatic; a familiar example is yellow sodium light with a wavelength close to 590 nm.

Ordinary light is considered to be vibrating in *all* directions perpendicular to the propagation direction (Figure 3.1b). *Plane-polarized light* vibrates in only *one* direction (Figure 3.1c), and is produced by the polarizer and analyzer, both of which in modern microscopes consist of a sheet of plastic (polaroid) which absorbs all light except that vibrating in one direction. Natural tourmaline crystals also strongly absorb light vibrating in all but one direction, and may be used as simple polarizers. Older microscopes employed an ingenious combination of calcite prisms to produce polarized light (described first by William Nicol in 1829, and known as *Nicol prisms*).

Figure 3.2 Colors of the visible spectrum.

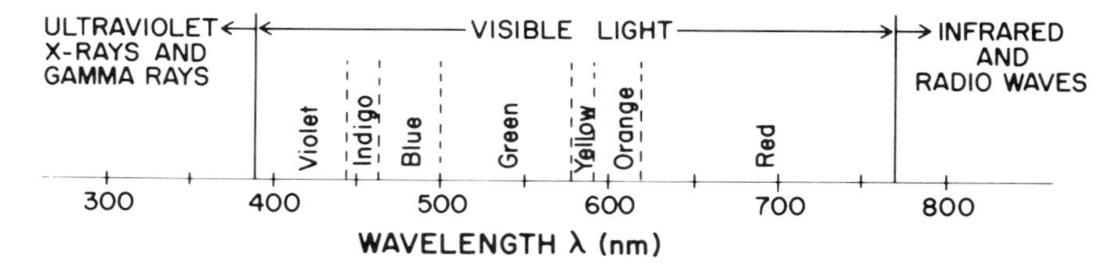

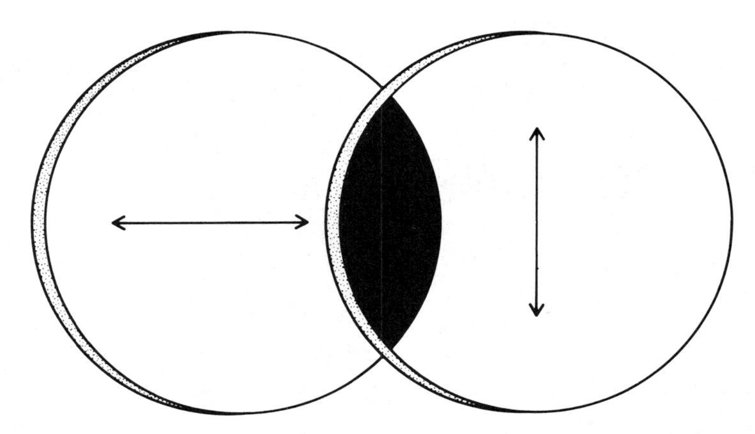

Figure 3.3 Two sheets of polaroid with mutually perpendicular vibration directions absorb all transmitted light when superimposed.

Observations with the petrographic microscope may be:

1. *In ordinary light*. Such a system is seldom used.
2. *In plane-polarized light*. Light is polarized by the polarizer (below the stage) before passing through the specimen on the stage.
3. *In crossed-polarized light*. In addition to the substage polarizer, the analyzer above the microscope stage is inserted. Light is then polarized after, as well as before, passing through the specimen. The two sheets of polaroid are oriented so that they transmit light vibrating in directions at right angles to each other (E–W and N–S). The effect of superimposing one upon the other is to absorb all light passing through the microscope (Figure 3.3). However, many minerals, when placed between the two sheets of polaroid so oriented, affect the light so that it once more reaches the eye.

Isotropic and Anisotropic Minerals

The majority of minerals polarize the light that passes through them. In general, two rays of light are produced and their vibration directions are both at right angles to the direction of propagation of the light and at right angles to each other when the rays emerge from a mineral section (however, within the crystal itself, the vibration directions are not always exactly perpendicular to the light rays, nor do the two rays follow precisely parallel paths). Minerals that affect light in such a way are termed *anisotropic*. All crystals except those that belong to the cubic system are anisotropic (though a few anisotropic minerals may in practice appear to be isotropic, e.g., perovskite).

Cubic minerals do not polarize light passing through them, and they do not vary directionally in their effect on light. Cubic minerals are therefore termed *isotropic*. Glass and amorphous substances are also isotropic.

Isotropism directly reflects the high degree of regularity in the atomic structure of cubic minerals. The specific details of anisotropic optical properties also reflect the particular symmetry of the crystals, as will be discussed later.

Refractive Index and the Velocity of Light

A ray of light is usually bent when passing from one substance to another (Figure 3.4), a phenomenon known as refraction. The RI of a substance is given by the ratio of the sines of the angles of incidence and refraction of light passing from air into the substance (the RI of air is taken to be 1). In more general terms, if light is incident from a substance which has a refractive index other than 1, the angles of refraction and incidence are related by Snell's law:

$$\frac{\text{RI r}}{\text{RI i}} = \frac{\sin \text{i}}{\sin \text{r}}.$$

RI r and RI i are the refractive indices of the media through which the refracted and incident rays travel. Snell's law requires modification for some rays passing through anisotropic crystals, but the details need not concern us here.

Light travels most quickly in a vacuum and is slowed when passing through any substance. The denser the substance, the slower the light is

Figure 3.4 The phenomenon of refraction.

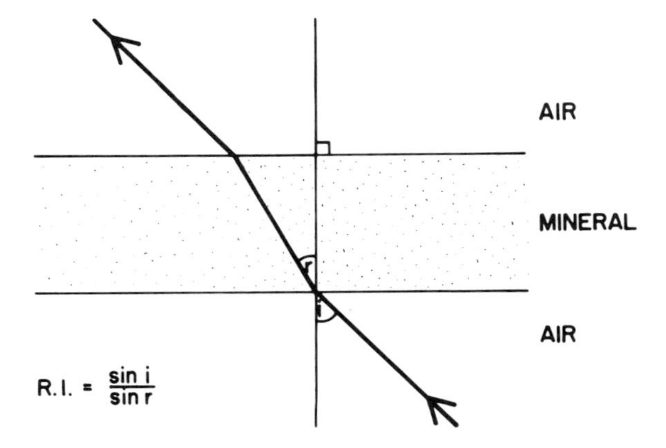

in traveling through it. It can be shown simply that the RI of a crystal is equal to the ratio of the velocities of light in air and in the crystal:

$$\mathrm{RI} = \frac{\text{Velocity of light in air}}{\text{Velocity of light in crystal}}.$$

In other words, the higher the refractive index of a crystal, the slower the light is in passing through it; in general, the density is higher too.

The RI of common minerals ranges from 1.43 to 3.22, and methods for its determination, either qualitatively or quantitatively, are described in Chapters 4, 5, and 6. In practice we will also learn to determine the speed of various light rays as relatively fast or relatively slow.

Isotropic minerals can be given a single value of refractive index. Anisotropic minerals, however, display the phenomenon of double refraction.

Double Refraction and Birefringence

If a clear rhomb of calcite is placed over a small illuminated hole in a metal plate, two images of the hole will be seen (Figure 3.5). The two images represent the two separately polarized rays of light produced by the anisotropic calcite. The vibration directions of the two rays are perpendicular to each other, and this can be demonstrated by rotating a sheet of polaroid above the rhomb. Every 90° of rotation, one of the images is successively extinguished as its associated vibration direction becomes perpendicular to the polaroid vibration direction (cf. Fig. 3.3). The fact

Figure 3.5 The two separately polarized and refracted light rays produced by a calcite prism.

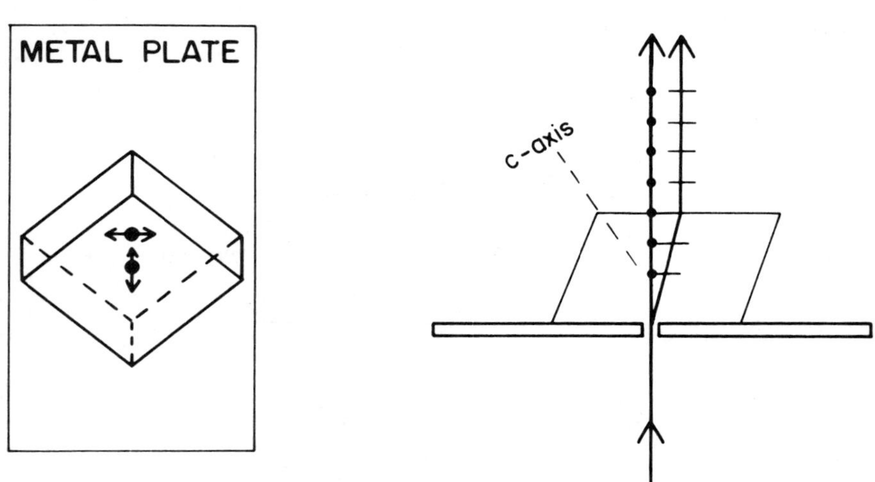

that two images are seen demonstrates that each ray represents a different refractive index for the crystal (and also a different velocity of light). This phenomenon is known as *double refraction,* and is characteristic of all anisotropic minerals, although the amount of double refraction is unusually large in calcite. The refractive index of all anisotropic minerals varies continuously depending on the vibration direction of the light within the crystal, and the difference between the largest and smallest possible values is called the *birefringence* (δ). Less than the maximum possible difference is called the *partial birefringence* (δ').

Isotropic Minerals in Crossed-Polarized Light

Light is not polarized by cubic minerals. Therefore when light travels through such a mineral placed on the microscope stage, the vibration direction of light produced by the polarizer is not changed (Figure 3.6), regardless of how we rotate the stage. The light that reaches the analyzer is thus polarized at right angles to the vibration direction of the analyzer (Figure 3.6), and is completely absorbed. In other words, all cubic minerals (and glass) will appear black whenever observed in crossed-polarized light. Only special sections of anisotropic minerals have this property.

Anisotropic Minerals in Crossed-Polarized Light

Anisotropic minerals generally polarize light into two rays vibrating at right angles to each other. In the general situation, where these vibration directions are not parallel to those of the polarizer and analyzer, light from the polarizer is resolved into two rays by the mineral (Figure 3.7). These two rays are resolved by the analyzer so that the light reaching the eye is only vibrating in the one direction allowed by the analyzer.

If the stage is rotated through 360°, there are four positions in which the two vibration directions of the mineral are parallel to those of the polarizer and analyzer (Figure 3.8). In these positions, light from the polarizer cannot be resolved into two rays and passes through the mineral un-

Figure 3.6 Behavior of isotropic minerals in crossed-polarized light.

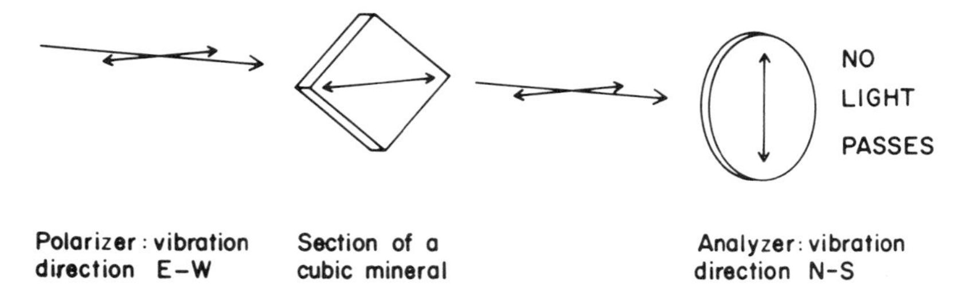

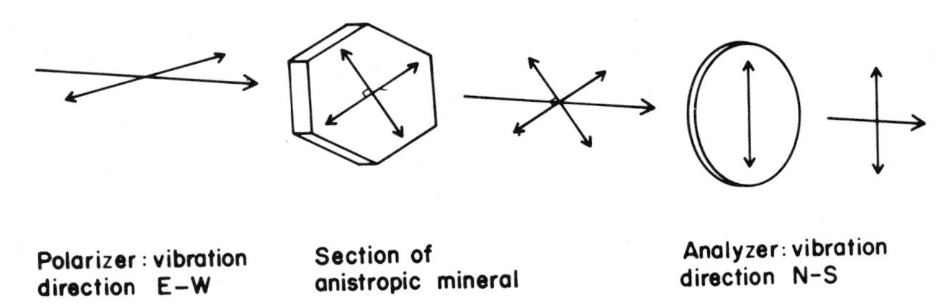

Figure 3.7 Anisotropic mineral section in crossed-polarized light with its vibration directions not parallel to those of the polarizer and analyzer.

changed in vibration direction. This light is vibrating at right angles to the direction allowed by the analyzer, which completely absorbs it. In general, therefore, all anisotropic minerals will go black (or will *extinguish*) in these four positions when the stage is rotated through 360°.

As we shall see later, there are special sections of anisotropic minerals that remain black or very dark when the stage is rotated, and it is then necessary to make further tests to distinguish them from isotropic mineral sections.

Retardation and Interference Colors

Within an anisotropic mineral the two polarized rays behave independently, and for much of our work it is normal practice to visualize the situation as illustrated in Figures 3.7 and 3.8. We will learn later to determine the orientations of the vibration directions and the related RIs and speeds of light. However, when white light is transmitted through an anisotropic mineral section between crossed polars, we observe an *interference color,* not white light, and this results from the interference of the two light rays when they pass through the analyzer. To understand interference colors, it is necessary to consider in some detail what happens to the various components of white light when traveling through a crystal.

Figure 3.8 Anisotropic mineral section in crossed-polarized light with its vibration directions parallel to those of the polarizer and analyzer.

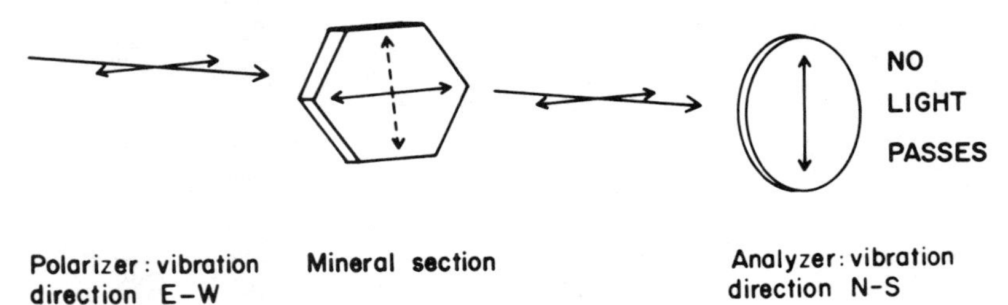

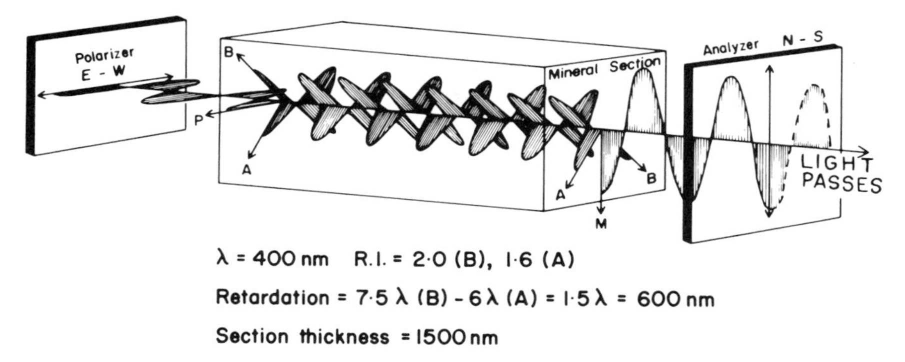

Figure 3.9 An example showing how polarized monochromatic light may be transmitted by the analyzer after passing through an anisotropic mineral.

Figure 3.9 illustrates a mineral section 1500 nm thick, which polarizes light into two rays A and B representing two refractive indices 1.6 and 2.0, respectively. We shall examine what happens to the violet part of the spectrum ($\lambda = 400$ nm) as it travels through this section. Not only does the speed of the light change within the crystal (ray B being slowed more than ray A) but the wavelength is shortened too. The relationship between wavelength and RI is a simple one:

$$\frac{\lambda(A)}{\lambda(B)} = \frac{\text{RI }(B)}{\text{RI }(A)}.$$

In other words, the higher the RI, the shorter the wavelength.

Applying this to Figure 3.9 we find that ray A completes 6.0 wavelengths ($\lambda = 250$ nm) and ray B 7.5 wavelengths ($\lambda = 200$ nm) within the crystal.* Because ray B travels more slowly through the crystal it emerges after ray A, though with a continuous stream of light we can imagine rays like A and B emerging together. The difference in number of wavelengths required by each ray to pass through the crystal is called the path difference or *retardation* which in this case of violet light is 1.5λ. On emergence into air the rays may be considered to interact (by vectorial addition) to produce plane-polarized light (M) vibrating parallel to the analyzer vibration direction through which the light can be transmitted without any loss in intensity. Because the two rays do not interact until emerging from the crystal, the retardation is measured in terms of the wavelength of light in air. In this case the retardation is 1.5×400 nm (= 600 nm).

---

* For a RI of 1 (air) $\lambda = 400$ nm. Therefore for a RI of 1.6, $\lambda/400 = 1/1.6$ or $\lambda = 250$ nm. Similarly, one can see that for a RI of 2.0, $\lambda/400 = 1/2.0$ or $\lambda = 200$ nm, or alternatively, that $\lambda/250 = 1.6/2.0$ or $\lambda = 200$ nm.

The nature of the interaction differs for different wavelengths. Consider the situation illustrated in Figure 3.10 for the orange part of the spectrum ($\lambda = 600$ nm). The crystal chosen is the same, and again has a thickness of 1500 nm; it resolves light to vibrate in two directions *A* and *B* related to two refractive indices 1.6 and 2.0, respectively. Using the formula given above we find that ray *A* completes 4.0 wavelengths and ray *B* 5.0 wavelengths within the mineral section. The retardation is 1.0λ. On emergence into the air the rays add vectorially to produce plane-polarized light, but now the vibration direction (*M*) is parallel to the polarizer vibration direction. In that this is 90° from the analyzer vibration direction, the light is totally extinguished.

In general, we can state that any light with a wavelength that results in retardations of 1.0λ, 2.0λ, 3.0λ, etc. will be totally extinguished by the analyzer. On the other hand, light with a wavelength that results in retardations of 0.5λ, 1.5λ, 2.5λ, etc. will be fully transmitted by the analyzer.

In the special case when both of the crystal's vibration directions are 45° from the polarizer vibration direction, and for retardations of 0.25λ, 0.75λ, 1.25λ, etc., the emerging light rays combine to produce a ray that vibrates successively with equal intensity in all directions perpendicular to the travel direction, a phenomenon termed *circular polarization*. Such light will be resolved by the analyzer with a 50% loss in intensity. In general, the emerging light rays combine to produce a ray vibrating successively with variable intensity in various directions perpendicular to the travel direction, a phenomenon termed *elliptical polarization*. The intensity of light resolved and transmitted by the analyzer will vary from close to 100% for retardations close to 0.5λ, 1.5λ, 2.5λ, etc. to almost zero for retardations close to 1.0λ, 2.0λ, 3.0λ, etc.

Thus an interference color transmitted through the analyzer represents white light from which some parts of the spectrum have been totally removed, other parts having been retained though with varying intensity.

Figure 3.10 An example showing how polarized monochromatic light may be extinguished by the analyzer after passing through an anisotropic mineral.

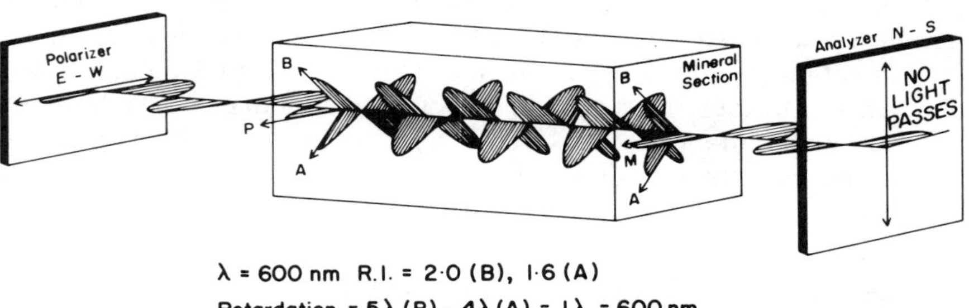

If we return to the mineral section illustrated in Figures 3.9 and 3.10 and consider what happens to white light, we can see that the orange part of the spectrum (600 nm) is totally removed whereas the violet part (400 nm) is fully transmitted. Intermediate wavelengths will be transmitted with variable intensities as described above. The resulting color observed through the analyzer is a deep blue-purple.

Some important relationships exist between retardation, birefringence, and the thickness of a mineral section. As shown in Figures 3.9 and 3.10, the retardation for a particular mineral section is constant regardless of the wavelength of light. In one case the retardation is 1.5λ, in the other 1.0λ, but in nanometers these values are the same, namely 1.5×400 or 1.0×600 or 600 nm. It is easy to see that retardation increases in simple proportion with thickness. Thus a doubling of the thickness of the mineral section would produce retardations of 3.0×400 nm or 2.0×600 nm (= 1200 nm). The retardation is similarly dependent on the difference in RI represented by the two rays A and B. The greater the difference, the greater the retardation. These interrelationships can be simply expressed as follows:

$$\text{Retardation} = \text{thickness} \times \text{birefringence (or partial birefringence).}$$

Thus for Figures 3.9 and 3.10, the retardation is 1500 nm $\times$ 0.4 (= 600 nm).

This relationship forms the basis of the *Michel–Lévy Color Chart* (at the back of the book) on which retardation is plotted against thickness, thus allowing lines of equal birefringence (or partial birefringence) to be drawn radiating outwards from the origin of the chart. The chart is used in practice to determine the birefringence of minerals in thin section; the technique is explained in the next chapter. The interference colors produced for various retardations are usually divided into *orders* at intervals of approximately 550 nm. Retardations higher than 2300 nm result in a sequence of paler and paler reds and greens. For very high retardations the interference color appears white.

Anomalous Interference Colors

Some minerals exhibit interference colors that are not represented on the Michel–Lévy chart, especially within the range of first order colors. These anomalous colors may result variously from: selective absorption of a particular wavelength by the mineral; widely different refractive indices for the mineral for different wavelengths of light; different preferred vibration directions (for some biaxial minerals) for different wavelengths of light. Such *dispersion* of refractive indices and principal vibration directions will be discussed more fully at the end of this chapter. Common anomalous interference colors are noted in the mineral descriptions of Chapter 9.

Uniaxial and Biaxial Minerals

We have shown that retardation depends in part on the double refraction within a mineral section. The amount of double refraction depends in turn on two factors: (1) the birefringence of the mineral; and (2) the orientation of the mineral section with respect to the crystal lattice. In order to understand this dependence on orientation of the section, we have to look in further detail at the variation of refractive index within anisotropic minerals.

There are two main groups of anisotropic minerals—uniaxial and biaxial. There is a close relationship between these two groups and crystallography, uniaxial minerals belonging to the hexagonal, trigonal, or tetragonal systems, biaxial minerals belonging to the orthorhombic, monoclinic, or triclinic systems.

Uniaxial Minerals

Hexagonal, trigonal, and tetragonal minerals are all characterized by two or three equal crystal axes (*a*-axes) that lie in a plane at right angles to an axis (*c*-axis) of different length (see Chapter 1). The RIs within such minerals reflect this characteristic, and are equal for light vibrating in all directions in the plane of the *a*-axes, but vary towards a maximum or minimum value for light vibrating parallel to the *c*-axis. This RI variation can be represented by what is termed the uniaxial indicatrix (Figure 3.11).

The indicatrix is an ellipsoid whose radii are directly proportional in length to the RIs of the mineral for light vibrating in those directions. It is most important to understand from the outset that the indicatrix represents the RI of the mineral for vibration directions, and not for travel

Figure 3.11 The uniaxial indicatrix.

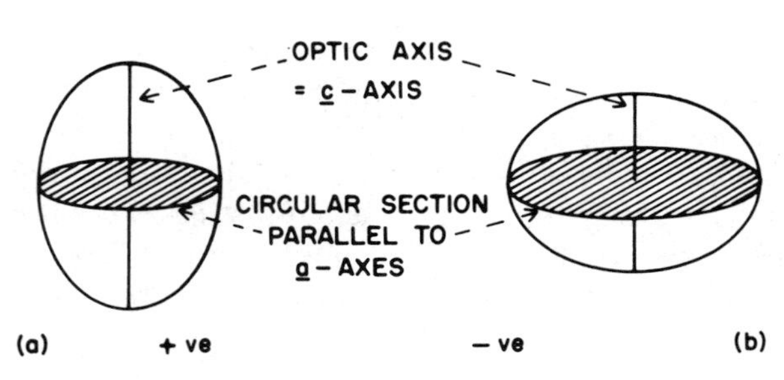

directions of light. Because all light rays that vibrate within the plane of the *a*-axes represent equal RIs, this plane is represented in the indicatrix by a circular section. In other words, the indicatrix can be described as an ellipsoid of revolution, the axis of revolution being parallel to the *c*-axis, and the ellipsoid being identical for all angles of rotation about that axis.

Some uniaxial minerals have a maximum refractive index for light vibrating parallel to the *c*-axis and these are called *uniaxial positive* (Figure 3.11a). Those with a minimum RI for light vibrating parallel to the *c*-axis are called *uniaxial negative* (Figure 3.11b).

There is an important rule governing the vibration directions of the two light rays that emerge from any section of a uniaxial mineral. We already know that the two vibration directions are at right angles to each other and at right angles to the direction of propagation (the microscope tube). It is also found that one of the vibration directions always lies in the circular cross section of the indicatrix, that is, in the plane of the *a*-axes of the crystal. The light ray with this vibration direction clearly represents a constant value of RI for any one mineral, and is called the *ordinary ray* (abbreviated ω). The other ray represents a variable RI, which reaches a maximum (uniaxial positive) or minimum (uniaxial negative) value when light vibrates parallel to the *c*-axis. It is called the *extraordinary ray* (abbreviated ε).

To illustrate these points more clearly, let us examine three types of section of uniaxial minerals which will be encountered in laboratory work. Uniaxial +ve quartz is used as an example.

(1) *With the c-axis parallel to the mineral section* (Figure 3.12). In this case light is polarized into the following two rays: one (ω) that has a vibration direction along the line where the circular section of the indicatrix intersects the plane of the mineral section; another (ε) that has a vibration direction, at right angles to ω and parallel to the mineral section, and in this special case parallel to the *c*-axis thus representing the maximum RI possible (in a +ve mineral). The double refraction in this section is the maximum possible and defines the birefringence of the mineral ($\delta = \varepsilon - \omega$).

Figure 3.12 The uniaxial indicatrix and vibration directions in quartz cut parallel to the *c*-axis.

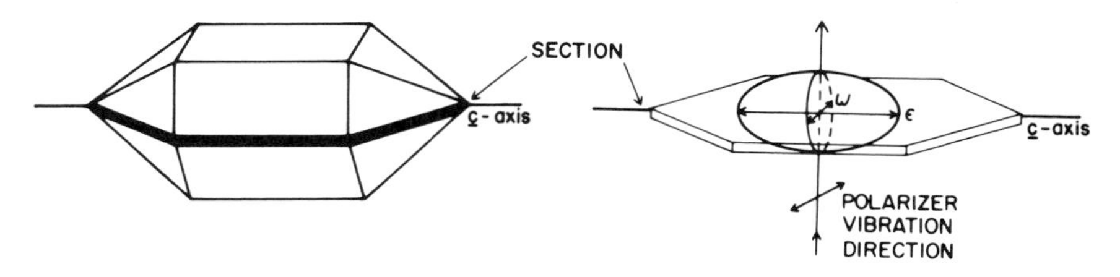

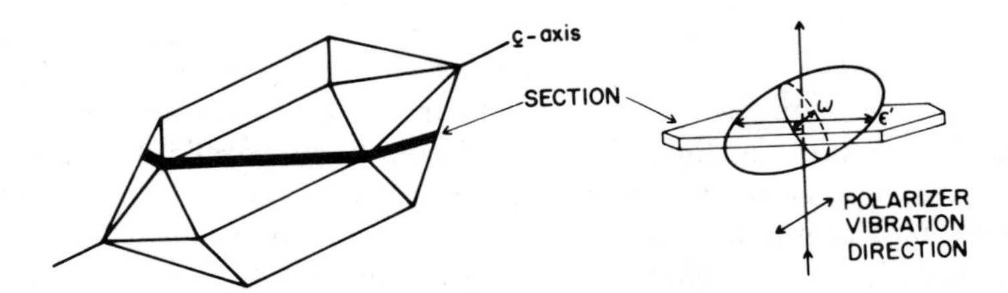

Figure 3.13 The uniaxial indicatrix and vibration directions in quartz cut oblique to the *c*-axis.

(2) *With the c-axis at some angle other than 90° to the mineral section* (Figure 3.13). In this case, light is polarized into the following two rays: one (ω) that has a vibration direction along the line where the circular section of the indicatrix intersects the plane of the mineral section; another (ε') that has a vibration direction at right angles to ω, but which also lies in the mineral section, and represents an RI value intermediate between ε and ω. Only a *partial birefringence* is displayed by this section ($\delta' = \varepsilon' - \omega$).

(3) *With the c-axis perpendicular to the mineral section* (Figure 3.14). In this special case, light can only vibrate in the circular section of the indicatrix, and all possible vibration directions represent equal refractive indices. The result is that light is not doubly refracted, and passes through the mineral with the vibration direction of the polarizer unchanged. Inserting the analyzer causes total absorption of the light, and the section will appear black in any position of rotation of the microscope stage. In other words the partial birefringence of this section is zero. The unique direction perpendicular to this section and parallel to the *c*-axis is called

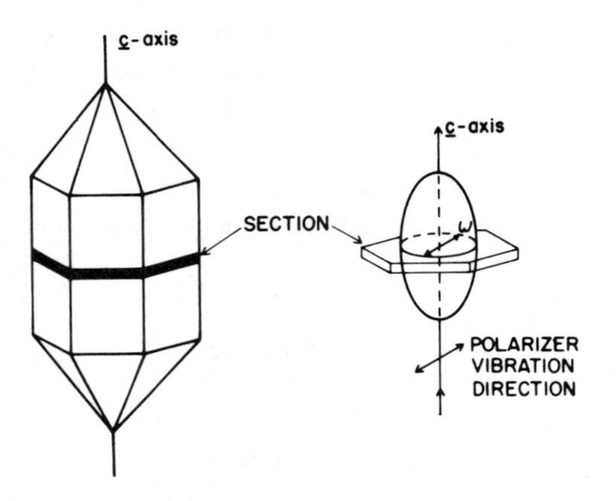

Figure 3.14 The uniaxial indicatrix and single vibration direction in quartz cut perpendicular to the *c*-axis.

the *optic axis*. The fact that there is only one such direction gives rise to the name uniaxial. A uniaxial mineral section with a zero partial birefringence can be distinguished from isotropic mineral sections by means of interference figures (see later).

One further aspect of routine terminology can be introduced at this point. Because the velocity of a ray is always inversely proportional to the RI with which it is associated, we can talk in terms of relative speeds of light rays as well as relative values of RI. Thus in a positive unaxial mineral, ω is the faster ray and ε the slower ray. In a negative mineral, ω is the slower, ε the faster ray.

Biaxial Minerals

Orthorhombic, monoclinic, and triclinic minerals are all characterized by three unequal crystal axes (Chapter 1). The RIs within such minerals reflect this characteristic and can be discussed in terms of three unequal values in three principal directions. Unlike the crystal axes, the three optical directions are always at right angles to each other.

We can represent the three optical directions by means of the biaxial indicatrix, which is an ellipsoid with three unequal principal radii (Figure 3.15). The shortest and longest radii are constructed parallel to the vibration directions of light representing the smallest and largest RIs (or the fastest and slowest rays, respectively). These vibration directions are termed X and Z, respectively. The intermediate radius at right angles to both X and Z is termed Y. As with the uniaxial indicatrix, it is essential to understand that the directions X, Y, and Z of the indicatrix are proportional in length to the refractive index of light vibrating in those directions. The light rays emerging from a mineral section actually travel at right angles to their vibration directions.

The refractive indices of the crystal when light vibrates parallel to X, Y, and Z are termed α, β, and γ, respectively; light rays passing through biaxial minerals do not necessarily vibrate in any of the directions X, Y, or Z.

There are the following relationships between crystal axes and X, Y, and Z, for the various crystal systems. *Orthorhombic minerals:* X, Y, and Z are parallel to a, b, and c, but in any order. Thus for one mineral a may be parallel to X, but for other minerals b or c may be parallel to X. *Monoclinic minerals:* the *ac*-plane contains two optical directions, but neither a nor c is necessarily parallel to either optical directions; b is parallel to the third optical direction. *Triclinic minerals:* a, b, and c are not necessarily parallel to any of the optical directions. In general, the slowest optical direction Z (representing the largest RI) tends to lie along the most strongly bonded or densely packed direction in a mineral, whereas the

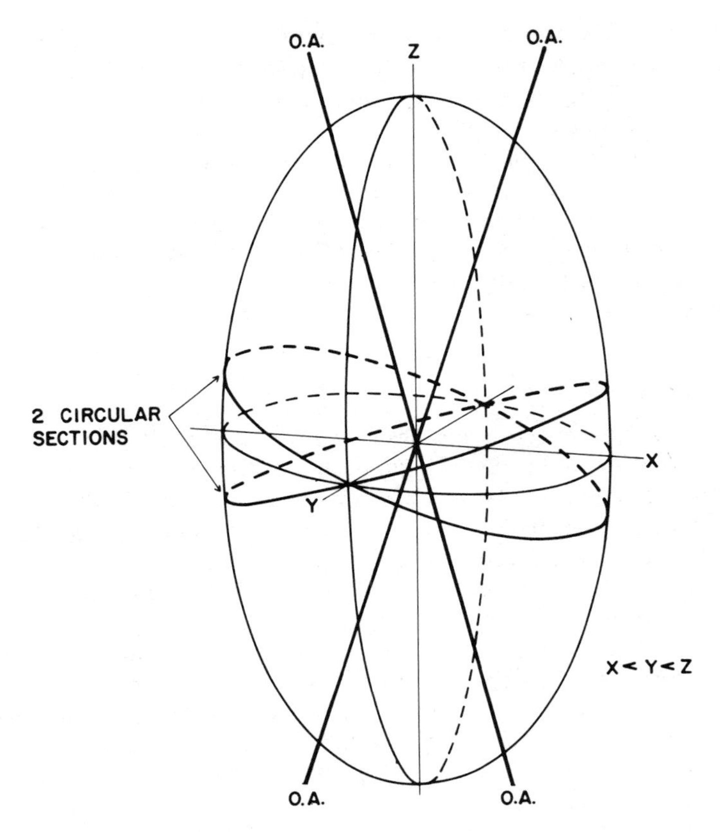

Figure 3.15 The biaxial indicatrix.

fastest direction *X* (the smallest RI) tends to lie along the most weakly bonded direction. Thus in micas, *X* is always at a high angle to the strongly bonded sheet structure.

There are always two sections of the biaxial indicatrix that are circular, and these represent planes in the crystal along which light has no preferred vibration direction. Both planes contain and intersect in *Y*, are perpendicular to the *XZ* plane, and are bisected by *X* and *Z* at an angle that depends on the particular mineral. More details of these two sections are given in (5) below.

To illustrate the general properties of biaxial minerals, let us examine the types of section likely to be encountered in laboratory work. The orthorhombic mineral olivine is used as an example.

(1) *Mineral section parallel to X and Z of the indicatrix* (Figure 3.16). In this case light is traveling parallel to *Y* and is polarized into two rays

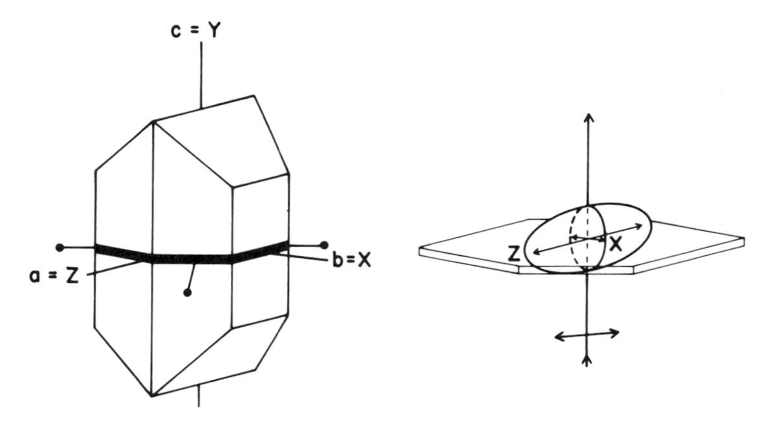

Figure 3.16 The biaxial indicatrix and vibration directions in olivine cut parallel to X and Z.

that vibrate parallel to the fastest and slowest directions X and Z. The double refraction in this section is the maximum possible and defines the birefringence of the mineral ($\delta = \gamma - \alpha$).

(2) *Mineral section parallel to X and Y of the indicatrix* (Figure 3.17). In this case light travels parallel to Z and vibrates parallel to the fastest and intermediate directions X and Y. Only a partial birefringence is displayed by the section ($\delta' = \beta - \alpha$).

(3) *Mineral section parallel to Y and Z of the indicatrix* (Figure 3.18). In this case light travels parallel to X but vibrates parallel to the intermediate and slowest directions Y and Z. Only a partial birefringence is displayed by the section ($\delta' = \gamma - \beta$).

Figure 3.17 The biaxial indicatrix and vibration directions in olivine cut parallel to X and Y.

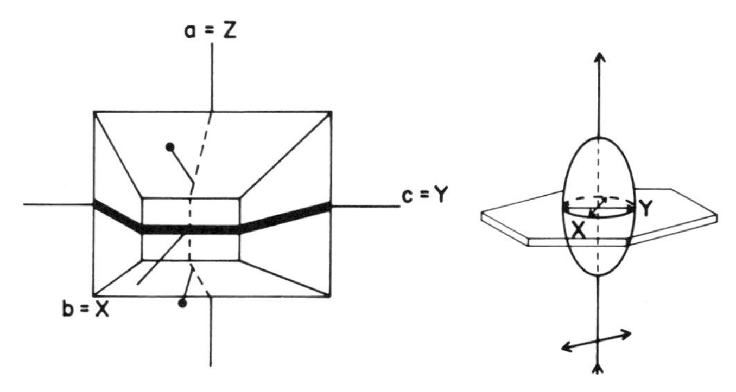

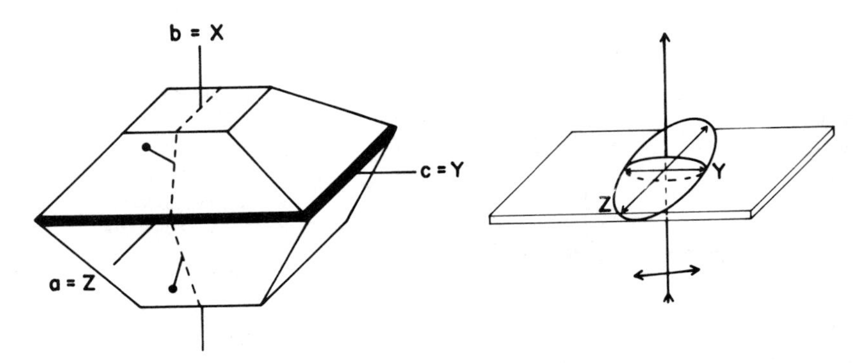

Figure 3.18 The biaxial indicatrix and vibration directions in olivine cut parallel to Y and Z.

(4) *Mineral section not parallel to the circular sections and not containing the directions X, Y, or Z.* Any general section of the biaxial indicatrix is an ellipse, and the orientations of vibration directions in such a section are parallel to the major and minor axes of that ellipse. Another way of determining the orientation of vibration directions is by the Biot–Fresnel law, which states that the vibration directions bisect the angles between the two planes that are normal to the section and contain one optic axis each. We shall find this law useful in understanding interference figures, but it need not concern us during routine laboratory work. A general section of the indicatrix displays a partial birefringence that is the difference between two refractive indices α' and γ', which have the general relationship:

$$\alpha < \alpha' < \beta < \gamma' < \gamma.$$

(5) *Mineral section parallel to either of the two circular sections of the indicatrix* (Figure 3.19). The angle between the two circular sections var-

Figure 3.19 The biaxial indicatrix and single vibration direction in olivine cut parallel to a circular section.

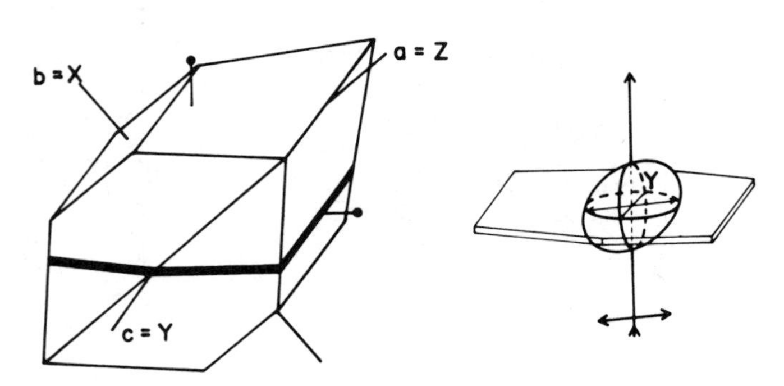

ies from mineral to mineral. We shall continue our examples with olivine in which the two sections commonly make an angle of approximately 90° with each other.

All possible vibration directions of light traveling perpendicular to one of the circular sections lie in that section, and represent the same refractive index. There are no preferred vibration directions. The vibration direction of light from the polarizer is therefore unaffected on passing through the mineral and is totally absorbed by the analyzer. The section appears black (or in practice often very dark grey due to the effects of dispersion or a lack of parallelism of the light rays), and remains so on rotation of the stage. The partial birefringence of the section is zero.

Optic Axial Angle ($2V$)

The direction perpendicular to the circular section is called the optic axis, and there are two such directions for all biaxial minerals (hence biaxial). The angle between the two optic axes is called the *optic axial angle* or $2V$. When this angle is acute about Z, the mineral is *positive;* when it is acute about X the mineral is *negative*. The optic axes always lie in the XZ-plane which is therefore called the *optic axial plane* (OAP). If the mineral is positive, Z is called the *acute bisectrix,* and X the *obtuse bisectrix*. If the mineral is negative, X is the acute and Z the obtuse bisectrix. An optic axial angle of, say, 20° about the Z direction may be recorded in the following ways: $2V_z = 20°$; $2V_x = 160°$; or +ve $2V = 20°$. If the optic axial angle is 90°, the mineral is neither positive nor negative and $2V_z = 2V_x = 90°$. If the optic axial angle is 0° (or close to zero) the mineral is *pseudo-uniaxial*.

The three-dimensional geometry of the biaxial indicatrix is such that there is the following relationship between the optic axial angle and the magnitudes of the refractive indices α, β, γ:

$$\cos^2 V_z = \frac{\alpha^2(\gamma^2 - \beta^2)}{\beta^2(\gamma^2 - \alpha^2)}.$$

Note that, in this formula, the angle determined is V, half the optic axial angle, and that $V_z = 90° - V_x$.

It can be seen from this formula that if β has a value closer to α than γ the mineral is positive, and vice versa. If β has a value closely approaching α or γ the mineral is pseudo-uniaxial +ve or −ve, respectively.

Interference Figures

Normally, we view a mineral section with so-called orthoscopic illumination, or a bundle of approximately parallel light rays. Microscopes are provided with a condenser system (Chapter 2), which when fully raised produces strongly convergent light (called *conoscopic illumination*). The more expensive microscopes are fitted with a removable condenser which can be swung in or out of position whenever necessary. Interference figures are produced by convergent light between crossed polars, and may be focused using high-power objectives. To view the figures it is necessary to either use the accessory Bertrand lens, or to remove the eyepiece and insert a pinhole stop in the top of the microscope tube.

Interference figures are used to determine whether a mineral is uniaxial or biaxial positive or negative; details of their appearance and use are explained in Chapter 4. In this chapter, we shall examine in principle how they are produced.

Uniaxial Interference Figures

Figure 3.20 illustrates a thin section of quartz cut at right angles to its *c*-axis and placed in convergent polarized light. The rays of light *A, B,* and *C* pass through the section at oblique angles and are polarized into two rays. The ordinary rays must vibrate in the plane of the section (the circular section of the indicatrix), and since the extraordinary rays vibrate at right angles to the ω ray, they must lie in vertical planes which contain the *c*-axis. The vibration directions of the ε' rays dip in towards the *c*-axis. All other rays of light passing obliquely through the section are similarly polarized into two rays, and in every case, the ε' ray must vibrate in a vertical plane containing the *c*-axis. Therefore in plan view (Figure 3.20c), the vibration directions of the ε' rays form a radiating pattern at right angles to which there is a concentric pattern of ω-ray vibration directions. Where these vibration directions are parallel to those of the polarizer and analyzer (E–W and N–S), light is extinguished completely, and a black cross (called an *isogyre*) is formed (Figure 3.20d). The black cross will not change in position when the stage is rotated. Outwards from the center of the figure, the dip of the ε' rays increases, and with it the partial birefringence. In normal thickness sections a low birefringence mineral such as quartz displays first-order grey and white interference colors in the four quadrants of the cross, but moderate or high birefringence minerals display a sequence of colors (as on the Michel–Lévy chart) in concentric rings around the center of the cross. The higher the birefringence the greater the number of rings displayed. Similarly, the thicker a section is the more color rings seen.

If the section is cut at a high angle but not exactly perpendicular to the *c*-axis, an off-centered uniaxial-cross figure is obtained, the center of which rotates in a circular path as the stage is rotated (Figure 3.21a). For sections at a moderate angle to the *c*-axis, one can imagine the center of the cross to lie outside the field of view. On rotation of the stage, each of the four arms of the cross successively enter and exit the field of view (Figure 3.21b).

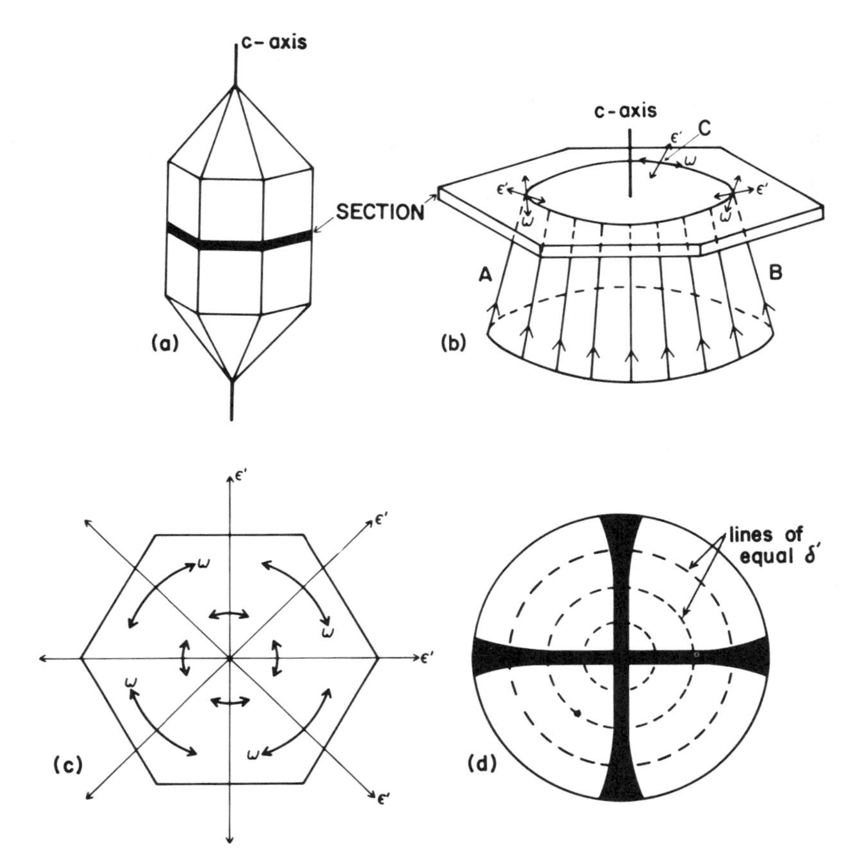

Figure 3.20 Production of the uniaxial cross-interference figure. See text for explanation.

For sections at smaller angles to the *c*-axis the arms of the cross enter and exit the field of view quickly, and they no longer retain regular E–W or N–S orientations but sweep across the field of view. When the *c*-axis is parallel to the section, a centered *flash figure* is produced with isogyres similar in nature to some biaxial obtuse bisectrix figures (see later), and so called because the isogyres flash or move very rapidly in and out of the field of view.

Biaxial Interference Figures

Figure 3.22 illustrates a section of a positive biaxial mineral with a small $2V$ cut perpendicular to Z (the acute bisectrix in positive minerals). The ray of light A passes through the section obliquely and is polarized into two

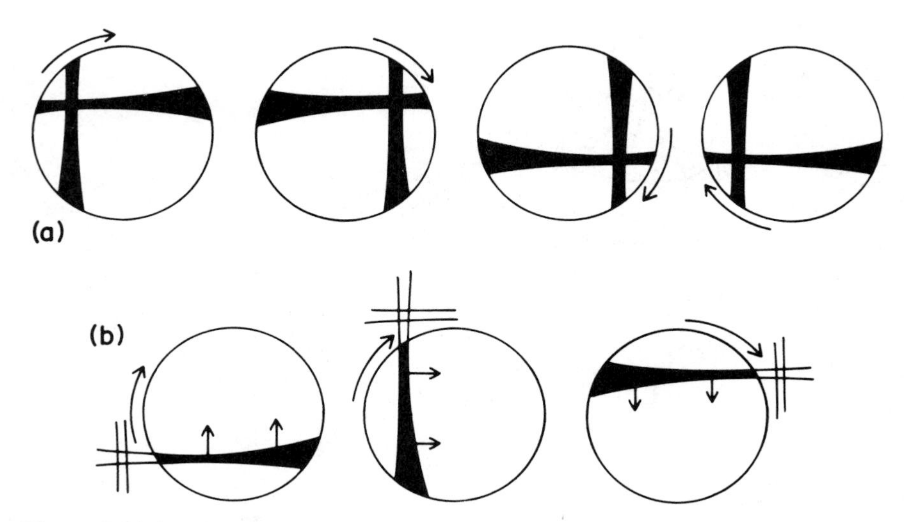

Figure 3.21 Moving pattern of isogyres for off-centered uniaxial cross figures when c-axis is **(a)** at a small angle to microscope tube, and **(b)** at a moderate angle.

Figure 3.22 Production of the biaxial acute bisectrix interference figure. See text for explanation.

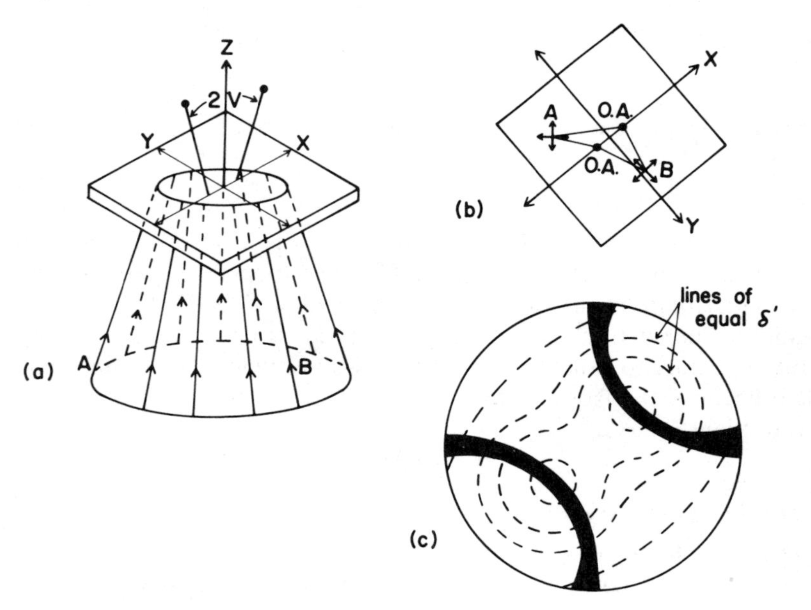

rays, the vibration directions of which can be determined by the Biot–Fresnel law as follows. On the plan view (Figure 3.22b), draw the traces of the two planes, each of which contains an optic axis, and the normal to the effective section (the effective section contains the two vibration directions and its normal, of course, is parallel to ray A). The two vibration directions bisect the angles between these two planes, and in this case are parallel to the polarizer and analyzer vibration directions, and light is extinguished between crossed-polars. The ray of light B, on the other hand, is polarized into two rays vibrating at an angle to the polarizer and analyzer directions, and light is not extinguished. If we determine the vibration directions of all other convergent rays of light, we find that there are two curved bands of extinction (isogyres) passing through the optic axes (Figure 3.22c). δ' increases outwards from the optic axes, and if the birefringence or thickness of section are sufficiently high for interference color rings to be present, they form a figure-of-eight pattern (Figure 3.22c).

Unlike the uniaxial cross-interference figure, the pattern of isogyres changes as the stage is rotated (Figure 3.23). The isogyres join to form a cross when the optical directions X and Y are parallel to the vibration directions of the polarizer and analyzer, and open up again in the other quadrants on further rotation. The positions of the optic axes are always marked by the figure-of-eight pattern of interference color rings (if these are present). The degree of opening between the isogyres when Y is in a 45° position as in Figure 3.22c depends on the value of $2V$ which can vary from $2V_z = 0°$ to 180°. Figure 3.23 illustrates the movement of isogyres for three different $2V_z$ values: very small (ca. 10°); small (ca. 25°); and moderate (ca. 50°). In each case, Z is perpendicular to the section and parallel to the microscope tube. The opening of the isogyres is almost imperceptible (Figure 3.23a) for a $2V_z$ which approaches zero. In fact, if the $2V_z$ *is* zero (pseudo-uniaxial), it is usually impossible to distinguish the figure from a uniaxial cross. The degree of opening increases progressively for increasing $2V_z$, until for a value of 50° or so (Figure 3.23c), the isogyres will approach the edge of the field of view. The precise amount of opening depends on the NA of the objective lens and the RI of the mineral (see Figure 4.15). For larger optic axial angles the isogyres may move completely out of the field of view and the rate of movement becomes progressively faster.

It is worth remembering that a $2V_z$ of 150°, for example, is the same as a $2V_x$ of 30°; in such a case the isogyres leave the field of view quickly when looking down Z, the obtuse bisectrix, but stay in the field of view and move slowly when looking down X, the acute bisectrix. Depending on whether the mineral section is cut perpendicular to the acute or obtuse bisectrix, the figure is called an *acute bisectrix figure* or *obtuse bisectrix*

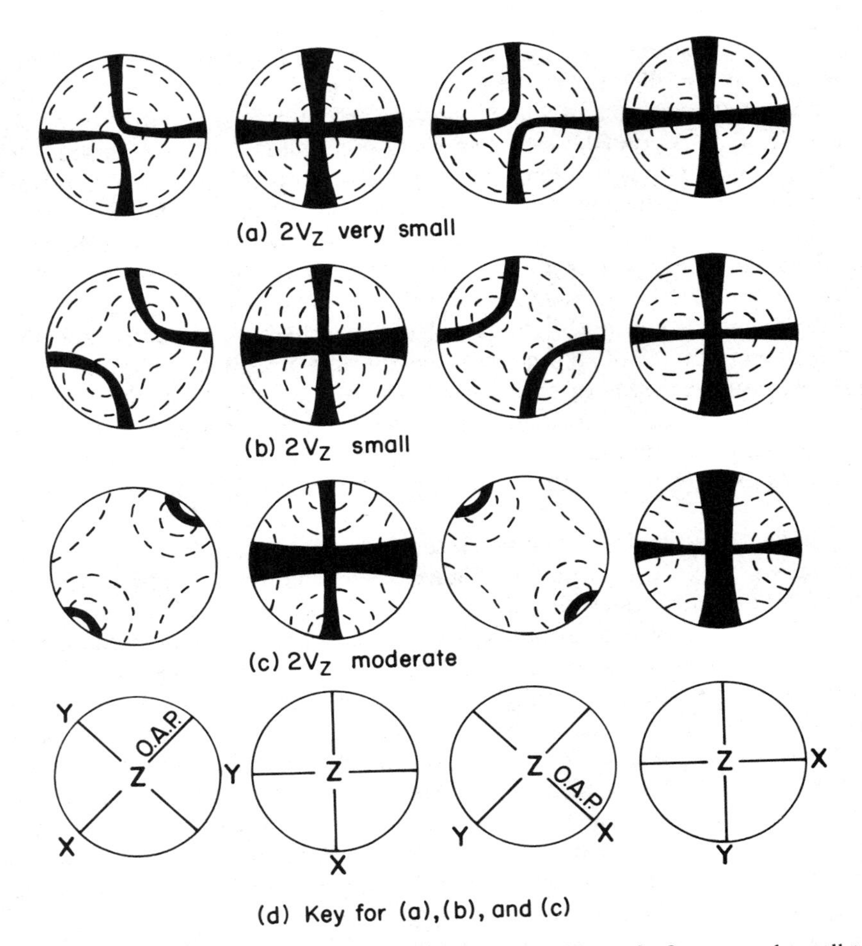

Figure 3.23 Changing pattern of isogyres for acute bisectrix figures and small to moderate $2V$. See text for full explanation. The key (d) shows direction of stage rotation.

figure. It should be clear from the above that, for a moderate or small $2V$, the acute bisectrix figure is easily identified by the slow-moving isogyres that stay in the field of view. It is often difficult or impossible to distinguish acute bisectrix figures for a higher $2V$ from obtuse bisectrix figures.

Sections at high angles other than 90° to the bisectrices provide off-centered bisectrix figures in which the isogyres join and part at a point which traces a circular path about the center of the field of view on rotation of the stage. Very off-centered figures display isogyres that lack any readily discernible pattern and sweep in and out of the field of view.

Sections cut parallel to X and Z (the optic axial plane) provide centered flash figures for which the isogyres usually have the appearance of a very rapidly moving obtuse bisectrix figure.

In routine work, the most useful and easily obtained biaxial figure is that provided by a section cut perpendicular to either of the optic axes, the so-called *optic-axis figure*. An isogyre will be present at the center of the field of view for such a figure, regardless of how the stage is rotated. For a $2V$ of 90° (Figure 3.24a), the isogyre is straight and rotates about the central position rather like a propeller blade as the stage is rotated. With smaller angles of $2V$, the isogyre becomes progressively more and more

Figure 3.24 Changing patterns of optic-axis figures for various values of $2V$. The key (d) shows direction of stage rotation.

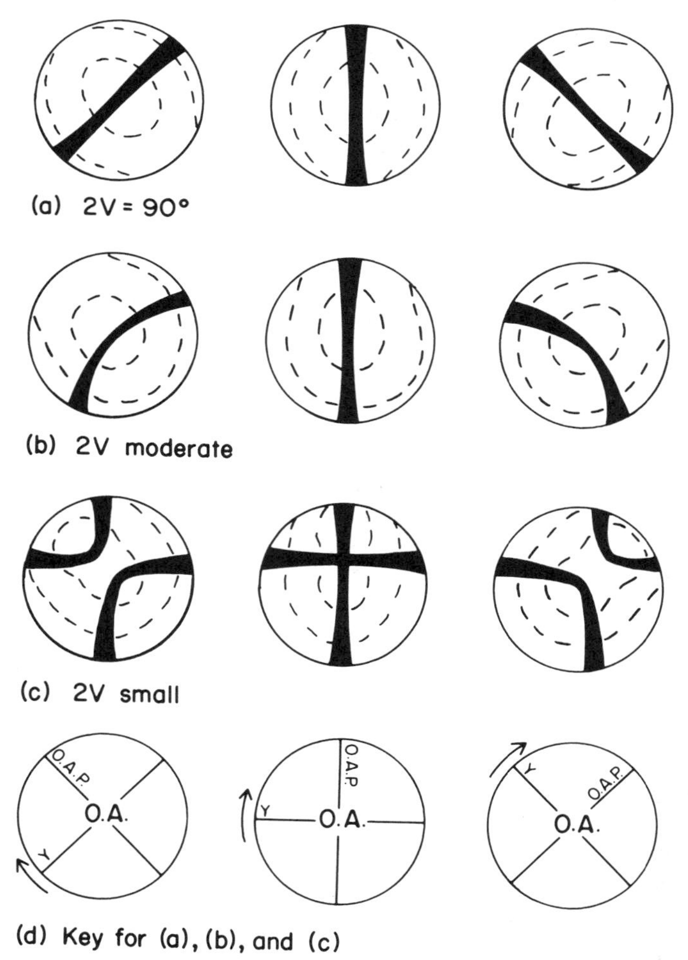

curved when Y and the OAP are in the 45° position (Figure 3.24b); the isogyre always straightens when Y is N–S or E–W. For a small $2V$ both optic axes emerge in the field of view and two isogyres are visible (Figure 3.24c). Their behavior on rotation of the stage is similar to an acute bisectrix figure for a small $2V$ though the point where the isogyres join (the acute bisectrix) is not in the center of the field of view. Of course, if the $2V$ is zero, a pseudo-uniaxial cross will be observed. In all these figures, interference color rings, if present, will be centered about the points of emergence of the optic axes (Figure 3.24).

Dispersion

The well-known experiment of splitting white light into its component colors by passing it through a triangular glass prism (Figure 3.25) serves to demonstrate that the RI of a substance varies for different wavelengths of light, a phenomenon known as dispersion. The spectrum of colors produced by the refraction of sunlight is not continuous, there being a number of dark lines (or absent wavelengths). These are known as Fraunhofer lines, and lettered from A in the red part of the spectrum to K in the violet part. In precise optical work, it has been common practice to report the refractive indices of minerals for wavelengths corresponding to the Fraunhofer lines C (656 nm), D (589 nm), and F (486 nm). Because the D-line wavelength corresponds to the relatively intense and easily obtained sodium light, the most commonly recorded RI measurement is that for 589 nm (RI_D).

The amount of dispersion varies from mineral to mineral and is commonly recorded in the form $RI_F - RI_C$, or the coefficient of dispersion. Such precise measurements of dispersion are seldom necessary, and for most purposes RI determinations in white or yellow sodium light suffice.

Figure 3.25 Differential refraction of sunlight by a glass prism to form the visible spectrum.

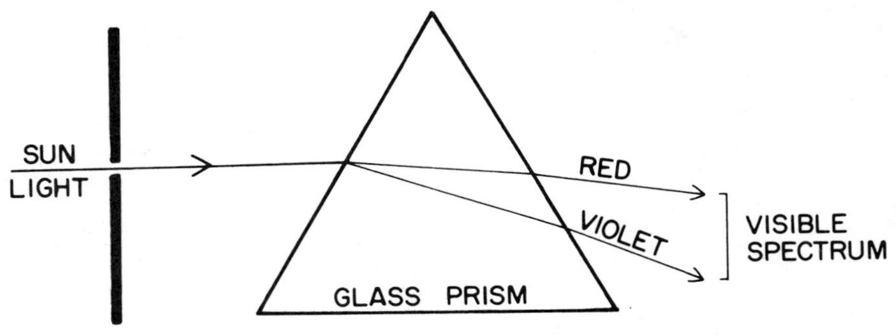

However, dispersion may produce visible effects in a number of ways, and these are noted briefly below.

Isotropic Minerals

The phenomenon of dispersion affects only the precise values of RIs.

Uniaxial Minerals

The coefficients of dispersion for each of ω and ε may differ significantly, in which case the retardation produced by a mineral section will differ for different wavelengths. Anomalous interference colors may result. In rare cases, a uniaxial mineral may be +ve for some wavelengths and −ve for others (with a δ of zero for some part of the spectrum).

Biaxial Minerals

Biaxial minerals may produce anomalous interference colors if the coefficients of dispersion for α, β, and γ differ significantly. Because $2V$ is dependent on the relative values of α, β, and γ (as given in the formula on page 43), $2V$ will vary with wavelength in sympathy with α, β, and γ. The effect is to produce color fringes to the isogyres of biaxial interference figures, and this effect may help in mineral identification. It is discussed again in Chapter 4.

The various symmetries of biaxial crystals control the ways in which dispersion may be effected. In orthorhombic minerals, *a, b,* and *c* are always parallel to *X, Y,* and *Z,* but not necessarily in that order. In extreme cases, not only will dispersion result in a variable $2V$ for different wavelengths but a swapping of the position of optical directions. Thus, in brookite, *X* is parallel to *c* for wavelengths less than 550 nm, whereas *Y* is parallel to *c* for longer wavelengths. In other words the optic axial plane switches in position from (100) to (001), a phenomenon known as *crossed axial-plane dispersion.* The resulting interference figure is complex and lacks the normal black isogyre patterns; further consequences are that many sections of brookite do not completely extinguish, nor do they display normal interference colors.

In monoclinic crystals, only *b* is exactly parallel to *X, Y,* or *Z*. If $b = Y$, a mineral may display *inclined dispersion* in which the positions of the acute and obtuse bisectrices and the optic axes may all vary for different wavelengths. The dispersion lies entirely within the *ac* plane perpendicular to the *b*-axis; hence color fringes to isogyres will have a mirror symmetry on either side of the trace of the optic axial plane (= *ac* plane). If $b =$ the acute bisectrix, a mineral will display *crossed dispersion;* if $b =$ the obtuse bisectrix, *parallel* or *horizontal dispersion*. In principle these are similar in that the dispersion results in a rotation of the OAP about the *b*-axis. The result is an asymmetric disposition of color fringes with respect to the isogyres.

The lack of symmetry in triclinic minerals may lead to correspondingly asymmetrical dispersion patterns.

More details are given on the effects of dispersion in the next three chapters, both with regard to interference figures and precise RI measurement. It is worthwhile commenting again that, for routine work, the only significant effects of dispersion are seen in (a) the production of anomalous interference colors in a small number of common minerals, and (b) the incomplete extinction of some biaxial minerals, particularly in sections more or less perpendicular to an optic axis.

Color and Pleochroism

The vast majority of the common rock-forming minerals are transparent in thin sections or as small grains. The intensity of color is naturally less than in hand specimen, and we find that many "colored" minerals are colorless in thin section. A few minerals are opaque, even in very thin sections, and require special methods of study using reflected light.

The color of an isotropic mineral is uniform regardless of its orientation on a microscope stage. Many anisotropic minerals, however, differentially absorb light passing through them depending on the vibration direction of the light, a phenomenon known as *pleochroism* (sometimes called dichroism in uniaxial minerals).

A good example of a strongly pleochroic mineral is biotite in which the X optical direction absorbs light weakly and the Y and Z directions absorb light strongly (such a pleochroic scheme is abbreviated $Z = Y > X$). The effect of this is illustrated in Figure 3.26. In plane-polarized light in Figure 3.26a, light can only vibrate parallel to X and we see a pale-brown color. When the crystal section is oriented as in Figure 3.26b, light only vibrates parallel to Z and we see a dark-brown color. In intermediate positions (Figure 3.26c), light from the polarizer is resolved to vibrate in both the X

Figure 3.26 Pleochroic biotite in plane-polarized light: $Z > X$.

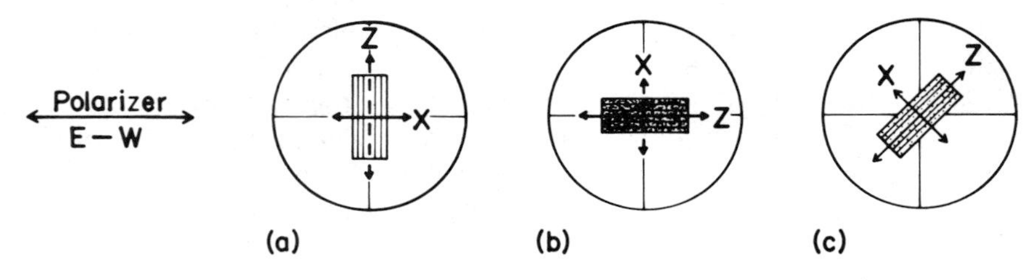

and Z directions, and we see an intermediate shade of brown. Pleochroic biotite is illustrated in Plates 7 and 9.

Another good example of a pleochroic mineral is uniaxial tourmaline, which very strongly absorbs light when ω is set parallel to the polarizer vibration direction, but which weakly absorbs light vibrating parallel to ε (its pleochroic scheme is abbreviated $\omega > \varepsilon$).

Some minerals have quite different colors for different vibration directions. Thus, glaucophane has a pleochroic scheme in which X = colorless, Y = violet, and Z = blue (Plate 19), and orthopyroxene has X = pink, Y = yellow, and Z = green (Plate 3).

Chapter 4

Flat-Stage Techniques—Thin Sections

It is common practice to examine rock and mineral samples under the microscope as very thin sections (approximately 0.03 mm thick). There are several good reasons for this: many minerals are not transparent except as very thin grains; a standardized 0.03-mm thickness allows us to examine almost the entire range of grain sizes in rocks without any serious effects of overlapping or superimposed grains; the important property of birefringence is most easily determined by observing the low-order interference colors produced in sections of known thickness close to 0.03 mm (using the Michel–Lévy color chart); grain sizes and volume percentages of minerals in a rock can be determined, and their textural relationships examined. Thin sections usually enable the RIs and the nature of the indicatrix (uniaxial +ve or −ve, biaxial +ve or −ve, value of $2V$) to be estimated, and with experience, the vast majority of the common rock-forming minerals can be determined with a great measure of confidence. The addition to the microscope of the universal stage (Chapter 6) allows the nature of the indicatrix to be determined with accuracy. Precise chemical analyses of minerals, using the electron microprobe, can be made on specially prepared, polished thin sections. Thin sections are permanent and easily stored.

Sample Preparation

The method of thin-section preparation depends on the nature of the sample and equipment available. Most laboratories have diamond saws and various grinding machines, which speed the process up, but they are not absolutely essential. A wide range of igneous and metamorphic rocks as well as lithified sediments present little difficulty to the skilled technician, and may be prepared in the following way. For a more detailed account of generally used methods, refer to Reed and Mergner (1953).

1. A small chip of rock or mineral is sampled, or a thin slice (a few millimeters thick) is cut from the specimen using a diamond saw.
2. The chip or slice is ground flat and smooth on one surface using progressively finer abrasive (carborundum powder, starting with 80 grade and finishing with 600 grade). The abrasive is mixed to a paste with water and the grinding done on thick glass plates or on a mechanical lap.
3. The smooth surface is cemented to a glass slide (usually 3″ × 1″, but 2″ × 1″ for universal-stage work). Sticks of Lakeside 70 cement (RI = 1.54) are generally used at this stage. The mounting is done on a hot plate at 100°C, which allows the cement to be spread to a thin, even layer. The heating should not be prolonged lest too many bubbles form.
4. After cooling, the other surface is ground down until the rock section is 0.03 mm thick. Progressively finer abrasives mixed with water are used and the finishing touches made with 600 grade carborundum powder. The thickness is gauged by observing the interference colors of common minerals such as quartz or feldspar (the section on birefringence in this chapter explains how to do this).
5. After washing thoroughly, the section is first coated with cellulose acetate to prevent breaking up, then heated and transferred to a fresh slide coated with Canada balsam. The balsam is usually thinned with xylol to ease spreading, and it should be cooked at 85°C to harden it. This is judged by using a glass rod to draw up a fine thread of balsam which should snap on cooling.
6. More balsam is spread on top of the section and a cover slip added. During the cooking process, bubbles should be gently squeezed out as the cover slip is eased into position.
7. After cooling, the slide is washed in alcohol to remove any excess balsam.

Some modern semiautomatic machines rely on mounting a specimen on a glass slide with a very thin film of strong epoxy resin. Thin sections of the correct thickness can then be made in a purely mechanical way, quickly, and in large numbers. For standard work it is best to use an epoxy such as Araldite AY 105 (Wright, 1964), which has a RI between 1.535 and 1.55, close to that of Canada balsam (1.537) and Lakeside 70 (1.54).

Delicate specimens may need to be handled with special care during all stages of thin-section preparation. Very friable sediments and metamorphic rocks with a pronounced fissility may require impregnating with a cement before a section can be made. Heating the specimen in Canada balsam is sometimes sufficient, but soaking in an epoxy resin in a vacuum (Hutchison, 1974) is a better way to impregnate the sample thoroughly.

Water is used during all the cutting and grinding stages of normal thin-section preparation. Special methods are required for the water-soluble salts, and a suitable procedure has been described by Bennett (1958). Similarly, rocks extensively altered to swelling clays may require special preparation either using kerosene instead of water during grinding of the section, or strong epoxy resins rather than Canada balsam as a cement (Wright, 1964).

Polished and Ultrathin Sections

Polished sections allow observations to be made in reflected as well as transmitted light, and are essential if individual grains are to be analyzed using the electron microprobe. The polishing is usually achieved with diamond or fine alumina powder; details of the techniques may be found in Hutchison (1974), Taggart (1977), and Lister (1978). Ultrathin sections (as thin as 0.002 mm) are useful when studying highly birefringent minerals, such as the carbonates, or very fine-grained minerals, such as the sheet silicates in slates. They are made in much the same way as polished sections (e.g., Lindholm and Dean, 1973).

Refractive Index Determination

Relief

A standard thin section is highly irregular in detail (see enlarged part of Figure 4.1). Boundaries between adjacent grains are not generally vertical and may be stepped or curved. The top and bottom surfaces of the section are not polished but pitted and grooved to some degree.

If there is a difference in RI between the mineral and the cement, these irregularities concentrate or scatter light by reflection and refraction. The

Figure 4.1 Thin-section surface irregularities that produce the effect of relief.

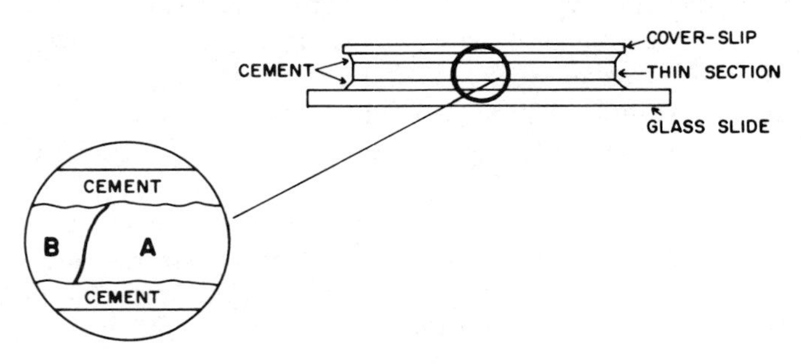

Table 4.1 Relative Relief in Thin Sections

RI	Relief	Mineral example
>1.90	extreme	rutile
1.78–1.90	very high	almandine garnet
1.68–1.78	high	epidote
1.57–1.68	moderate	actinolite
1.49–1.57	low	quartz
<1.49	moderate (less than CB)	fluorite

effect is to give an impression of three-dimensional relief. If the difference in RI is small, the irregularities will be barely visible and the mineral has *low* relief. A large difference results in *high* relief.

Relief should be observed in plane-polarized light with a low or medium-power objective. In practice, it will be found that a high or very high relief (e.g., epidote, Plate 27, or garnet, Plate 25) is clearly discernible, even when the lower diaphragm of the microscope is completely open. A moderate relief will become discernible by lowering the substage condenser system and/or partially closing the lower diaphragm. Low-relief minerals will retain their low-relief appearance even when the lower diaphragm is closed. The description of relief is nevertheless somewhat subjective, and the apparent relief increases with both the thickness of a section and the degree of irregularity of its surfaces.

Exercise

Estimating RI from Relief

Practice estimating RI from relief, using at first a group of reference thin sections. Acquaint yourself with the amount of diaphragm closure (or lowering of substage system) necessary to bring out various reliefs. Suggested minerals for examination are listed in Table 4.1. Canada balsam (RI = 1.537) or Lakeside 70 (RI = 1.54) is presumed to be the cement. This is generally the case, but it is as well for the student to inquire in the laboratory what cement has been used.

Becke Lines

If a mineral has a high or very high relief, it is safe* to say its RI is higher than that of Canada balsam. However, the RI of a mineral with low or moderate relief may be either higher or lower than that of Canada balsam, and it is necessary to make a Becke line test.

---

* The lowest RI of a common rock-forming mineral is 1.43 (fluorite). There are, however, various rare minerals (not described in this book) with much lower RI. A well-known example is cryolite with an RI of 1.34, approximately the same as that of water.

In plane-polarized light, the irregular edge of a mineral grain against another substance of different RI will concentrate light as a thin, bright white line along its margin. This line, known as the Becke line, will move inwards or outwards as the mineral grain is brought slowly into or out of focus. It is best viewed with the medium-power objective (×10) of the microscope, and with the lower diaphragm at least partially closed. The following rule regarding the direction of movement of the Becke line enables us to compare the RI of a mineral and Canada balsam (e.g., at the edge of the thin section), or the RI of two adjacent minerals: the Becke line moves into the substance of higher RI when the distance between objective and section is increased (Figure 4.2). Becke lines cannot be successfully observed with very high-relief minerals.

The Becke line can be explained by the concentration of light rays above the substance with higher RI through a combination of total reflection at nearly vertical boundaries and refraction of light rays towards the substance with higher RI. "False" Becke lines moving in the opposite direction to the true lines are sometimes observed, though they are seldom a problem if the lower diaphragm is properly closed. False lines may be seen in thin sections if grain boundaries have been penetrated by

Figure 4.2 To illustrate the movement of Becke lines as microscope tube is raised (or stage lowered) for minerals with **(a)** RI > medium or **(b)** RI < medium.

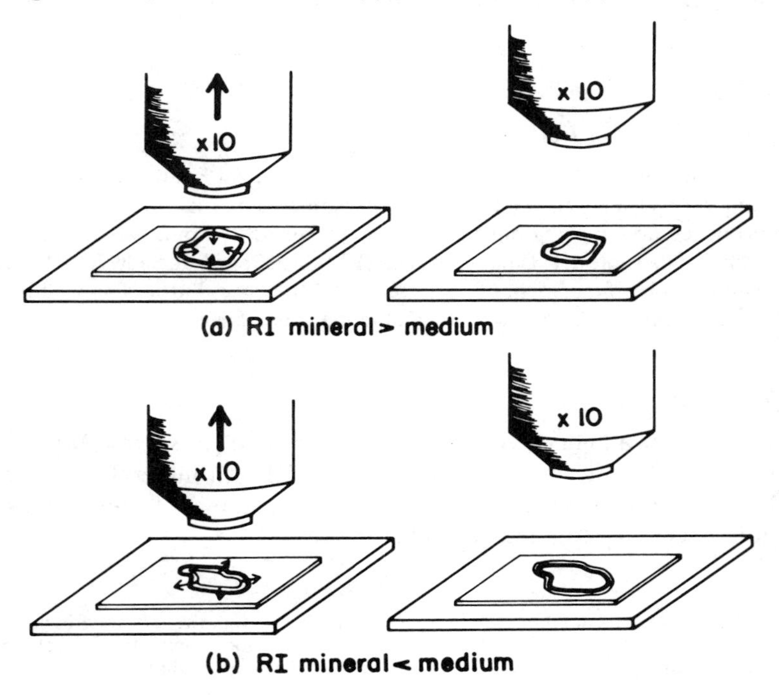

Canada balsam; thus not only are the refractive indices of two adjacent grains being compared but their relative refractive indices with balsam too.

Exercises

Movement Directions of Becke Lines

Observe the movement directions of Becke lines for microcline and quartz in a granite, and calcic-plagioclase in a gabbro, as follows:

(1) In crossed-polarized light, select fresh grains which lie against Canada balsam at the edge of each section. Microcline is easily identified by its cross-hatch twinning (Plate 8), the quartz by its lack of cleavage, twinning, and alteration (Plate 8), and the plagioclase by its lamellar twinning (Plate 4).

(2) Examine each mineral in turn using plane-polarized light and a ×10 objective. The substage system should be lowered and/or the substage diaphragm should be well closed.

(3) Focus up and down (use the coarse focusing adjustment but not too vigorously), and note the type of Becke line movement. When the distance between the stage and objective is increased, the line moves out of the microcline but into the quartz and plagioclase. The line is weak and difficult to observe for quartz, which has a very low relief.

(4) Find a boundary between quartz and microcline in the granite. Observe the strong Becke line, which moves from microcline into the quartz when the distance between objective and stage is increased.

(5) These observations demonstrate that the relative RIs are microcline < CB < quartz < calcic-plagioclase.

Twinkling

Some minerals with a high birefringence have one RI close to that of Canada balsam and the other one considerably different. Such minerals change markedly in relief on rotation of the stage in plane-polarized light, a phenomenon known as twinkling. Calcite and dolomite are good examples of minerals that twinkle, and the effect is shown in Plate 39.

Accurate Determination of Refractive Index (RI)

Thin-section work generally allows only a rough estimation of refractive index (RI). In some special situations where the RIs of a mineral overlap that of Canada balsam, or those of a mineral with known indices such as quartz, a semiquantitative determination may be possible (see, for example, methods of plagioclase determination, page 255). If precise RI measurement is desired, a separate crushed-grain mount is usually prepared and examined in various immersion oils (Chapter 5). Sometimes it may be necessary to measure the RI of a particular grain in a thin section, and there are two ways of doing this. In either case, the cover slip must first be removed from the thin section, and the balsam cleaned away. If it is possible to prise a mineral grain out of the section, it can then be mounted on a spindle stage for measurement of RI (Chapter 6). On the other hand, accurate measurement may be made in situ following the method described by Laskowski et al. (1979).

Determination of Birefringence

The birefringence or partial birefringence of a mineral section may be determined by observing interference colors in crossed-polarized light with the diaphragm below the stage open, and using a low- or moderate-power objective. Figure 4.3 diagrammatically represents the Michel–Lévy color chart (at the back of book). Values of birefringence are represented by lines radiating from the lower left-hand corner of the diagram. Thickness of the mineral section is represented by horizontal lines, and the interference colors are vertical lines. Basically, the technique involves determining: (1) the thickness of the section, and (2) the interference color. For example, a mineral section with interference color *A* and thickness *B* (Figure 4.3) has a birefringence or partial birefringence *C*. To identify an unknown mineral, we are normally concerned with determining the true birefringence, that is, the difference between ω and ε, or α and γ.

Choosing a Grain for Determination of True Birefringence

The several grains of a mineral in a thin section will normally be cut parallel to different crystallographic directions, and hence intersect the

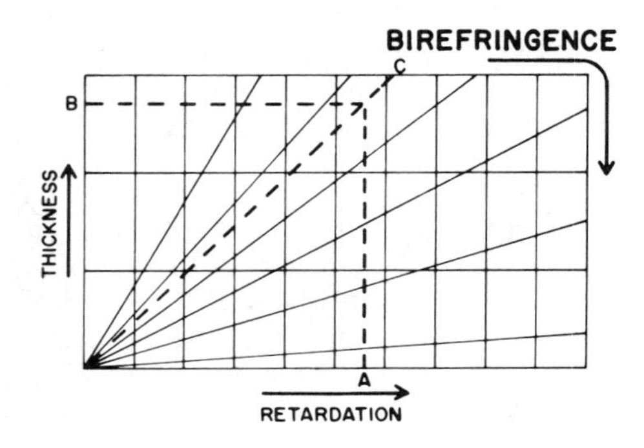

Figure 4.3 How to use the Michel–Lévy chart.

indicatrix in different orientations. The interference color will vary from black (or very dark grey), for those cut parallel to a circular section of the indicatrix, to a maximum for those cut parallel to ω and ε, or X and Z. Thus for a mineral with birefringence E in a section thickness D, we may see a number of grains with interference colors ranging from 1 to 7 (Figure 4.4). Plates 2 and 28 show several grains of olivine and epidote, respectively, displaying a variety of interference colors. To measure the true birefringence we must therefore search for the grain with the highest interference color (e.g., grain 7 in Figure 4.4). The search must be thorough, and it is worth noting that the grain selected should yield a centered flash figure. A problem arises if the mineral has a preferred orientation (as is often the case in metamorphic rocks), when suitably oriented sections may not be present.

Identifying the Interference Color

The interference colors are normally divided into orders: first, second, third, fourth, etc., corresponding to multiples of 550-nm retardation. It will be noticed that some colors repeat themselves, so that there is a second- and third-order red, for example. Higher-order colors not shown on the Michel–Lévy chart consist of a repeated sequence of paler and paler pinks and greens. Although these colors do become paler with increased order, the student must beware of determining the order of color from the color intensity, or by direct comparison with the chart.

The simplest method to determine the order of color is to look at the edge of mineral grains. Grains are normally wedge-shaped at their margins, and as the thickness reduces, so will the interference color. For example, if a section with birefringence F reduces in thickness at its edge from G to H, the interference color will change from I to J as a series of color rings (Figure 4.5). It is nearly always possible to observe the unique and distinctive first-order white or grey at the edge of a grain that displays color rings (using a moderate- or high-power objective). By counting the

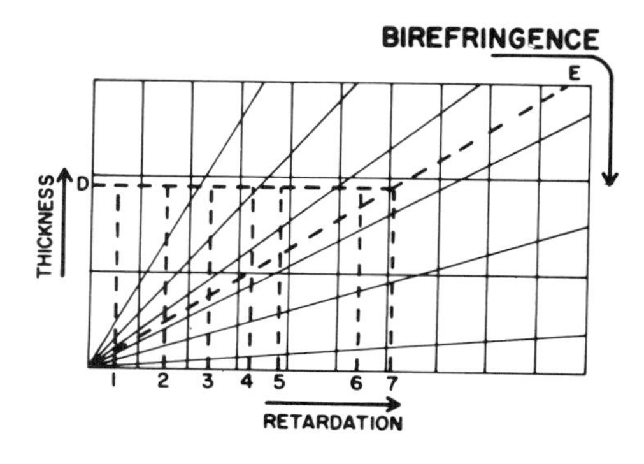

Figure 4.4 Some possible interference colors produced by a mineral with birefringence E in a section of thickness D but with a number of orientations.

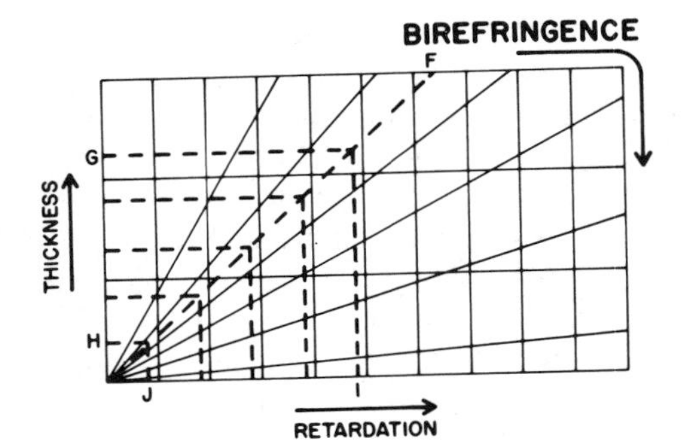

Figure 4.5 Change of interference colors with change of thickness in a section with birefringence *F*.

color rings, the order of color in the main part of the mineral section may be determined. Plates 18 and 28 show examples of a number of grains with color rings at their edges, thus enabling the main interference color of each grain to be determined.

Some difficulty may be experienced with very-high-order colors. These become so pale that they may be confused with first-order white. The distinction is easily made using a sensitive-tint plate which, as explained in the next section, will cause first-order white to change to an obvious second-order blue or first-order yellow; a high-order "white" will not change perceptibly on insertion of the tint plate. Color rings may, of course, be visible at the edges of very highly birefringent grains; even though it will not be possible to count the number of rings exactly, the very fact of an extreme birefringence is often diagnostic.

Accurate Measurement of Retardation

The retardation of a mineral grain can be measured precisely by using a quartz wedge or rotary compensator. This is explained in the next section on compensators, although one should note that there is little purpose in measuring retardation with precision unless the thickness and optical orientation of the mineral grain are known precisely, too. In practice this means that (a) the mineral grain should provide a centered interference figure (see later under orientation diagrams), and (b) an adjacent quartz grain must display a centered flash figure (i.e., the section is cut parallel to the optic axis), so that its retardation, and hence thickness, can be determined precisely with a rotary compensator.

Determination of Thin-Section Thickness

Standard modern thin sections are close to 0.03 mm in thickness. It is easy for the student to confirm this provided a mineral of known birefringence has already been identified in the section. Quartz is ideal for the purpose because it is common, easily identified, and of constant birefringence. Feldspar and other minerals may be used but with less accuracy.

Quartz has a birefringence of 0.009. By reference to the Michel–Lévy chart, it can be seen that, in a section of 0.03-mm thickness, there should be a few grains with first-order pale-yellow interference colors. If there are none, the section is thinner than normal; if there are grains with orange interference colors, it is thicker than normal. Plate 32 illustrates the range of interference colors for quartz in a standard thickness section. With some experience, it is possible to determine the thickness within 0.01 mm (note, however, that a section may vary in thickness, usually being thinner at its margins).

The precise thickness of any quartz grain which provides a centered flash figure (i.e., cut parallel to its optic axis) can be determined if we first make an exact measurement of the retardation using a rotary compensator (see next section). Inasmuch as the birefringence of quartz is known (0.009), the thickness can then be read from the Michel–Lévy chart.

Exercises

Determining Birefringence

Obtain thin sections of a muscovite-rich granite, an olivine-augite-gabbro, and a calcite-cemented quartz sandstone, and determine the birefringence of muscovite, augite, olivine, and calcite, as follows:

(1) Practice discerning the entire population of grains of a particular mineral species by general examination, first in plane-polarized, then crossed-polarized light. Muscovite (Plates 29 and 30) will be evident from its moderate relief (close the diaphragm), tabular form, good cleavage, and generally bright interference colors. Augite and olivine (Plates 1 and 2) in the gabbro both show a high relief; augite is usually slightly colored and displays cleavages. Much of the calcite in the sandstone will twinkle.

(2) Examine numerous grains of quartz (in the sandstone and granite) and plagioclase (in the gabbro) using crossed-polarized light. For a standard thickness (0.03 mm), some grains should display first-order white and creamy-white interference colors. Use the Michel–Lévy chart to estimate thickness if only lower colors or some higher colors are observed.

(3) In crossed-polarized light for each mineral in turn, carefully search for the grain displaying the highest interference color. Use a ×10 objective. The entire population must be surveyed. In a standard thickness section, muscovite typically displays second- or third-order red or green; look at grain edges for color rings which should be counted and correlated with colors on the Michel–Lévy chart. Augite normally displays a range of colors up to a maximum of second-order blue (Plate 2), whereas the olivine will show colors up into the third order (Plate 2); count the color rings for olivine.

Calcite displays very-high-order colors not represented on the Michel–Lévy chart. Look for closely spaced rings at grain edges, and compare first-order white of quartz with the very pale high-order color of calcite.

(4) For each mineral in turn, use the value of retardation represented by the highest interference color and the value of section thickness to determine birefringence with the Michel–Lévy chart. Check your determination against the reported values for each mineral in Chapter 9.

Anomalous Interference Colors

A number of minerals including vesuvianite, melilite, the epidote group of minerals, and some members of the chlorite family display abnormal first-order colors instead of grey and white. Common anomalous colors are various shades of blue, purple, and brown (Plates 26 and 30), or in the case of epidote simply the absence of white (Plate 28). These result from the various effects of dispersion as discussed in principle in Chapter 3. There should be no difficulty in distinguishing anomalous first-order blues and purples from normal second- and third-order colors, because the normal sequence of color rings will be absent at the edges of grains. The presence of anomalous interference colors may be a diagnostic property.

An Alternative Method of Birefringence Determination Using Interference Figures

It often happens that, in any one section, an insufficient number of grains exists to determine the birefringence by the method of searching for the highest interference color. Sometimes the highest interference color grains are not found because of a strong mineral-preferred orientation. In such cases, the birefringence may be estimated from interference figures.

For a standard thickness section, we can make the following generalizations about interference figures. Isogyres are wide, ill-defined, and almost fill the entire field of view for very-low-birefringence minerals; they appear progressively thinner and more sharply defined with increasing birefringence. The interference colors observed between isogyres vary, too, so that first-order grey, white, and yellow will be seen with very-low to low-birefringence minerals, one or two color rings with moderate-birefringence minerals, and many rings with high-birefringence minerals.

A more precise method of birefringence measurement using centered optic-axis figures has been described by Winchell (1965). First, the isogyre(s) of the figure must be placed in the 45° position (as in the left- or right-hand diagrams of Figure 3.24). The retardation (R) at the edge of the field of view is measured on the concave side of the isogyre and birefringence determined from the relationship $\delta = K(R/t)$, where t is the thick-

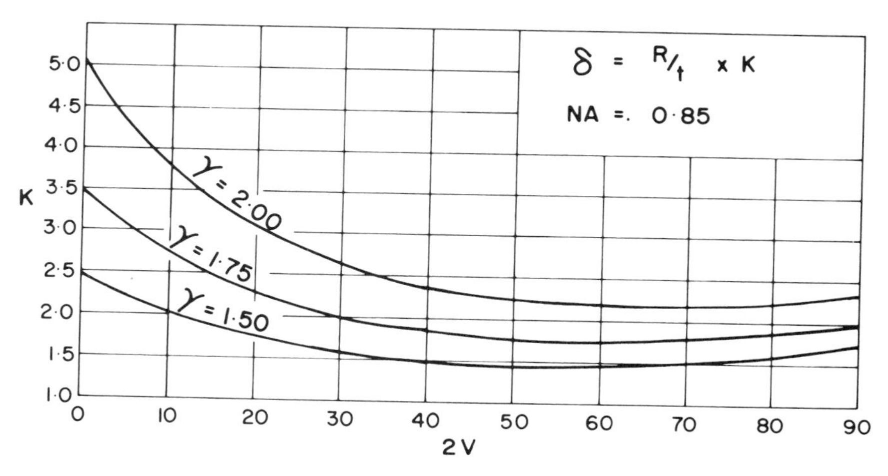

Figure 4.6 Determining K and hence birefringence from optic-axis figures. See text for full explanation.

ness of the section and K a figure dependent on $2V$, RI, and the numerical aperture of the objective lens. Values of K are given in Figure 4.6 for NA = 0.85 and values of γ between 1.5 and 2.0.

Fast and Slow Directions and the Use of Accessory Plates (Compensators)

All petrographic microscopes are provided with one or more accessory plates. They are made of mineral sections cut with thicknesses and orientations to produce retardations of known amount. When inserted in the accessory slot of the microscope tube, the plates have two vibration directions in the 45° positions, and these are marked as fast and/or slow (or X and Z or α and γ, respectively). Accessory plates are known as compensators.

The most useful of the compensators is the sensitive-tint plate, also known as a gypsum or 1-wavelength plate. It consists of a section of quartz or gypsum that produces between crossed-polars an interference color of first- to second-order pink (corresponding to a retardation close to 560 nm). The color can be observed simply by inserting the plate and using crossed-polarized light. The mica or quarter-wavelength plate produces a first-order grey interference color corresponding to a retardation of approximately 140 nm. Rather more expensive is the quartz wedge, which has a variable thickness producing retardations from zero to as much as 3500 nm. Viewed in crossed-polarized light, it displays a gradational series of interference colors up into the sixth and seventh orders. More expensive again are various graduated rotary compensators. These consist of a crystal plate, which can be inserted into the microscope tube and rotated so that a gradational series of interference colors are seen. The precise value of retardation can be determined from the degree of rotation and by reference to a set of tables accompanying the compensator. An example is the Ehringhaus rotary quartz plate manufactured by Zeiss which produces up to 4230-nm retardation.

Determination of Fast and Slow Directions

Referring back to Chapter 3, we can recall that the interference color produced by a mineral section is a measure of retardation which is a consequence of two light rays passing through an anisotropic mineral with different speeds. The two light rays have specific vibration directions perpendicular to each other, and the positions of these directions are readily determined by rotating a mineral grain until it extinguishes. The vibration directions of the two rays are then E–W and N–S, parallel to the cross-hairs of the microscope (Figure 4.7a; cf. Figure 3.8). By rotating the stage 45° from extinction, we can place the two vibration directions NE–SW and NW–SE (*A* and *B* in Figure 4.7b).

The accessory plates described above similarly have two vibration directions in the NE–SW and NW–SE positions when inserted into the accessory slot of the microscope. For most microscopes, the slot and simple nonrotary plates are oriented so that the slow direction of the plate is in the NE–SW position. The student is, however, cautioned to be careful because the Swift microscope illustrated in Figure 2.2 has a NE–SW slot and a length-slow sensitive-tint plate whereas the Zeiss microscope of Figure 2.1 has a NW–SE slot and a length-fast plate; both therefore have the slow direction NE–SW, but swapping plates between microscopes could result in some confusion!

The effect of the accessory plate is (a) to *increase the retardation* produced by the mineral section if the slow vibration directions of min-

Figure 4.7 Use of an accessory plate to locate fast and slow directions. See text for explanation.

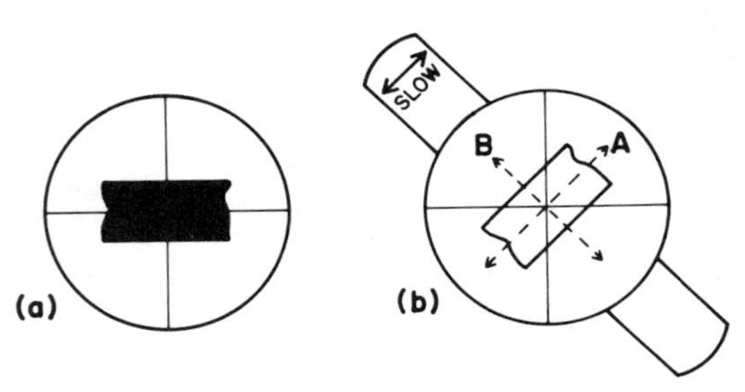

eral and plate coincide, or (b) to *compensate* for the retardation produced by the mineral section if the slow direction of the plate is parallel to the fast direction of the mineral. The compensation (subtraction) or increase (addition) in the amount of retardation will be precisely equal to the retardation of the plate. Examples of the use of each type of compensator to determine fast and slow directions follow.

Sensitive-Tint Plate

Let us imagine the mineral section in Figure 4.7b displays a second-order blue interference color corresponding to a retardation of 700 nm. If *A* is the slow vibration direction, insertion of the plate with its slow direction NE–SW parallel to *A* will result in an increase in retardation by 560 nm to 1260 nm; the interference color changes to third-order green. If, on the other hand, *A* is the fast direction, compensation takes place and the retardation is reduced to 140 nm or first-order grey.

Many mineral sections display first-order colors, and it is worthwhile examining the effect of the tint plate on these. Imagine the mineral section in Figure 4.7b displays a first-order white (a retardation of 200 nm). If *A* is the slow direction, insertion of the plate results in an increase in retardation to 760 nm, or a second-order blue-green. If, on the other hand, *A* is the fast direction, the retardations of plate and mineral compensate for each other leaving a retardation of 360 nm, a first-order yellow.

The changes in color for mineral sections displaying higher-order colors may not be so clear. Thus a fourth-order green will change either to a third-order or a fifth-order green, and whereas the fifth-order color is paler, the distinction is not always easy to discern. The simplest solution to this problem is to observe the changes in the lower-order color rings at the thin edge of a grain. If this is not possible, a quartz wedge or rotary compensator may be required.

Mica Plate

The addition or subtraction of retardation when using the mica plate is about 140 nm. Thus in Figure 4.7b, if *A* is slow, the interference color will change, for example, from a first-order white (200 nm) to yellow (340 nm), or from a second-order blue (700 nm) to yellow green (840 nm). If *A* is fast, first-order white will change to dark grey and second-order blue to the sensitive-tint color.

There is little advantage in using the mica plate rather than the sensitive-tint plate. In my experience, students prefer the latter mainly because they use it anyway in preference to the mica plate for interpreting interference figures.

Quartz Wedge and Rotary Compensators

The quartz wedge works in principle in exactly the same way as the sensitive-tint and mica plates, but for the wedge, addition or subtraction of retardation can be accomplished for up to 3500 nm. It is therefore particularly useful in determining fast and slow directions for mineral sections displaying up to sixth- and seventh-order colors. If the mineral illustrated in Figure 4.7b displays such a color, and if *A* is the slow direction, insertion of the plate (check again that the slow direction of the wedge is NE–SW) will cause an increase in total retardation resulting in even paler higher-order colors. If, however, *A* is the fast direction, insertion of the plate will result in a progressive decrease in total retardation, and, at some position during insertion, compensation will become total and the section will go black. By observing carefully the position on the quartz wedge for which total compensation occurs, the original retardation of a highly birefringent grain can be more precisely measured.

A rotary compensator is simply a more sophisticated means of determining the precise amount of retardation. Using Figure 4.7 as an example, place whichever of *A* and *B* is the fast direction in the 45° position exactly parallel to the slow direction of the rotary compensator; then rotate the accessory plate until total compensation occurs. The degree of rotation can be read off and the retardation determined from the tables provided with the compensator.

The use of compensators to determine the positions of fast and slow directions with respect to distinctive crystal faces or good cleavages is an essential part of identifying many minerals. We shall learn how to do this and specify the nature of the optical directions with more precision in the later section on orientation diagrams. Compensators, particularly the sensitive-tint plate and quartz wedge, are also essential for interpreting interference figures.

Exercises

Using Compensators

(1) Insert each plate in turn into the accessory slot (without a thin section on the stage), and examine the interference colors produced in crossed-polarized light.

(2) Obtain thin sections of a muscovite-rich granite, an olivine-augite-gabbro, and a calcite-cemented quartz sandstone, as for the last exercise.

(3) Select crystals of quartz, augite, olivine, muscovite, and calcite displaying first-order white, second-order blue, third-order colors, and very-high-order colors, respectively.

(4) For each crystal in turn, place the grain in extinction, make a sketch of it, and superimpose on the sketch the traces of the cross-hairs as shown in Figure 4.7a. Label the two vibration directions *A* and *B*.

(5) Rotate successively clockwise and anticlockwise so that both *A* and *B* are placed in turn in the NE–SW position.

(6) Observe the changes in interference colors for each position and for each accessory plate. The account immediately preceding this exercise gives a detailed explanation of the changes expected. For olivine, examine the changes to the color rings at grain edges, too.

(7) Label each of the vibration directions *A* and *B* fast and slow as a result of your observations.

(8) For each mineral, place the fast direction exactly in the NE–SW position and insert the quartz wedge (slow direction NE–SW) until complete compensation occurs (i.e., the mineral goes black). Remove the thin section so that the color of the quartz wedge is clearly visible; then withdraw the wedge slowly noting the color changes precisely. In this way the position on the wedge for complete compensation can be determined and the retardation of that mineral section measured. Complete compensation may not be possible with some calcite sections.

Interference Figures, Determination of Optic Sign, and the Measurement of 2*V*

Interference figures are obtained as follows: focus on a grain with a high-power objective (×40 or greater); make sure the grain is central in the field of view, and that the microscope is centered perfectly (see Chapter 2); raise the substage condenser, insert the accessory condenser (if the microscope has one) and open the substage diaphragm; use crossed-polarized light; insert the Bertrand lens, or remove the eyepiece and insert a pinhole stop; if there is a diaphragm in the microscope tube, partially close it. The stage should be rotated when observing interference figures so that the moving pattern of isogyres can be noted.

The primary use of interference figures is to give information on the indicatrix of the mineral. A secondary use is to identify more precisely the position of principal optical directions. Every nonopaque anisotropic grain in a thin section will produce some sort of interference figure, but often this provides little useful information. Luckily the most useful figures are also the easiest to find. These are the optic-axis figures (including the uniaxial cross), which can always be obtained from grains cut parallel to a circular section of the indicatrix; such grains display a zero partial birefringence (black or at least very dark grey in crossed-polarized light).

Optic-axis figures enable one to determine whether the mineral is uniaxial (or pseudo-uniaxial) or biaxial, measure the $2V$ if biaxial, and with the aid of accessory plates, determine whether it is +ve or −ve. Some acute bisectrix figures also provide complete information on the indicatrix.

We must also learn to interpret the interference figures from other grains. Thus, a centered flash figure characterizes sections cut parallel to X and Z (the optic-axial plane) of a biaxial mineral or cut parallel to the optic axis of a uniaxial mineral, and centered bisectrix figures enable us to specify the positions of Y and X or Z. This information is essential for understanding orientation diagrams (see later) and for the precise measurement of principal refractive indices (Chapters 5 and 6).

In standard thin sections, there is a practical limitation to assessing the nature of the indicatrix for very-low-birefringence minerals. The isogyres are so broad for such minerals that, for a birefringence of less than approximately 0.005, they essentially fill the field of view. This is one of the major drawbacks of thin sections 0.03 mm thick because thicker sections would provide discernible interference figures for such minerals. Isotropic (cubic) minerals are always black in cross-polarized light, and will not produce an interference figure.

Use of Accessory Plates with Interference Figures

It will be recalled that the ε' vibration directions radiate out from the center of a uniaxial cross-interference figure, whereas the ω directions are concentric (Figure 3.20). For a positive mineral, ε' is slow compared with ω; for a negative mineral, ε' is fast. Therefore, in the NE quadrant of the uniaxial cross, for a positive mineral, slow directions are more or less NE–SW; in the same quadrant for a negative mineral, slow directions are NW–SE (Figure 4.8). Insertion of an accessory plate with its slow direction NE–SW will therefore cause addition in the NE quadrant for the +ve mineral but subtraction for the −ve mineral. Because of the radiating and concentric patterns of ε' and ω, addition and subtraction will be produced in the alternate quadrants as shown in Figure 4.8.

The situation is a little more difficult to visualize for biaxial interference figures, but let us take as an example an acute bisectrix figure looking down the X optical direction (Figure 4.9). In (a) the stage has been rotated so that the isogyres have moved to the NW and SE quadrants and the OAP trace lies NW–SE. At the center of the field of view, the two vibration directions are Y perpendicular to the OAP and Z, corresponding to the refractive indices β and γ. Remembering that the figure represents an image produced by convergent light rays (essentially a three-dimensional picture of the indicatrix) we can see that, moving towards either optic axis, the value of birefringence decreases and the slower ray (Z') approaches Y in value. In this three-dimensional picture, the vibration direc-

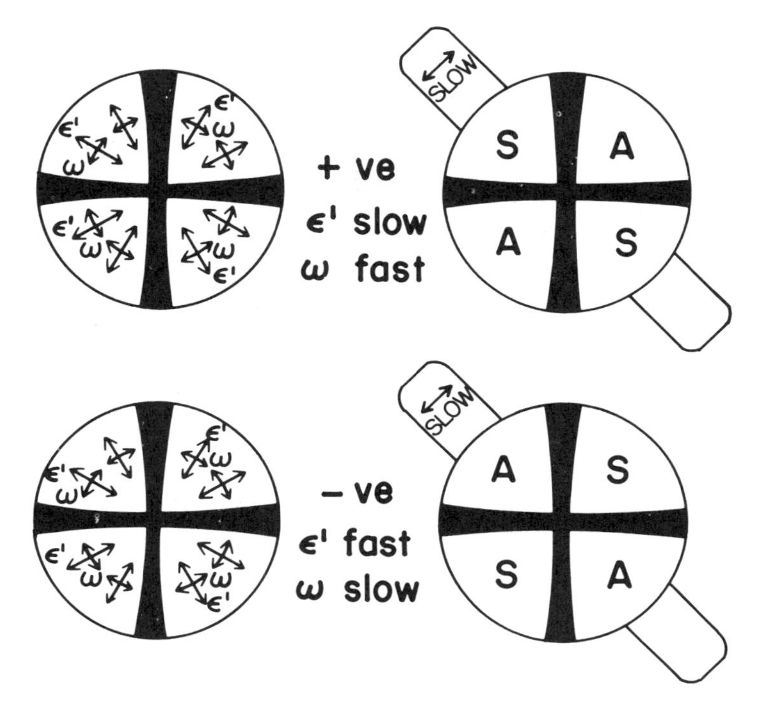

Figure 4.8 Patterns of addition (*A*) and subtraction (*S*) of retardations when an accessory plate is superimposed on +ve and −ve uniaxial cross figures.

tion perpendicular to *Y* on the other side of either optic axis becomes the faster ray or *X'*. Insertion of an accessory plate with its slow direction NE–SW therefore produces subtraction in the center of the field of view but addition on the other sides of the isogyres. When the stage is rotated 90° so that the OAP lies NE–SW as in Figure 4.9b, the vibration directions are similarly rotated 90°. Insertion of the plate now produces addition at the center of the field of view, but subtraction on the other sides of the isogyres. If we were to draw a bisectrix figure similar to Figure 4.9, but with *Z* vertical instead of *X*, we would find the areas of addition and subtraction reversed.

Use of the Sensitive-Tint Plate

A sensitive-tint plate is used to determine the patterns of addition and subtraction for low- to moderate-birefringence minerals when first-order white is the dominant or at least a conspicuous interference color between the isogyres of an interference figure. Even with several color rings present, a small area of white will be seen immediately adjacent to the

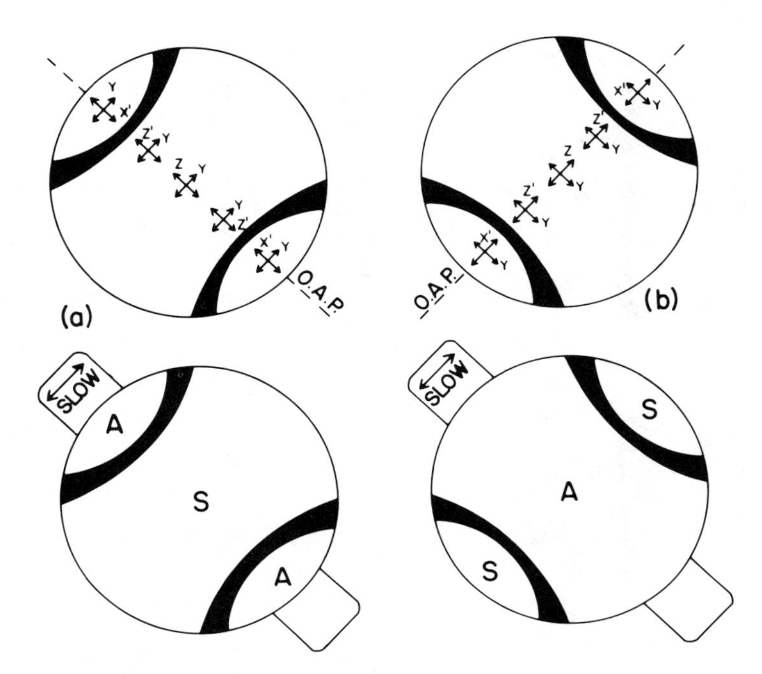

Figure 4.9 Patterns of addition (A) and subtraction (S) of retardations when an accessory plate is superimposed on a −ve acute bisectrix figure (looking down X).

position of an optic axis. The white represents a retardation of approximately 200 nm, and it will change to a blue or blue-green (760 nm) by addition, or to a yellow (360 nm) by subtraction, using the sensitive-tint plate (R = 560 nm). The black isogyres represent a zero retardation, and change to the sensitive-tint pink (560 nm) on insertion of the plate. If color rings are present they will increase or decrease by one order with addition or subtraction. Particularly distinctive is the change of the first color ring to black in the area of subtraction. The change from white to either blue (addition) or yellow (subtraction) is so easily observed that this will be used as the principal determinative guide on the diagrams that follow.

Use of the Quartz Wedge

For highly birefringent minerals, so many color rings may be present that first-order white is no longer conspicuous. In this situation, the quartz wedge is the most useful accessory plate. As the wedge is inserted (thin edge first), color rings will appear to move in towards an optic axis if there is progressive addition of retardation, but away from that optic axis if there is a progressive subtraction of retardation. As the wedge is removed, the rings move, of course, in the opposite direction.

Use of the Mica Plate

The mica plate may be used for moderate- to high-birefringence minerals, in which case the black isogyres disappear and small areas of black substitute for first-order white where subtraction takes place. For low-birefringence minerals such as quartz, the entire field of view changes to white or grey, and the pattern of addition and subtraction cannot be discerned. Although, in principle, the effects of the mica plate are readily analyzed, my experience is that students prefer the other plates. In the following, I have chosen therefore to omit illustrating its effects.

Interpretation of Interference Figures

In the following diagrams, *b* and *y* refer to the second-order blue and first-order yellow interference colors produced by addition and subtraction (from first-order white), respectively, on insertion of a sensitive-tint plate. The arrows refer to the movement of color rings when the quartz wedge is inserted. All diagrams assume the accessory plates have their slow directions NE–SW (or fast NW–SE). If plates are used with a slow direction NW–SE, all the color and arrow schemes should be reversed.

The isogyres appear red when the sensitive-tint plate is used.

Uniaxial Cross-Interference Figures (Figure 4.10)

Interpretations possible. Mineral is uniaxial or biaxial with 2*V* approximately 0°. The optic axis or axes are approximately vertical in the section. +ve or −ve character may be determined.

Figure 4.10 Distinction of uniaxial +ve from uniaxial −ve figures using a sensitive-tint plate or quartz wedge. *b* = blue; *y* = yellow; arrows indicate movement of color rings when wedge is inserted.

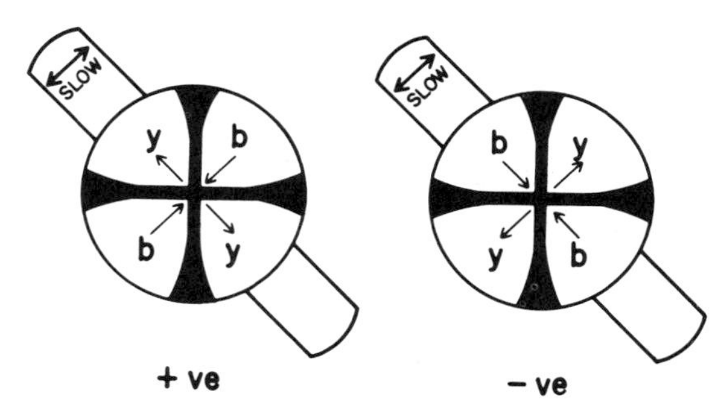

Notes. Off-centered figures can be used provided the center of the cross can always be seen (Figure 3.21a). In these cases the optic axis may be as much as 20–30° from vertical in the section. Off-centered figures in which straight isogyres move across the field of view but with the cross outside the field of view (Figure 3.21b) should not be used because they cannot be easily distinguished from some biaxial figures.

Uniaxial Flash Figures

These are similar to biaxial flash figures (page 78 and Figure 4.17). A centered flash figure from a uniaxial mineral indicates the section is cut parallel to the optic axis.

Biaxial Optic-Axis Figures—An isogyre remains centered in the field of view during rotation (Figure 4.11)

Interpretations possible. Mineral is biaxial, and an optic axis is vertical in the section. The degree of curvature of the isogyre when the OAP is in the 45° position enables $2V$ to be *estimated*. It is important to place the OAP precisely in this position before judging the curvature. This is best achieved by first rotating the stage until the isogyre is exactly straight and N–S or E–W, then rotating 45° from that position using the graduated scale of the stage as a measure. If $2V$ is 90°, the isogyre remains straight during rotation. The curvature becomes greater with decreasing $2V$ (Figures 4.11 and 3.24), and when the $2V$ is less than approximately 20°, an additional isogyre is seen towards the edge of the field of view. If the $2V$ is 0° or close to 0°, the figure is similar to or indistinguishable from the uniaxial cross. +ve or −ve character can always be determined (Figure 4.12). Remember that if the $2V$ is 90°, the mineral is neither +ve nor −ve.

The figure can be used to determine the positions of Y and an optic axis for orientation diagrams.

Figure 4.11 Measurement of $2V$ and location of Y and OAP using the biaxial optic-axis figure.

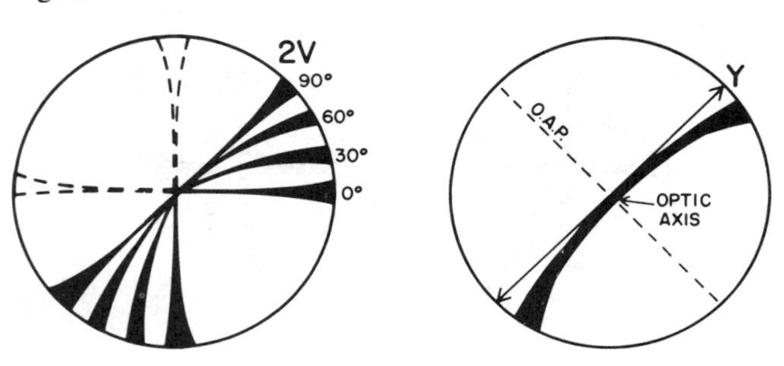

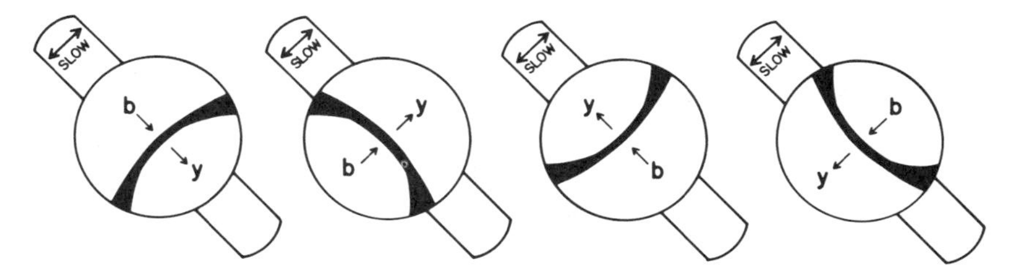

Figure 4.12 Optic-axis figure of a biaxial +ve mineral using a sensitive-tint plate or quartz wedge. *b* = blue; *y* = yellow; arrows indicate movement of color rings when wedge is inserted.

The field of view can be calibrated to enable a more accurate determination of 2*V*, as prescribed by Kamb (1958). The calibration requires that the RI β and the NA of the objective lens be known.

Acute Bisectrix Figures (Small to Moderate 2V)—Isogyres remain in field of view when stage is rotated (Figure 4.13)

Interpretations possible. Mineral is biaxial. The acute bisectrix is near vertical in the section. +ve or −ve character may be determined. 2*V* may be *estimated* from the maximum separation of the isogyres in the 45° position. The isogyres barely separate if the 2*V* is very small (less than 10°), whereas if the 2*V* is of the order of 50° or 60° the isogyres move to the edge of the field of view (Figure 3.23). The figure can be used to determine the position of *X, Y,* and *Z* for orientation diagrams. Off-centered figures may be used provided the isogyres clearly cross and always stay in the field of view.

A rough estimation of 2*V*, as described above, is often sufficient for routine work. However, Tobi (1956) has devised a method for more precise measurement using centered acute bisectrix figures; the NA of the objective lens and the RI β must be known. The traces of the optic axial plane and *Y* are placed in the 45° position (rotate exactly 45° from where the isogyres cross). A micrometer placed in the microscope eyepiece is used to measure (a) the diameter of the field of view (2*R*), and (b) the distance between the optic axes (2*D*), as shown in Figure 4.14. By using the ratio $2D/2R$ and the value of β, 2*V* can be read from a chart for any given NA. Tobi's chart for an NA of 0.85 is reproduced in Figure 4.15, and the dashed lines illustrate an example so that, for $\beta = 1.7$ and $2D/2R = 0.25$, $2V = 14°$. The chart is easily modified for objectives of other NA. If, for example, the NA is 0.65, simply multiply $2D/2R$ by 0.65/0.85 before using the chart. Off-centered figures may also be used provided the OAP is very close to vertical.

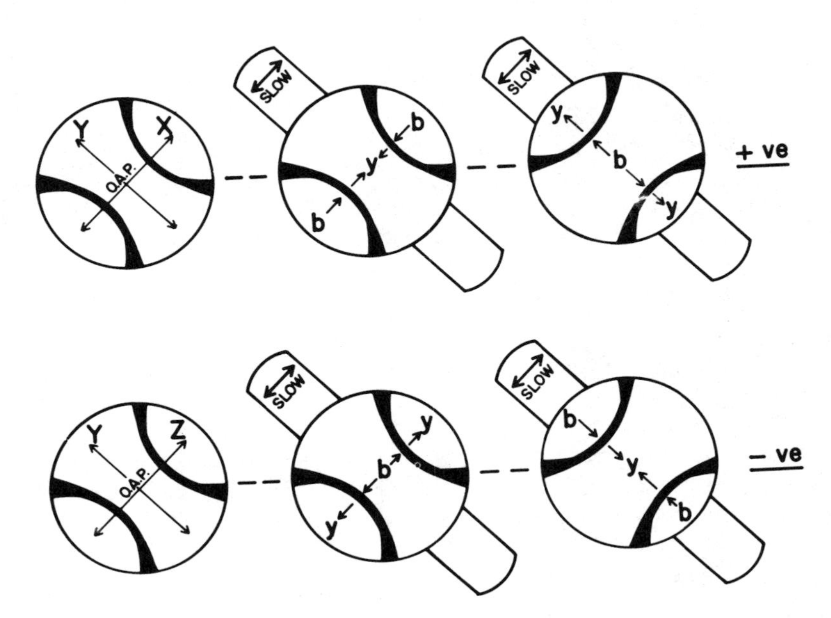

Figure 4.13 Distinction of biaxial +ve from biaxial −ve acute bisectrix interference figures using a sensitive-tint plate or quartz wedge. b = blue; y = yellow; arrows indicate movement of color rings when wedge is inserted.

Acute and Obtuse Bisectrix Figures (Moderate to Large 2V)—Isogyres move slowly out of field of view when stage is rotated (Figure 4.16)

Interpretations possible. Mineral is biaxial. The acute or obtuse bisectrix is near vertical, but they cannot be distinguished. Therefore, the +ve or −ve character of the mineral cannot be determined. The $2V$ is moderately large or large. Can be used to determine the positions of X, Y, and Z for orientation diagrams.

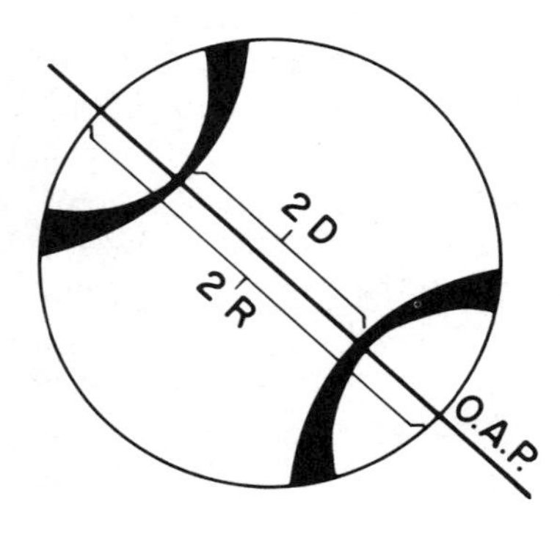

Figure 4.14 The distances $2D$ and $2R$ to be measured when using Tobi's chart (Figure 4.15) for determining $2V$ from acute bisectrix figures.

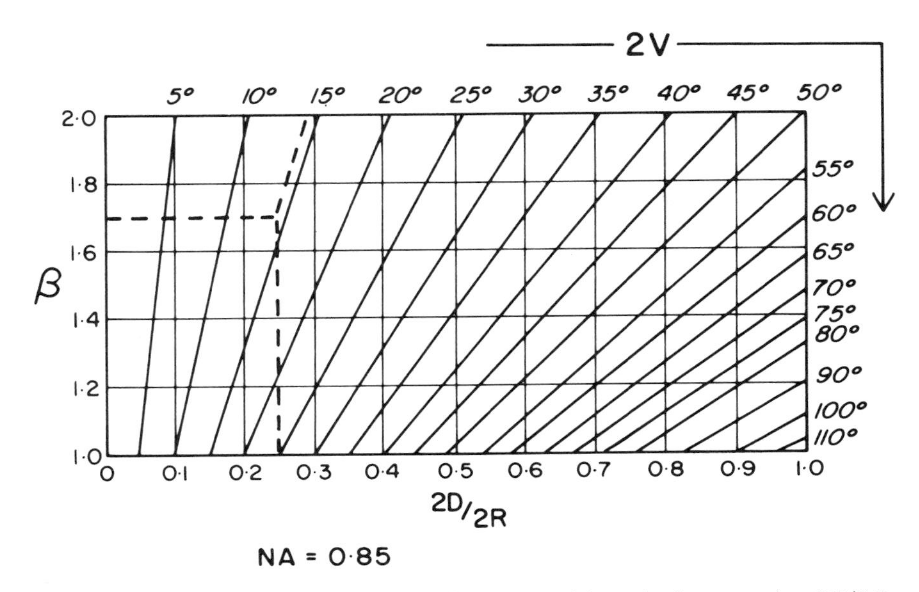

Figure 4.15 Chart for determining $2V$ from acute bisectrix figures using $2D/2R$ (Figure 4.14) and β (after Tobi, 1956). See text for full explanation.

Figure 4.16 Use of sensitive-tint plate or quartz wedge in bisectrix figures when isogyres move slowly out of field of view. b = blue; y = yellow; arrows indicate movement of color rings when wedge is inserted.

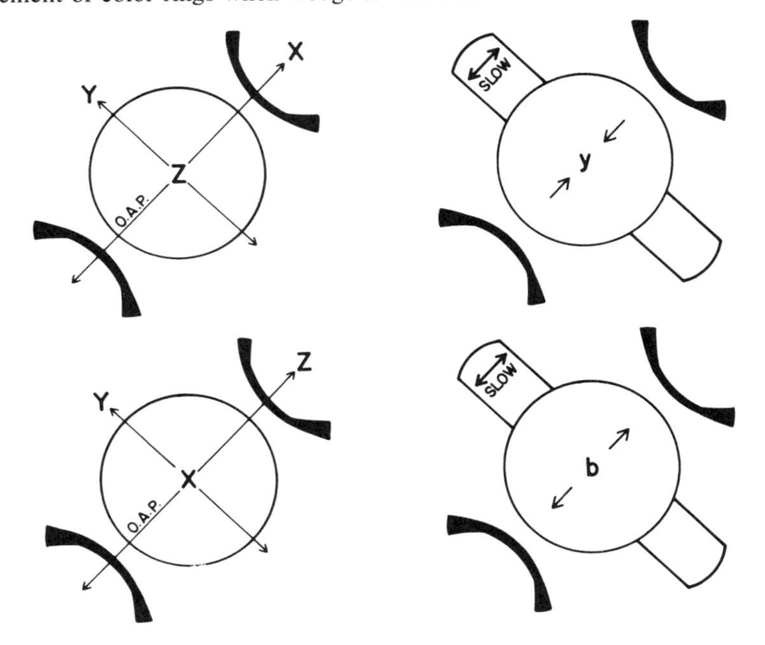

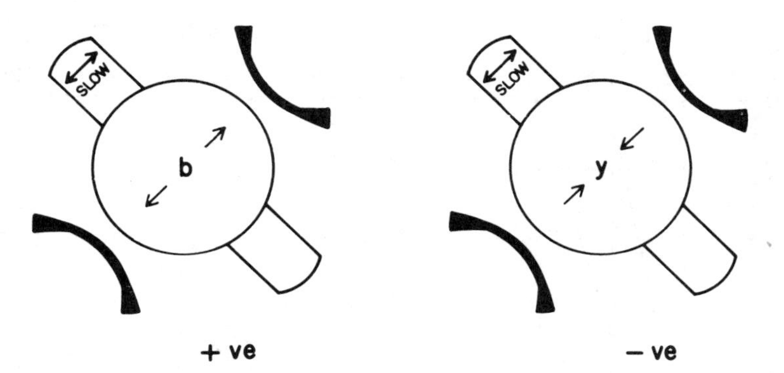

Figure 4.17 Flash figures from +ve and −ve minerals using a sensitive-tint plate or quartz wedge. b = blue; y = yellow; arrows indicate movement of color rings when wedge is inserted.

Notes. Off-centered figures may be used provided the isogyres clearly cross, and move slowly out of the field of view. It is necessary to obtain an optic-axis figure to determine $2V$ and the +ve or −ve character.

Obtuse Bisectrix Figure (Small to Moderate 2V) and Uniaxial Flash Figures—Isogyres move quickly out of field of view when stage is rotated (Figure 4.17)

These figures are called flash figures because the isogyres move out of the field of view very quickly (within approximately 30° rotation of the stage).

Interpretations possible. May be used to determine whether mineral is +ve or −ve, although it is advisable to check the determination with an optic-axis figure. Optic axes lie in or close to the plane of the section.

Notes. The figure does not distinguish biaxial and uniaxial minerals. It is necessary to obtain an optic axis or an acute bisectrix figure to determine $2V$.

Figures Produced When X and Z Lie Parallel to the Section

If the $2V$ is small, a flash figure similar to that described above is seen. If the $2V$ is large, a flash figure is produced in which the direction of separation of the isogyres cannot be seen clearly. Such centered figures may be useful in confirming the presence of an XZ section.

Exercises

(1) For success in obtaining interference figures, it is essential that the microscope be manipulated exactly as described on page 69. Start with large mineral grains and progress gradually to smaller grains, learning through experience what is the minimum practical grain size for obtaining figures.

(2) Using again thin sections of the muscovite-rich granite, olivine-augite-gabbro, and calcite-cemented quartz sandstone, search for zero or nearly zero partial birefringence grains of quartz, calcite, muscovite, augite, and olivine. The search usually requires a survey of the entire population of grains using both plane-polarized and crossed-polarized light. Remember that suitable grains are easily overlooked because of their atypical grey-black interference color.

(3) For each mineral in turn, obtain figures from the grains selected in (2) above. Persevere until a more or less centered optic-axis figure is obtained in each case.

(4) Note that suitable grains of muscovite are quite atypical in appearance, and do not display the platy habit or good cleavage. Note also that the birefringence of calcite is so high that only the most ideal grains display recognizably low-interference colors, and even those grains in conoscopic illumination may appear highly birefringent because of the convergent light. Note that suitable grains of calcite do not twinkle.

(5) Using a sensitive-tint plate (or for calcite, the quartz wedge), determine whether the minerals are +ve or −ve using Figures 4.10 and 4.12. Estimate 2V from Figure 4.11. The minerals selected provide examples of uniaxial +ve and −ve, biaxial +ve and −ve, as well as a range of 2V values.

(6) From each optic-axis figure, estimate the birefringence of each mineral using the method described on page 64.

(7) Optic-axis figures are always useful, and with some practice the easiest to obtain. Once these can be obtained and interpreted with confidence, examine interference figures from any other grain. Unless the movement pattern of color rings and isogyres is sensibly symmetrical, figures are difficult to interpret. Be very careful only to interpret them to the extent indicated in the preceding descriptions.

(8) XY and $\omega\varepsilon$ sections are characterized by centered flash figures. For each mineral they can be obtained from grains displaying the highest interference color.

The Effects of Dispersion on Biaxial Interference Figures

In Chapter 3 it was shown in principle how dispersion will result in differing values of 2V for different wavelengths of light, and for monoclinic and triclinic crystals, a variation in the positions of X, Y, and Z. In standard thin sections the visible effects are seldom marked except for the production of anomalous interference colors in a number of common minerals.

Similarly, the effects on biaxial interference figures are seldom marked; in some cases, however, color fringes will be seen along isogyres, and their relative disposition may assist identification. The effects are more marked in thick sections.

Let us take as an example the orthorhombic mineral sillimanite. $2V_z$ is larger for red light than for violet light (abbreviated $r > v$). Therefore, in an acute bisectrix figure, the isogyres (where light is extinguished) for red light are more widely spaced than those for violet light (Figure 4.18). Where red light is preferentially extinguished we see blue-violet, and vice versa. In other words, we see a red fringe on the inside and a blue fringe on the outside of the isogyres. The situation is reversed if the $2V$ is larger for violet light ($v > r$).

The amount of dispersion may be classified as weak, moderate (distinct), or strong, and is best observed in acute bisectrix or optic-axis figures.

In orthorhombic crystals, the positions of X, Y, and Z are fixed (though in rare cases such as brookite they may swap for one another). Therefore the distribution of color fringes in the figure is always symmetrical about the trace of the OAP and Y (Figure 4.19a and b).

For monoclinic minerals the dispersion may be inclined if $Y = b$ (Figure 4.19c), horizontal or parallel if b = obtuse bisectrix (Figure 4.19d and e), or crossed if b = acute bisectrix (Figure 4.19f and g).

For triclinic minerals, the patterns of dispersion may be quite irregular.

Although the dispersive effects are weak for many minerals and not readily determined, a few minerals such as brookite and some of the soda-amphiboles, for example, display such a strong dispersion that isogyre patterns may no longer be discernible in white light.

Figure 4.18 Acute bisectrix figure with $2V$ larger for red than violet light. The result (b) is a red fringe on the inside curve of the isogyre.

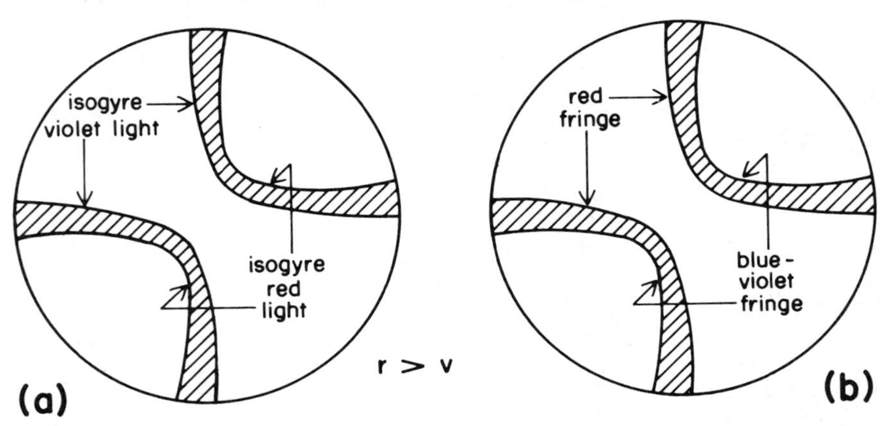

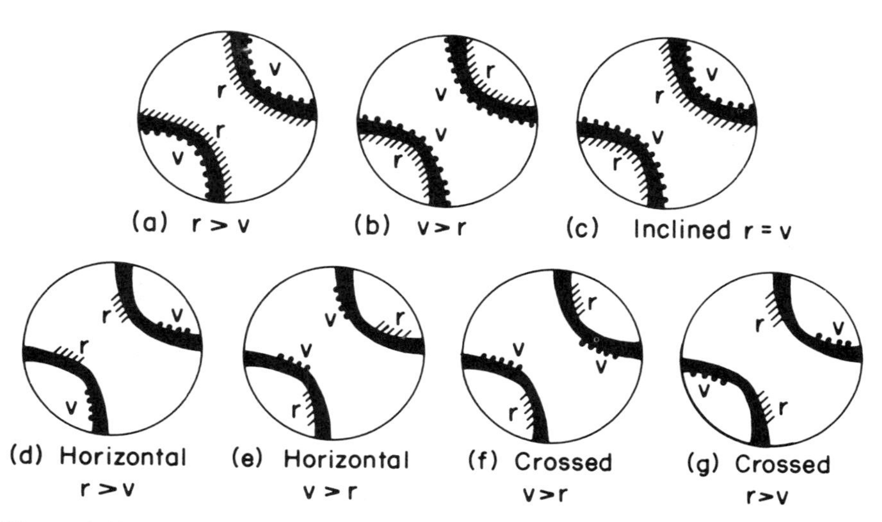

Figure 4.19 A variety of possible dispersion patterns in acute bisectrix figures. See text for discussion.

Pseudo-Uniaxiality and Pseudo-Biaxiality

Many biaxial minerals have a $2V$ of zero (or close to zero), and it may then be impossible to distinguish their "biaxial" interference figures from the uniaxial cross. Any visible dispersion of the optic axes will, of course, indicate a biaxial mineral, and in practice it will be found that some pseudo-uniaxial monoclinic and triclinic minerals have inclined extinction, thus establishing their biaxial character (see later section on orientation diagrams).

On the other hand, some cubic and uniaxial minerals display biaxial figures (pseudo-biaxiality). This effect can usually be ascribed to strain of the crystal lattice (as commonly found in quartz and calcite) or to the superposition of fine twin lamellae (as for example in calcite). Recent literature on this subject includes Foord and Mills (1978) and Turner (1975a and b).

Crystal Shape and Cleavage

Great care must be exercised in determining the shape of crystals from thin-section studies. For example, the two-dimensional appearance of an oblique section of a square prism is a rhomb (Figure 4.20).

Neither should it be forgotten that in many rocks the crystal boundaries are not crystal faces but mutual growth boundaries with adjacent grains. This is particularly true in metamorphic rocks where the shape of grains in a mosaic may bear no relationship to crystallography.

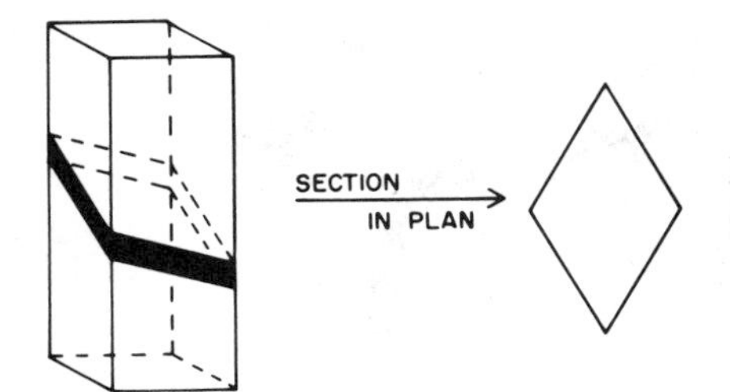

Figure 4.20 Rhomb-shaped section from a square prism.

Nevertheless, euhedral crystal shapes and habits can be judged in an *approximate* way if sufficient care is exercised, and they may be useful in identification of the mineral.

Cleavage is commonly observed as a set of cracks or bands in a mineral section. Cleavage vertical to the section appears as fine cracks, but cleavage oblique to the section will appear as dark broad bands. Cleavages more than 30° from vertical to the section are not normally seen at all. Information on the number of cleavages and their relationship to optical directions and crystal shapes is of considerable importance in the identification of minerals. It is useful to remember that cleavages are nearly alway parallel to simple crystal faces, e.g. (100), (110), etc., and that the presence or lack of cleavage in certain directions may reflect crystal symmetry (see Chapter 1).

To make the best use of crystal shape and cleavage information, orientation diagrams should be constructed, as described in the next section.

Orientation Diagrams and Extinction Angles

Extinction Angles

The angle between a vibration direction and a cleavage or prominent crystal face is called the extinction angle. If the angle is zero, the mineral is said to have straight extinction. If the angle is not zero, the extinction is inclined; sometimes extinction is symmetrical to the cleavages or crystal faces. The angle is determined by putting the grain in extinction and then rotating the stage until the cleavage or crystal face is parallel to the cross-hairs (polarizer and analyzer vibration directions), and noting the angle of rotation. This is a most useful property especially if it is properly related to precise optical directions. This can be achieved by constructing orientation diagrams. First, however, some general statements can be made with regard to extinction angles in minerals of the various crystal systems (Figure 4.21).

Figure 4.21 Typical extinction angles in various types of uniaxial and biaxial crystals.

TYPICAL EXTINCTION ANGLES

UNIAXIAL

(a) Prismatic Crystals

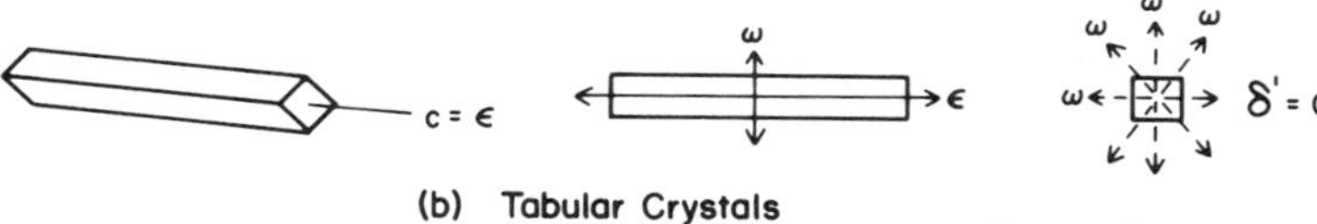

(b) Tabular Crystals

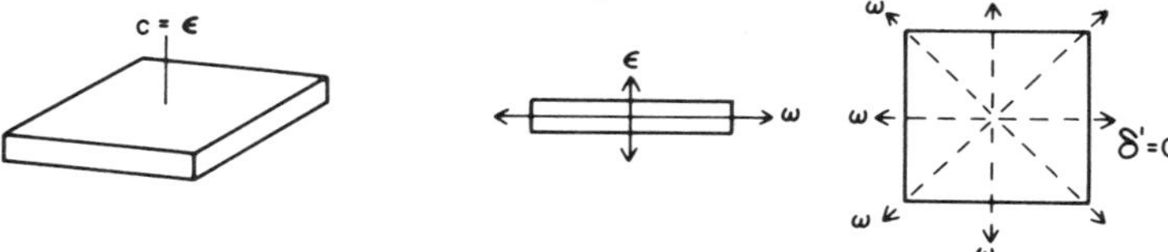

BIAXIAL

(c) Orthorhombic

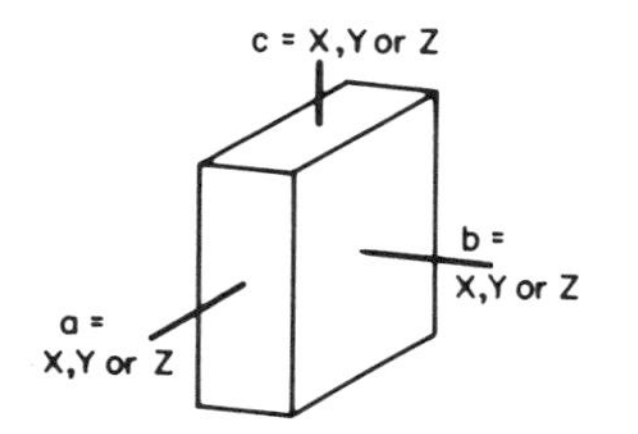

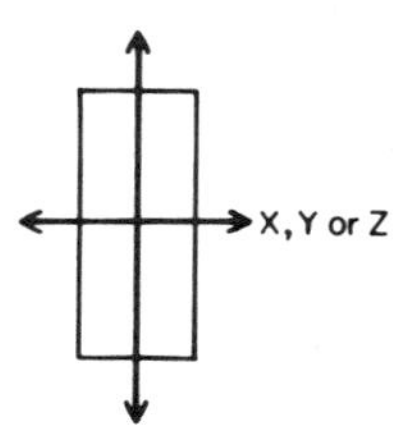

(d) Monoclinic

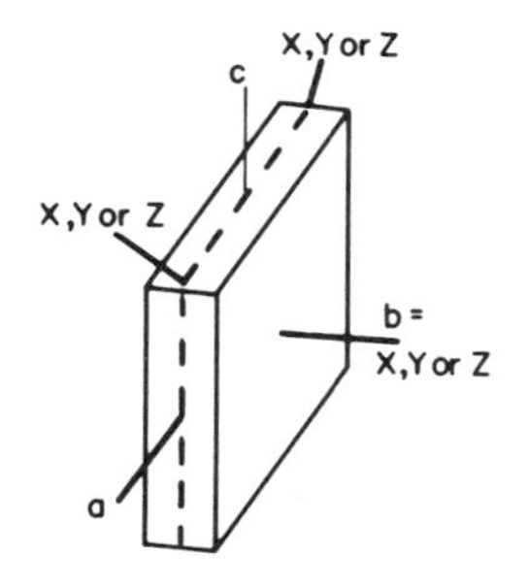

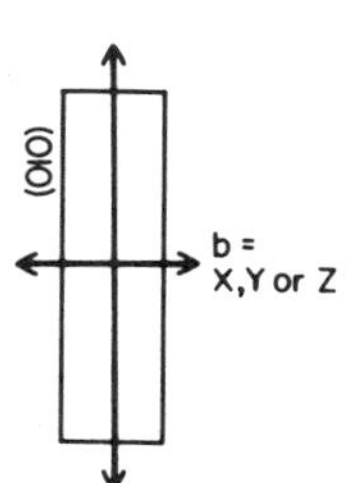

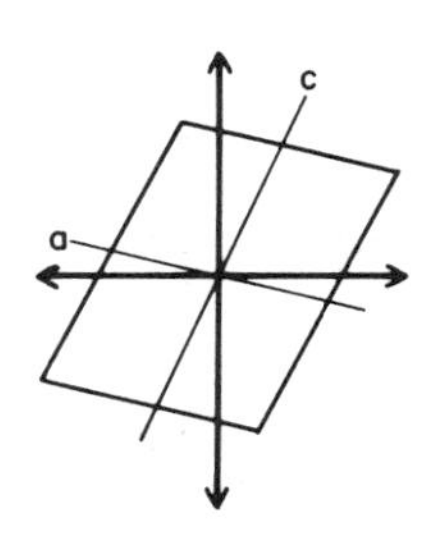

(e) Triclinic

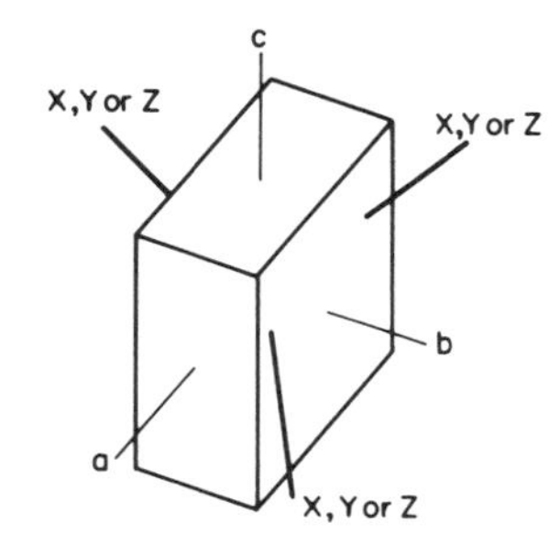

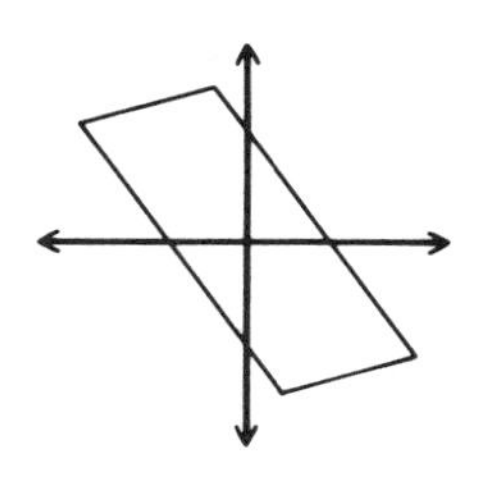

Hexagonal, trigonal, and tetragonal minerals are uniaxial; in other words the *c*-axis is always parallel to the optic axis and ε-vibration direction, and the *a*-axes lie in the circular section of the indicatrix. This relationship means that such minerals generally display straight or symmetrical extinction (Figure 4.21a and b), although there may be inclined extinction to nonvertical cleavage traces. Although orthorhombic minerals are biaxial, *X*, *Y*, and *Z* are parallel to *a*, *b*, and *c* (in any order), and straight or symmetrical extinction is similarly general (Figure 4.21c). In monoclinic minerals, however, only $b = X$, *Y*, or *Z*, and only in sections containing the *b*-axis is straight or symmetrical extinction to be expected; other sections generally display inclined extinction (Figure 4.21d). Triclinic minerals lack any systematic relationship between optical directions and crystal axes and therefore generally display inclined extinction (Figure 4.21e).

These are useful general rules, but it must be remembered that some monoclinic and triclinic minerals have a coincidental parallelism of crystal axes and optical directions that can be misleading. For example, many of the micas display perfectly straight extinction because *X* is exactly per-

Figure 4.22 Relationships between sign of elongation and the shape and optical sign of uniaxial minerals, and the optical directions of biaxial minerals.

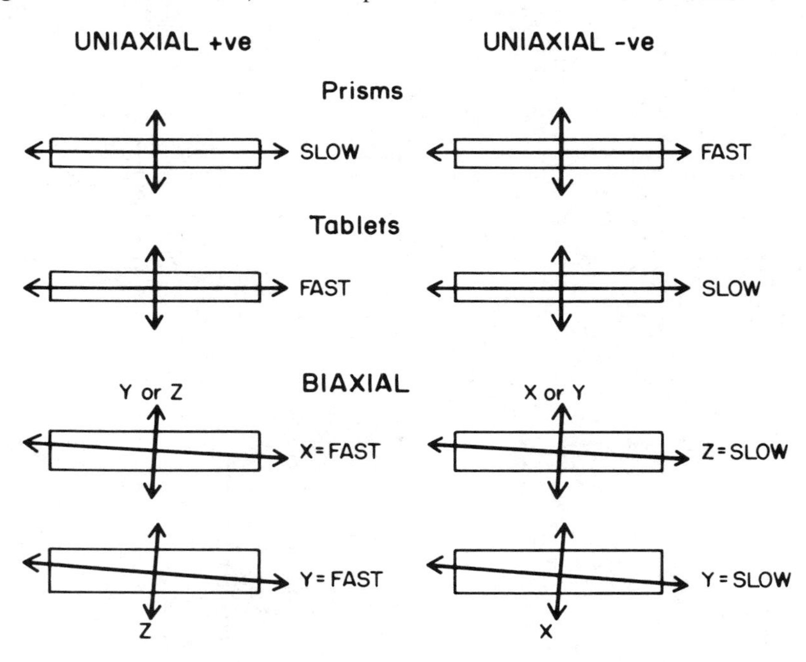

pendicular to (001), despite being monoclinic; furthermore, some micas are pseudo-uniaxial.

Sign of Elongation

Elongate mineral sections are described as *length slow* (possessing a *positive elongation*) if the vibration direction parallel or close to that elongation is the slow direction; they are described as *length fast* (possessing a *negative elongation*) if it is the faster direction. The method of determination of fast and slow directions has been described on page 66.

It is worth noting that *prismatic* uniaxial minerals (Figure 4.22) are +ve if length slow and −ve if length fast, whereas *tabular* uniaxial minerals (Figure 4.22) are −ve if length slow, and +ve if length fast. No such simple relationship exists with biaxial minerals. However, note that, if X is the optical direction most closely parallel to the length of a prismatic biaxial mineral, it will always be length fast; if it is Z, the grains will always be length slow; but, if it is Y, some will be length fast, others length slow (Figure 4.22).

Exercises

(1) Obtain thin sections containing the minerals sillimanite, tremolite, tourmaline, and wollastonite. All of these (except perhaps some tourmaline) are strongly prismatic or even fibrous.

(2) For each mineral, examine several sections showing the well-developed prismatic shape. Put each grain into extinction with its length more or less E–W, make a sketch of it, and superimpose the positions of the cross-hairs.

(3) Note that sillimanite (orthorhombic) and tourmaline (uniaxial) display straight extinction, but that tremolite (monoclinic) and wollastonite (triclinic) generally display inclined extinction, although the angle is very small for the latter.

(4) Measure the extinction angles for tremolite and wollastonite by rotating from extinction so that the prism length is exactly E–W. For tremolite they will vary from grain to grain between 0° and 20°; for wollastonite the maximum angle will be 5°.

(5) Rotate from extinction 45° anticlockwise so that prisms lie more or less NE–SW.

(6) Insert the sensitive-tint plate, observe the changes in interference color, and determine whether the crystals are length slow or length fast. Sillimanite and tremolite are always length slow, tourmaline always length fast, but because Y is along the length of wollastonite prisms, some of these will be length slow and the others length fast.

Orientation Diagrams

Throughout the systematic description of minerals in this (and other) books, orientation diagrams are employed to illustrate optical properties. It is essential that the student learn to use them, and that he be capable of constructing them for himself. There are two basic approaches to their construction.

1. Using Characteristic Optical Sections

A search is made for the sections that show the highest and lowest interference colors for the mineral. These are sought in any case for the determination of birefringence and the nature of the indicatrix.

(A) *Lowest interference color section.* Such a section is cut parallel to one of the circular sections of the biaxial indicatrix or perpendicular to the *c*-axis of a uniaxial mineral. Proceed as follows:

1. Obtain an interference figure. Determine whether uniaxial +ve or −ve or biaxial +ve or −ve (this will always be possible if the correct section is chosen). If biaxial, estimate $2V$.
2. If uniaxial, remove Bertrand lens, and view in plane-polarized light. Draw the section, noting any crystal shape (if euhedral, the section should be a basal one reflecting the hexagonal, trigonal, or tetragonal symmetry) and/or cleavages present.
3. If biaxial, rotate the interference figure into the 45° position. In such a position the orientations of the OAP and Y can be specified (Figure 4.23). Without rotating the stage, remove the Bertrand lens and view the section in plane-polarized light. Draw the section noting any crystal shape and position of cleavages if present. Superimpose on the drawing the position of Y and the OAP as determined from the interference figure.

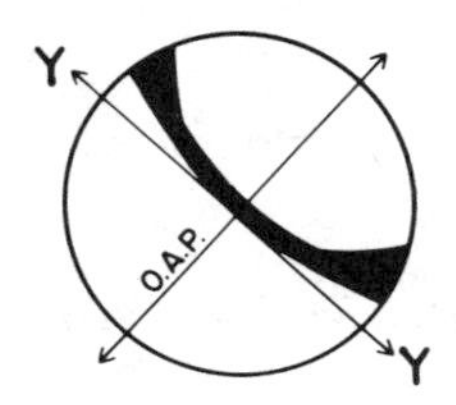

Figure 4.23 Use of biaxial optic-axis figures for specifying Y and the position of OAP.

(B) *Highest interference color section.* Such a section is parallel to X and Z of a biaxial mineral or parallel to the c-axis of a uniaxial mineral. Proceed as follows:

1. Put the grain into extinction, remove the analyzer and draw the section, noting any crystal shape and/or cleavages. The two optical directions can be marked as being parallel to the cross-hairs.
2. If the mineral is uniaxial we should already know from section A (above) whether the mineral is +ve or −ve. If uniaxial +ve, then the direction parallel to the c-axis (ε) is slow, if −ve it is fast. Replace the analyzer, rotate the stage 45°, and using an accessory plate determine the position of the c-axis by noting the fast and slow directions. Remember that uniaxial minerals *always* have straight or symmetrical extinction with respect to crystal shapes and cleavages in this section.
3. If the mineral is biaxial, replace the analyzer, rotate the stage 45°, and using accessory plates determine which direction is fast and which is slow. Mark these on the drawing as X and Z, respectively. Remember that euhedral minerals belonging to the orthorhombic system will have straight or symmetrical extinction with respect to crystal shapes and cleavages in this section. There will normally be inclined extinction in triclinic minerals. If X or $Z = b$ of a monoclinic mineral, there will be straight or symmetrical extinction, but if $Y = b$, there will normally be inclined extinction.

2. *Using Characteristic Crystallographic Sections*

A search is made for sections which are parallel or perpendicular to prominent crystal faces or cleavages. If the mineral is uniaxial, these sections will usually be the same as the two characteristic optical sections already studied under 1 (above). Because optical and crystallographic directions have more complex interrelationships in biaxial minerals, their characteristic crystallographic sections will not necessarily be the same as those studied under 1 (above).

Therefore, for biaxial minerals proceed as follows:

1. Look for sections that may be judged as parallel to important crystal faces, or that display cleavages perfectly vertical to the section. (Note that, if a mineral has one good cleavage, there should be sections parallel to it which do not appear to have a cleavage. The presence of such sections should be anticipated, and a careful search made for them. Similarly, minerals with two cleavages should be observed not only in sections displaying the two cleavages, but also in those displaying only one.)

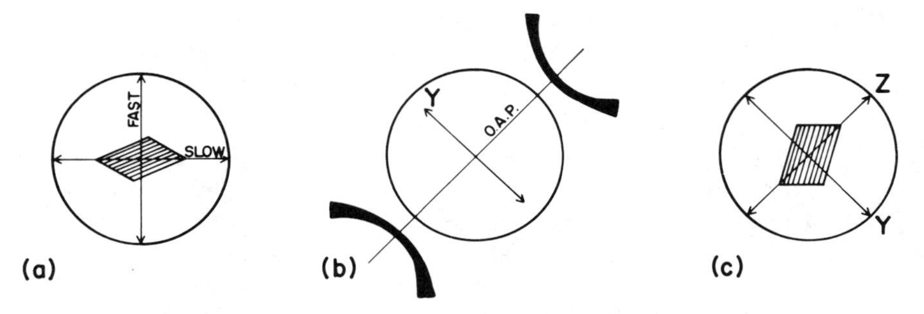

Figure 4.24 How to construct an orientation diagram.

2. Put the mineral into extinction, remove the analyzer, and draw the grain illustrating the crystal faces (if any) and cleavages (if any), and superimpose the orientation of the cross-hairs (which indicate the positions of two vibration directions).
3. Replace the analyzer, rotate the stage 45°, and using an accessory plate determine the fast and slow directions and note them on the drawing.
4. Obtain an interference figure.

Sometimes one is fortunate enough to obtain a centered bisectrix figure, in which case the position of X, Y, and Z can immediately be noted on the drawing of the mineral section. An example of the procedure is given below.

a. The mineral was drawn in its extinction position (Figure 4.24a), and by later rotation and use of an accessory plate, the fast and slow directions were identified and noted.
b. A bisectrix interference figure was obtained in which the isogyres rotated out of the NE and SW quadrants. We can specify therefore the positions of the OAP and Y (always perpendicular to the OAP) (Figure 4.24b).
c. Without rotating from this position, the analyzer and Bertrand lens were removed and the grain was observed (Figure 4.24c).
d. The slow direction in this case lies in the OAP and must therefore be Z. The section also contains Y. X is therefore vertical in the section.

Even if a noncentered figure is obtained, by combining the determination of slow and fast directions and the partial birefringence of the section, sufficient data may be accumulated to allow comparison with the diagrams in the systematic section of this book. If noncentered figures are always obtained on characteristic crystallographic sections, this in itself is significant, and usually implies that the mineral has a triclinic symmetry.

Color and Pleochroism

Transparent Minerals

Color in thin section is observed in plane-polarized light, usually with the diaphragm below the stage fully open. It is one of the first properties observed when inspecting a section, and is an important and sometimes diagnostic one. Minerals that may be colored are listed in Determinative Table IV.

Color is seemingly a simple and obvious property to describe, but to do this properly for the pleochroism that many minerals display (Chapter 3 and Figure 3.26) requires a thorough knowledge of the principles and techniques of optical mineralogy (which explains why this section concludes the chapter).

To determine the pleochroic scheme of a mineral proceed as follows:

1. Draw orientation diagrams, as outlined in the previous section.
2. In plane-polarized light, rotate the grains so that ω or ε, or X, Y, or Z are in turn parallel to the polarizer vibration direction. Note the color and its intensity for each optical direction that has been identified.

Opaque Minerals

Ideally, a polished section should be made and the mineral then observed with a special reflected light microscope, such as are used by ore mineralogists (Galopin and Henry, 1972). With an ordinary microscope, a limited amount of information can be obtained by shining a light over the specimen. Determinative Table X should be referred to for the colors of the common opaque minerals.

Exercises

(1) Obtain a thin section containing relatively coarse and randomly oriented euhedral hornblende.

(2) Make a general examination of the slide noting the presence of pleochroic sections perpendicular to c that display two {110} cleavages at 126°, sections parallel to (100) that display little pleochroism and no visible cleavage (because {110} planes are at too low an angle in these sections), and pleochroic (010) sections with a single cleavage trace.

(3) Find a zero partial birefringence section that displays no cleavage, and obtain an optic-axis figure. From a position where the isogyre is exactly E–W or N–S rotate 45° until the OAP is NE–SW. Determine the sign and $2V$ using a sensitive-tint plate. Remove the Bertrand lens and, without rotating the stage, sketch the grain and superimpose the positions of the OAP and *Y*.

(4) Find a section perpendicular to *c* (i.e., displaying two cleavages at 126°), and place it in extinction. Sketch it and superimpose the cross-hairs. Determine which vibration directions are fast and slow; then obtain an interference figure. Note the positions of the OAP and *Y* on your sketch.

(5) Find an *XZ* section (highest interference color and centered flash figure). Sketch the grain in extinction, superimpose the cross-hairs, and then determine which vibration direction is fast (*X*) and slow (*Z*). Measure the extinction angle between *Z* and the trace of the cleavage (= *c*-axis).

(6) Compare your results with Fig. 9.40.

(7) Using the grains already examined, place in turn *X, Y,* and *Z* exactly parallel to the polarizer vibration direction, and observe the color in plane-polarized light. Describe the pleochroic scheme.

(8) Repeat the above exercise with glaucophane.

(9) It is worth remembering that crystallographic data, such as *c*-axis, (010), (100), etc., cannot be specified, strictly speaking, until a mineral has been identified from its optical (or other) properties.

Chapter 5
Flat-Stage Techniques—Grain Mounts

It is customary for the sedimentary petrologist to examine friable or disaggregated sediments as loose grain mounts. Often the grains have been cleaned and separated into size fractions by sieving, or density fractions by using heavy liquids, before mounting on a glass slide. In comparison with thin sections, such mounts have the advantages that RIs may be determined with greater precision by immersion in a succession of liquids of known RI, and original grain shapes and sizes can be described more accurately. A disadvantage is that birefringence is less readily measured from interference colors, either because the grains vary in thickness to an unknown extent, or because they rest with a preferred orientation on a crystal face or cleavage; furthermore, such a preferred orientation may not allow easy determination of the nature of the indicatrix. Unless grains are permanently mounted, specimens are less easy to store than thin sections.

Similar to sedimentary grain mounts are crushed-grain mounts. These often provide the petrologist or mineralogist with the quickest and easiest means of mineral identification. For example, scrapings from the crystals in a rock may be collected, or crystals from a vein may be crushed, and examined on a glass slide. The advantages and disadvantages over thin sections are the same as for sedimentary grain mounts: grain shapes, sizes, and textural information are lost in crushed-grain mounts.

For thin sections, the universal stage provides an extra tool for more detailed work. For grain mounts, the spindle stage more appropriately fills that position, and is described separately in Chapter 6.

In the following descriptions it will be assumed that the reader has already mastered the techniques described for thin sections.

Sample Preparation

It is outside the scope of this book to describe the many methods of disaggregating sediments, cleaning and sieving grains, and concentrating suites of heavy minerals. Recent guides to the literature include Müller (1967), Carver (1971), Hutchison (1974), and Lewis (1983).

Permanent mounts may be made using Canada balsam or a variety of epoxy resins. An epoxy such as Araldite AY 105 with a RI close to the conventional balsam is preferred for routine work, not only to maintain a standard, but to allow distinctions between the feldspars and quartz to be made easily. Mounting is done on a hot plate set to a temperature and for a time appropriate to the cement used.

Temporary mounts are made by spreading the grains on a glass slide, covering them with a glass cover slip, and introducing an immersion liquid by capillary action from the edge of the cover slip until the grains are totally immersed (Figure 5.1). Naturally this works most easily for fine grains; coarse grains may need to be placed on a glass slide which has been hollowed out. For routine work, a liquid with a RI close to Canada balsam is preferable. This may be a mixture of petroleum oil and α-chloronaphthalene (as described below under RI determination), though clove oil (RI = 1.53) may be used instead. Clove oil requires careful handling inasmuch as it is an effective paint stripper and will quickly remove the black finish of microscopes if spilled.

Heavy mineral suites are composed of high RI grains, and it may be convenient to use an immersion liquid of RI = 1.633 (α-chloronaphthalene). In such a liquid, tourmaline and apatite are readily spotted because of their low relief, the inevitable stray grains of quartz and feldspar stand out because of their lower RI, and the difference in relief of xenotime and monazite versus zircon is visible.

If grains are scarce or need to be preserved, a temporary mount in gelatin can be made allowing examination in a succession of immersion liquids. The method, which is described more fully in Müller (1967), for

Figure 5.1 Immersion liquid introduced at the edge of a cover slip which is placed over mineral fragments.

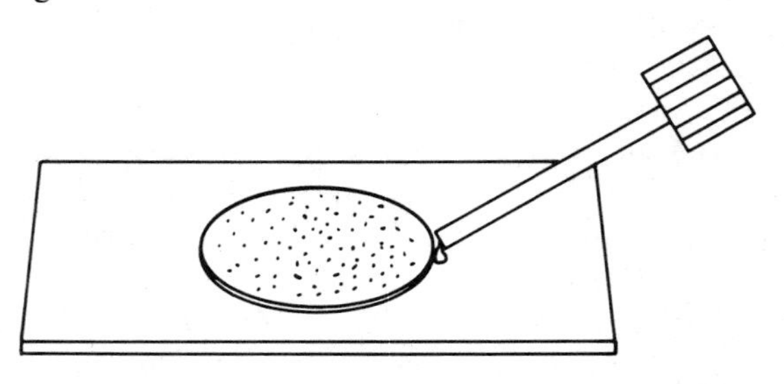

example, requires preparation of a 0.25 g/50 ml of water gelatin solution, which is dropped onto a glass slide and dried at 80°C. A second softening solution of 10 ml of water/5 ml of acetone and 2 ml of 2% formalin is applied before mounting grains, which are left projecting above the gelatin film. The slide is dried again at 80°C. A succession of liquids may now be introduced below a cover slip and washed away using acetone or carbon tetrachloride without disturbing the grains.

A thin section of a grain mount may be prepared, as described by Hutchison (1974) or Lewis (1983). In essence, the methods involve grinding one side of a slab of cemented grains until a large surface area of grain is exposed, remounting that side, and grinding the other until the standard 0.03-mm thickness is achieved. The section is processed in much the same way as an ordinary thin section.

Refractive Index Determination

A useful estimate of refractive index can be made by judging relief and using Becke lines, as for thin sections. However, the appearance of relief in grain mounts can be very misleading. Not only does variation in thickness influence relief (thicker grains show a higher relief), but the actual surface relief of grains in a mount is extremely variable. Thus, a smooth-topped grain will show less relief than a highly pitted or broken grain. The worker must, of course, know what the refractive index of the mounting medium is before judging relief and Becke lines.

Immersion Liquids

A more accurate means of determining RI is by immersing mineral grains in a series of liquids of known RI. The relief of the mineral will disappear when its RI matches that of the liquid.

It may be convenient to reduce the grain size of coarse material by crushing (some experience is necessary because too fine a powder cannot be easily observed whereas it is difficult to immerse too coarse a powder in the liquid). The grains are placed on a glass slide (a minute amount suffices), and covered with a small glass cover slip (between 8 and 18 mm in diameter is ideal). Drops of liquid are introduced by capillary action from the edge of the cover slip until the grains are totally immersed (Figure 5.1). By observing relief and Becke lines, the RI of the mineral is compared with that of the liquid. This procedure is followed with several liquids until one is found with the same RI as the mineral.

Suitable immersion liquids can be purchased from Cargille Laboratories Inc., Cedar Grove, N.J. 07009, but a cheap and effective series with a range 1.466–1.778 (with intervals of approximately 0.01 for routine work) can be made by mixing the following liquids:

1.466–1.633	petroleum oil (Nujol) + α-chloronaphthalene
1.633–1.739	α-chloronaphthalene + methylene iodide
1.739–1.778	methylene iodide + methylene iodide saturated with sulphur

α-Bromonaphthalene (RI = 1.655) may be used instead of α-chloronaphthalene.

Whatever liquids are used, they should be handled carefully so as not to dirty the microscope. Make a practice of washing hands after use because the skin may absorb α-chloro (or bromo) naphthalene with deleterious effects (although such effects have not been reported for mineralogists as far as I'm aware).

A variety of volatile liquids, including some of the alcohols and petroleum distillates, may be used for a lower range of RI (Harrington and Buerger, 1931; Larsen and Berman, 1934; Weaver and McVay, 1960). Liquids with RI higher than 1.778 are generally toxic, corrosive, and difficult to handle. Such liquids are readily available commercially from Cargille Laboratories. One high-index series can be made as follows:

1.655–1.814	α-bromonaphthalene + 10% S in $AsBr_3$
1.814–2.00	10% S in $AsBr_3$ + 20% S and 20% AsS_2 in 60% $AsBr_3$

Seldom are higher-index liquids required, but they may be prepared from various low-temperature melting compounds (Larsen and Berman, 1934; Meyrowitz, 1955).

The lower-RI liquid mixtures can be measured and periodically checked using a standard refractometer. Various Abbe types will measure liquids in the range 1.3–1.84 with an accuracy of ±0.0002, and the Leitz–Jelley type liquids in the range 1.3–1.9 with an accuracy of ±0.001.

This simple immersion method is often adequate for routine mineral identification. It is, however, a small step to measure the values of ω, ε, α, β, and γ separately within the limits of accuracy allowed by any given set of liquids.

Measurement of ω, ε, α, β, and γ

The method requires (a) identification of grains that contain one or two of the directions ω, ε, $X(=\alpha)$, $Y(=\beta)$, or $Z(=\gamma)$ in an orientation that can be set exactly horizontal in the slide and parallel to the polarizer vibration direction, and (b) a rather more careful matching of mineral and liquid RI. Identification of suitable grains is made as follows.

(a) *Uniaxial minerals.* ω, the ordinary ray, can be measured in any section of a uniaxial mineral. ω is the only direction present in sections that provide a centered uniaxial–cross-interference figure. In other sections, to determine ω it is first of all necessary to know whether the mineral is +ve or −ve, usually by means of an interference figure. ω is the fast direction in any section of a +ve mineral, the slow direction in any section of a −ve mineral; the fast or slow vibration direction, whichever is the case, is positioned exactly parallel to the polarizer vibration direction by rotating to the appropriate extinction position (the analyzer is then removed to compare the grain RI with the immersion liquid). Alternatively, ω can be determined as the direction of constant RI in several sections of varying orientation (and the sign indirectly established). ε can only be measured in grains with the *c*-axis horizontal. Such grains will show the highest interference colors for a particular thickness, and should provide a centered flash figure. Euhedral prismatic crystals will naturally lie with their *c*-axis (and hence ε) parallel to the slide. In the case of tabular grains or crystals with a perfect (001) cleavage, the *c*-axis naturally lies at a high angle to the slide, and it may be necessary to use a spindle stage to rotate ε into a suitable orientation. Similarly, the trigonal carbonates (e.g., calcite) have such a perfect rhombohedral cleavage that it is impossible to measure ε directly without the addition of a spindle stage.

(b) *Biaxial minerals.* β can be determined in any direction in a grain that lies with the circular section of the indicatrix parallel to the slide (i.e., a grain that provides a centered optic-axis figure). α and β, or γ and β, may be determined on grains providing bisectrix figures (see Chapter 4 on orientation diagrams on how to identify these directions). α and γ may be determined in a grain that shows the highest interference color for a particular thickness (α being the fast direction, γ the slow direction), provided the grain gives a centered flash figure. A preferred orientation resulting from a well-developed shape or cleavage may prevent the measurement of α, β, or γ without the use of a spindle stage. For example, micas, when crushed, form a mass of flakes that always lie with the {001} cleavage parallel to the slide making it impossible to measure α. More generally, if the principal RIs cannot be measured, the fast and slow directions of any grain will provide two refractive indices α' and γ', respectively, which always have the relationship $\alpha < \alpha' < \beta < \gamma' < \gamma$; such imprecise measurements at least give a minimum and maximum value for β.

The matching of mineral and liquid RIs may be made (a) in plane-polarized monochromatic light, usually direct sodium light or using an appropriate filter to produce the equivalent or (b) in white light. With a monochromatic light source, relief of the grain should completely disappear and no Becke line should be visible when a precise match is achieved. In white light, a good match for yellow light can still be made by observing the differing dispersive effects of mineral and liquid. These are evident as colored Becke lines, which move in opposing directions when mineral and liquid approach a match. Figure 5.2 illustrates typical dispersion curves so that for liquid *A* the mineral has a higher RI than the liquid for red light but a lower RI for blue light. The RIs match for yellow light, and the intensity of the red and blue Becke lines (which move in opposing directions) will be more or less equal. Liquid *B* (Figure 5.2) does not match the mineral in RI for yellow light, and the intensities of the colored Becke lines will differ.

By using a suitably graduated series of liquids, measurements with an accuracy of ±0.002 are attainable. The matching of liquid and mineral, and measurement of the liquid RI (on a refractometer) should be made at the same temperature because RI changes with temperature. Commercially available liquids provide a correction factor d*n*/d*t* (this is of the order of 0.0005 decrease in RI per 1°C rise in temperature) and a temperature at which the given RI was measured.

Even more accurate measurements (±0.0002) can be made if both the temperature and the wavelength of light are varied systematically (Emmons, 1943). If that degree of precision is required, the spindle stage

Figure 5.2 Intersecting liquid and mineral dispersion curves, which explain colored Becke lines and their opposing movement directions. See text for full explanation.

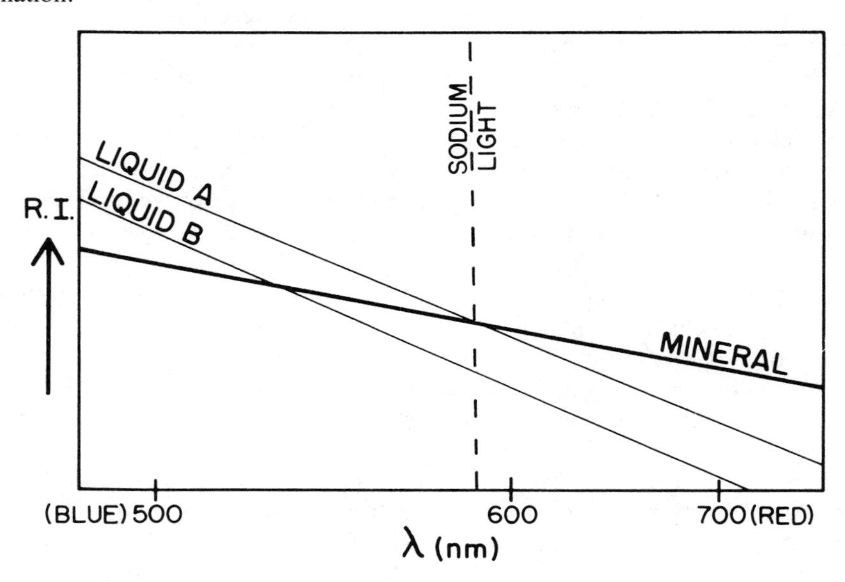

should be employed in order that the various optical directions be accurately positioned (Chapter 6).

Determination of Birefringence

If the RIs ω and ε, or α and γ have been measured using immersion liquids, these provide a direct determination of birefringence. If for some reason, such as a preferred grain orientation, only β and γ, or α and β could be determined, the other index and hence the birefringence can be calculated from the formula given on page 43, provided $2V$ has been measured too.

In permanent grain mounts, or for routine work generally, an attempt to estimate the birefringence of a mineral may be made by inspection of interference colors, as for thin sections. The method is, however, fraught with problems.

To use the Michel–Lévy chart, a requirement is that the thickness of the mineral under inspection be known. But seldom is this the case with grain mounts. Naturally a grain will rest on the glass slide so that its shortest dimension is approximately perpendicular to the slide; the thickness is therefore never likely to be greater than the smallest diameter (t') seen down the microscope (Figure 5.3). Because thicker grains of any one orientation produce a higher retardation, we must limit our search for the highest retardation among grains with a particular value of t'. Grains with a well-developed shape or cleavage rest on the slide with a preferred orientation and create a further problem. For example, micas always lie on their cleavage so that only a low partial birefringence can be seen. The best way to ensure that a grain suitably oriented for birefringence determination is present (i.e., an $\omega\varepsilon$ or XZ section) is by means of a centered flash figure. Any measure of δ made with this method must be regarded as a minimum value.

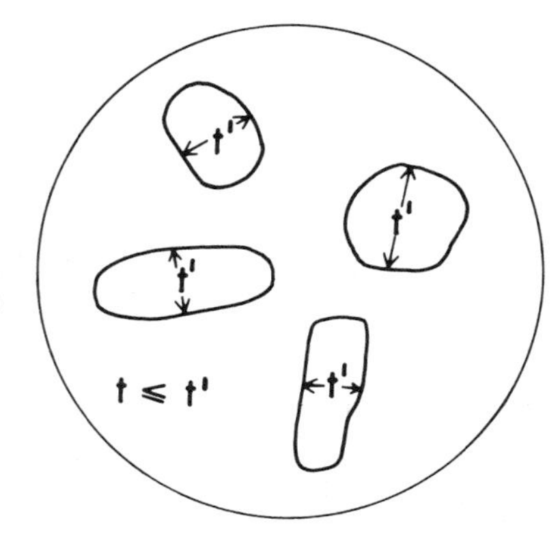

Figure 5.3 Dimension t' in grain mounts, which is always greater than or equal to the grain thickness.

Determination of birefringence from optic-axis figures (page 64) is similarly beset by uncertainty over the thickness of grains.

Crystal Shape, Cleavage, Extinction Angles, and Interference Figures

Many sedimentary grains retain vestiges of their original morphology or have shapes controlled by important cleavages. The photographs and drawings in Milner (1962) and Devismes (1978) provide excellent references.

Again it should be remembered that grains will naturally lie on a well-developed cleavage and have their shortest dimension perpendicular to the slide. If a set of grains, particularly crushed grains, consistently display one cleavage, it is probable that the grains have two cleavages and are resting on one of them; similarly, grains that have two cleavages visible are likely to possess three.

The presence of a particular shape or cleavage is in itself an important clue to mineral identification, but to be of most use, they should be related to optical directions using the techniques described on pages 82 and 95. Perhaps the most important points to remember are that well-shaped grains of uniaxial and biaxial orthorhombic minerals are characterized by straight or symmetrical extinction whereas monoclinic and triclinic minerals are more generally characterized by inclined extinction.

Interference figures should be obtained wherever possible, but obviously a strong shape factor or cleavage may prevent distinctive figures being obtained. With some experience, useful deductions can nevertheless be made. For example, prismatic grains, which are length slow, have inclined extinction, and consistently provide flash figures, can be deduced to be biaxial +ve. Prismatic grains, which are length slow, have straight extinction, and consistently provide flash figures, are either uniaxial +ve or biaxial +ve and probably orthorhombic.

Color

Because the fragments in a grain mount are commonly thicker than grains in a thin section, their color will be correspondingly more intense. Many minerals, which are colorless in thin section, display distinctive colors in thick grains. Thus sillimanite may be blue or brown, and cordierite blue. Pleochroic schemes should be determined as described on page 89.

Exercises

Crushed grains using a set of immersion liquids with refractive indices graduated in steps of 0.01

A suitable set of liquids with RIs from 1.46 to 1.78 can be readily made from the mixtures described on page 94. Use a standard refractometer at normal room temperature for calibration. Alternatively, liquid sets may be purchased from the Cargille Laboratories.

The following exercises serve to illustrate not only the principles of RI determination, but also give practice in a very simple and quick method of identifying common rock-forming minerals. Naturally, for precise mineralogical work, more closely calibrated sets of liquid are required, and the use of the spindle stage is recommended (Chapter 6). Nevertheless, for most routine work, the degree of accuracy that can be achieved as in the following exercises is adequate.

Although the exercises are done with known minerals, the student is advised to pretend, as it were, that the minerals are unknown; the Determinative Tables (Chapter 8) should be used to confirm their identity, once properties have been measured.

(1) Obtain crystals of quartz, calcite, gypsum, actinolite, and halite. Crush up each mineral to a grain size of the order of 0.1 mm (you will learn to judge this from experience), spread a few grains onto a separate glass slide for each mineral, and place small cover slips over the grains.

(2) Introduce an immersion liquid of RI = 1.54 under the cover slips by capillary action, as shown in Figure 5.1. Examine each mineral in turn as follows.

Quartz (Figure 5.4)

(A) Find the grains first in crossed-polarized light. Note their jagged shapes and the conchoidal fractures. No cleavage is evident, and there is a uniform

Figure 5.4 Crushed-grain mount of quartz. Crossed-polarized light. View measures 2.1 × 1.4 mm.

range of interference colors from black up into the second or third orders (depending of course on grain size).

(B) Find a black or dark-grey interference color grain, and obtain the uniaxial cross figure. Use a sensitive-tint plate to establish the +ve character. From this information, we can say that ω is the fast direction in any grain.

(C) Choose any large grain, find its fast direction, and then place this direction exactly parallel to the polarizer vibration direction by rotating to the appropriate extinction position.

(D) Close the substage diaphragm and/or lower the substage system, and examine the relief and Becke lines in plane-polarized light.

(E) Note that the relief is low, and the Becke line colored. Therefore ω is close to 1.54, but the only strong (orange) line movement indicates $\omega > 1.54$. Clearly for a +ve mineral $\varepsilon > 1.54$, too.

(F) Because ω is so close to 1.54, choose a liquid with RI = 1.55 to immerse another set of grains. Again, set the fast direction parallel to the polarizer vibration direction for any grain. The Becke line now moves in the other direction indicating $\omega < 1.55$.

(G) To measure ε, we must find a high interference color grain (for any specific thickness), which provides a centered flash figure. Several grains will probably need to be examined before a suitable one is found. Such a grain contains both ω (fast) and ε (slow) vibration directions. Place the slow direction parallel to the polarizer vibration direction (rotate into the appropriate extinction position), and examine the Becke line movement in plane-polarized light. We find that $\varepsilon > 1.55$, although the line is again colored.

(H) Progress to a liquid of RI = 1.56. Find another suitable grain for measurement of ε, as in G above. We find that $\varepsilon < 1.56$.

(I) Referring to Determinative Tables VI and VIII (it is necessary to check biaxial +ve minerals with $2V = 0°$) we find that quartz is the only mineral (described in this book) with the optical properties determined above.

Calcite (Figure 5.5)

(A) In crossed-polarized light, the extreme birefringence is immediately evident, and the rhomb-shaped fragments indicate the presence of three cleavages; two are visible and the grains lie on the third. Note the symmetrical extinction.

(B) The cleavage is so perfect that all grains lie with the *c*-axis (= optic axis) at an angle of approximately 44° to the slide, and a centered figure cannot be obtained. The figure is such that we cannot properly judge the mineral to be

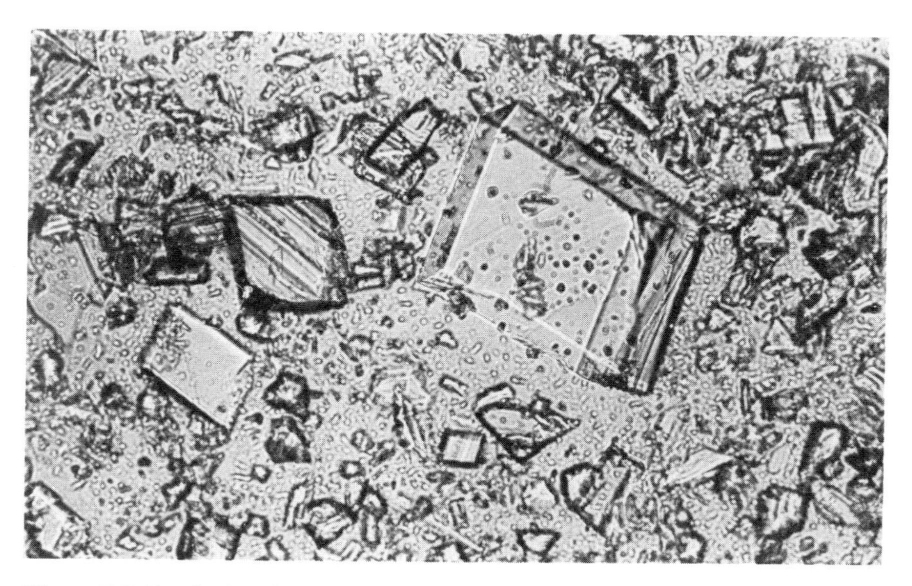

Figure 5.5 Crushed-grain mount of calcite. Plane-polarized light. View measures 0.9 × 0.6 mm.

uniaxial −ve; the possibility that it is biaxial −ve with a small $2V$ must be entertained (a quartz wedge may be required to determine sign).

(C) In plane-polarized light, note the change of relief (twinkling) when the stage is rotated. Both fast and slow directions have RIs > 1.54, although the fast direction is close to that value.

(D) If the mineral is uniaxial −ve, ω is the slow direction, and if biaxial −ve with a small $2V$, β will be close to γ and close to γ' in any grain. Therefore, inspect the grains in liquids with a RI much higher than 1.54 in order to match the slow direction. Because of the high birefringence and twinkling, the slow direction can be judged from relief; a sensitive-tint plate may also be used if low-order rings can be seen, otherwise use the quartz wedge.

(E) After making a number of trial matches, it will be found that the slow direction is very close to 1.66 (pure calcite has ω = 1.658), and colored Becke lines with $\omega > 1.66$ (red) and $\omega < 1.66$ (blue) will be seen (cf. Fig. 5.2).

(F) Inspection of Determinative Tables VII and IX shows that there is only one mineral with the above combination of properties, namely calcite.

(G) Because of the perfect cleavage, it is impossible to measure ε. Loupekine (1947) has constructed graphs of ε' for cleavage rhombs of the trigonal carbonates. Confirm that the RI for the fast direction (ε') is close to 1.57. Colored Becke lines will be seen with $\varepsilon' > 1.57$ (red) but < 1.57 (blue).

Gypsum (Figure 5.6)

(A) Like calcite, the presence of at least three cleavages is immediately obvious. Most fragments display two at 114°, but unlike calcite, the extinction is not symmetrical, thus indicating here a biaxial mineral.

(B) Fragments displaying two cleavages at 114° provide a centered flash figure and enable α (fast) and γ (slow) to be measured. Note that the angle between Z and one of the cleavages is 14°.

(C) Put X and Z in turn parallel to the polarizer vibration direction. Both α and $\gamma < 1.54$.

(D) Immerse another set of grains in a liquid of RI = 1.53. Selecting again a grain with 114° cleavages compare α and γ with the liquid. $\alpha < 1.53$ but γ is very close, and red and blue Becke lines are seen moving in opposite directions.

(E) Immerse another set of grains in liquid of RI = 1.52. In using the same type of cleavage fragment, note that α is now very close to the liquid with $\alpha >$ 1.52 for red light but $<$ 1.52 for blue light.

Figure 5.6 Crushed-grain mount of gypsum. Plane-polarized light. View measures 3.3 × 2.2 mm.

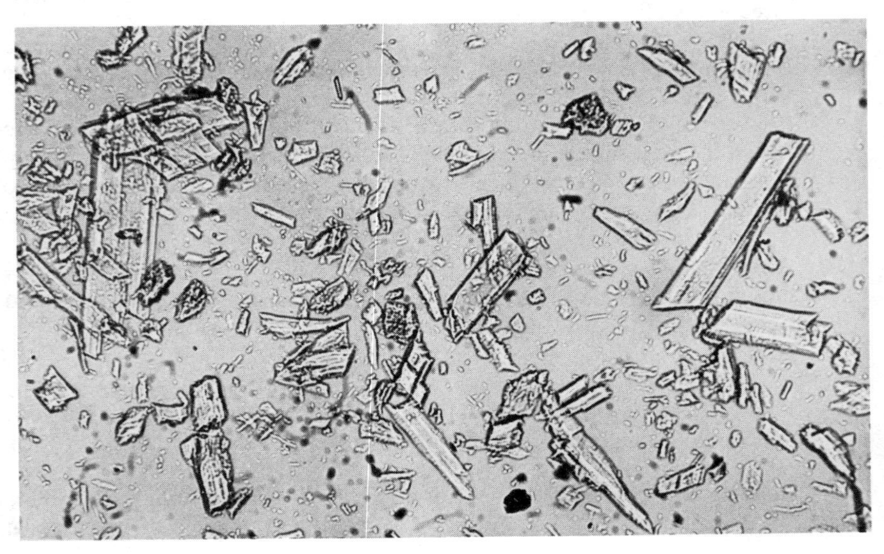

(F) From these measurements we can deduce that δ = ca. 0.01.

(G) It will have been noted that some fragments have straight extinction and display slightly off-centered bisectrix figures looking down X. A determination of β shows it to be much closer to 1.52 than 1.53, suggesting indirectly that the mineral is +ve. One is seldom lucky enough to find a fragment providing direct confirmation of this by way of an interference figure.

(H) Inspection of the Determinative Tables reveals that only gypsum has the above properties (distinguished from feldspars and cordierite, apart from its softness, by its cleavages).

Actinolite (Figure 5.7)

(A) The prismatic character is immediately evident, as is the presence of cleavages parallel to the length of prisms. In crossed-polarized light, inclined extinction demonstrates the mineral is biaxial.

(B) Note that extinction varies from straight to an angle of about 15°. All the grains are length slow.

(C) Grains with straight extinction provide off-centered acute bisectrix figures looking down X (note the mineral is −ve with a high $2V$). The OAP is vertical and parallel to the prism length, allowing β to be measured in such grains, as well as γ'.

Figure 5.7 Crushed-grain mount of actinolite. Plane-polarized light. View measures 3.3 × 2.2 mm.

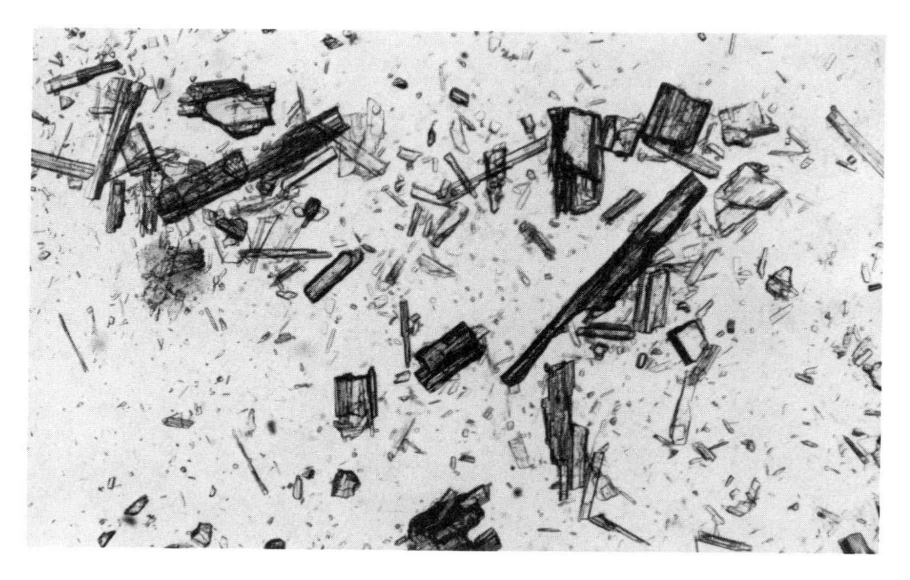

(D) Grains with inclined extinction provide consistently off-centered figures, allowing only α' (fast) and γ' (slow) to be measured.

(E) In plane-polarized light all RIs are clearly more than 1.54.

(F) With other sets of grains and several trial matches of RI, it will be found that RIs are somewhere in the range 1.62–1.66 (depending on the type of actinolite examined). δ is of the order of 0.02.

(G) Using Determinative Table IX, the only minerals with these properties, a prismatic habit and inclined extinction up to 15°, are the amphiboles, and actinolite can be quickly distinguished from the others.

(H) Referring to Fig. 9.40, note that the inability to measure α is a consequence of the perfect {110} cleavages. Grains naturally lie on (110) or $(1\bar{1}0)$ (off-centered figures and inclined extinction) or on a combination of the two (OAP vertical in what is effectively a (100) section with straight extinction).

Halite (Figure 5.8)

(A) The isotropic character is immediately evident in crossed-polarized light. Note also the perfect cubic cleavage.

(B) Isotropic minerals are characterized by a single refractive index n. It is not necessary therefore to place the grains in a specific position before matching RIs of mineral and liquid.

Figure 5.8 Crushed-grain mount of halite. Plane-polarized light. View measures 2.1 × 1.4 mm.

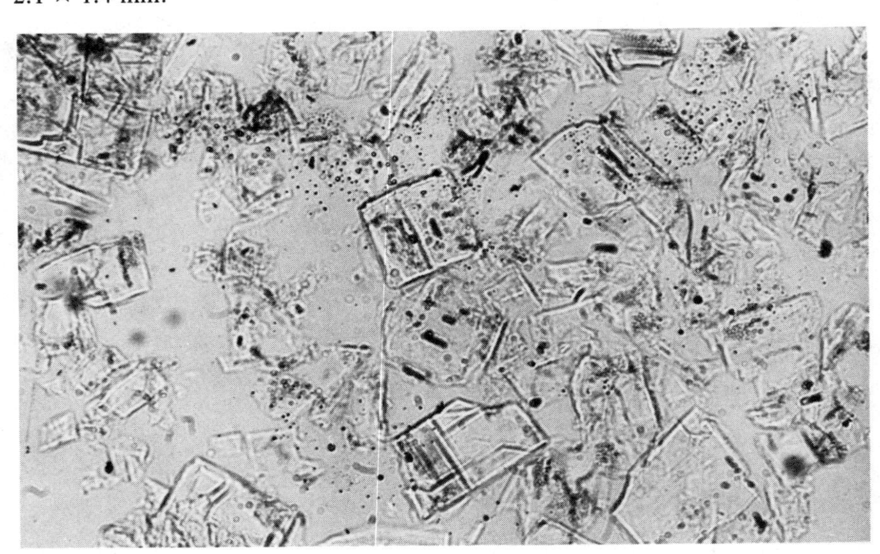

(C) In the liquid RI = 1.54, $n > 1.54$.

(D) Use a liquid RI = 1.55 with another set of grains to show that $n < 1.55$.

(E) Referring to Determinative Table V we find that halite is the only isotropic mineral with such a RI. We should also refer to the pseudo-isotropic minerals of Determinative Table III. The minerals apophyllite, dahlite, and kaolinite are easily eliminated as possibilities on the basis of their cleavage (or lack of cleavage) characteristics.

Chapter 6

Universal-Stage and Spindle-Stage Techniques

Although most of the common rock-forming minerals can be identified by using the flat-stage techniques described in Chapters 4 and 5, many of the properties are "estimated" rather than precisely measured. With experience and a certain amount of guile, the "estimates" may be very good, certainly good enough for routine work. Usually it is the limitation of only being able to rotate grains about the axis of the microscope stage that prevents more precise measurements from being made, and this can be overcome with either the universal or the spindle stage, both of which provide additional axes of rotation.

The universal stage (U-stage) is designed primarily for thin-section work, and its most important applications are in petrology. It enables the positions in space of the uniaxial optic axis and the biaxial optical directions X, Y, and Z to be located precisely and related to such crystallographic features as cleavage and twinning. The exact angle of $2V$ can be measured. This information may be useful for identifying problematical minerals, and is essential for reporting precise mineralogical data, although the U-stage has been largely superseded by the spindle stage for purely mineralogical work. The U-stage can be used for the determination of the anorthite content and twin laws of plagioclase (Slemmons, 1962), and for the distinction of the trigonal carbonates (Wolf et al., 1967). It is also used to determine the preferred orientation of mineral grains in a rock. The method involves determination of the orientation of various crystallographic features and/or the optical indicatrix for a large number of mineral grains in a single thin section; the results are plotted on a stereographic net to assess the degree of any preferred orientation. Examples of the use of the U-stage are given after the description of the equipment.

The spindle stage is designed for examining single mineral grains. Individual crystals can be oriented precisely for determination of the principal RIs and $2V$, and with temperature control and a variable wavelength monochromatic light source, indices can be measured to an accuracy of ± 0.0003 (Louisnathan et al., 1978).

Universal-Stage Methods

The Equipment

The U-stage described here (Figure 6.1) is that manufactured by Leitz, and has four axes of rotation. U-stages with three or five axes are also available. In fact, three axes suffice for most purposes.

The Axes of Rotation

With the central glass plate of the U-stage parallel to the microscope stage, the four axes can be designated as two "vertical" axes of rotation, A1 and A3, and two "horizontal" axes, A2 and A4.

A1	inner stage	initially vertical
A3	outer stage	
A2	N–S axis of tilt of inner ring	initially horizontal
A4	main E–W control axis	

The microscope stage axis is also vertical and is termed A5. A3 is not normally used, and it is best kept permanently clamped in zero position.

Figure 6.1 *Right:* four-axis universal stage manufactured by Leitz Wetzlar, Germany. *Center, from top to bottom:* circular glass plate; lower hemisphere; upper hemisphere. *Left:* upper hemisphere with a square mount and slide guide, used for petrofabric work.

The Hemispheres

Tilting a mineral section in a path of light soon results in a high degree of refraction and reflection which makes observations impossible. To avoid such reflection it is necessary to use glass hemispheres (Figure 6.1) both below and above the mineral section. If there is to be no deviation of the light ray on passing through the mineral and the hemispheres, the RI of the glass should equal that of the mineral. Glass hemispheres are manufactured with a variety of RIs, and they should be chosen to match that of the mineral to be studied. Corrections to readings necessary because of a difference in RI are easily made, and must always be made if the angle of tilt (between the normals to A1 and the microscope stage) is greater than 30–40° or the difference in RI is greater than 0.10. The study of minerals such as calcite, which have a very high birefringence, may necessitate corrections to be made to U-stage measurements. Errors due to small differences in RI of mineral and hemisphere are not normally large enough to cause concern in routine work, and are tolerated.

Corrections. These are simply made using the well-known Federow diagram (Figure 6.2). The way in which it is used is shown in the inset diagram. The observed angle is plotted along a radial line to the point where it reaches a concentric line representing the RI of the mineral. A vertical line is then drawn up to or down to the concentric line that represents the RI of the hemisphere, and the radial line intersecting this point gives the corrected angle.

Objectives

Unusually large working distances between the section and objectives are involved when working with the glass hemispheres. Special objectives must be used. As noted below, it is desirable that these objectives be fitted with individual diaphragms.

Illumination

Sharp extinction positions must be obtained to measure accurately the optical directions of minerals. An intense source of light is required. The diaphragms below the stage and within the special U-stage objectives are used to produce a good parallel beam of light, and to avoid reflection from the hemispheres.

Setting Up a Specimen

A small drop of glycerine is placed on the base of the circular glass plate supplied with the U-stage. The lower hemisphere is placed on the drop,

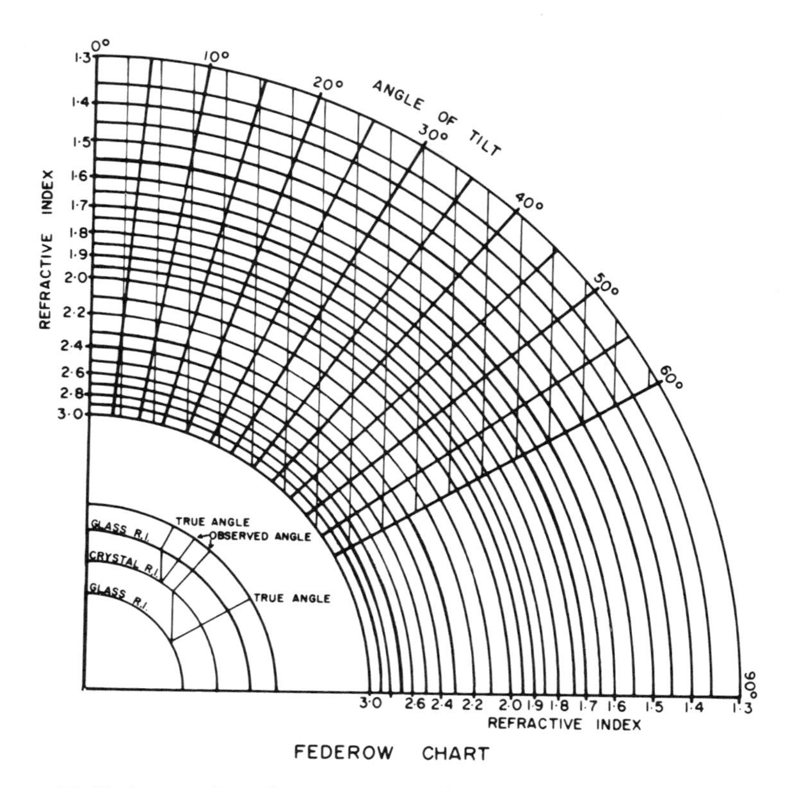

Figure 6.2 Federow chart for making corrections to universal-stage measurements.

thus sealing the contact area from air. The hemisphere and plate are then placed on the U-stage; a spring fitting prevents the hemisphere from falling. The thin section is placed on the glass plate and the upper hemisphere screwed into place on top, both contact surfaces of the thin section being sealed with glycerine. If the thin section has to be kept in a particular orientation (e.g., for petrofabric work), an upper hemisphere with a square edge mount is invaluable. This special mount may be fitted with a slide guide (Figure 6.1).

Adjusting the U-stage involves the following steps:

1. Center the microscope stage by rotating A5 and adjusting the centering screws of the microscope.
2. Clamp A5, then rotate A1 and center, using the centering screws on the U-stage.
3. Rotate A4 and A2. It is necessary that these rotation axes lie in the plane of the thin section. If they do not, the image will appear to rock backwards and forwards on rotation. Adjustment is made by turning

the threaded mount of the lower hemisphere (be careful not to mark the hemispheres with your thumb!).

4. Finally, focus on some irregularity or pitting on the top of the upper hemisphere (a special objective is usually provided for this). Adjust A5 to a position so that when A4 is rotated the surface irregularities move precisely parallel to the N–S cross-hair. Note the position of A5 and clamp.

Recording Measurements

Readings are recorded on tracing paper placed over a stereographic net. A small reference mark should be made on the tracing paper to coincide with the index mark on the A1 axis ring, which is the S pole on the Leitz stage. Data are normally recorded on the lower hemisphere of the net in petrological work. The methods of stereographic projection are described in Chapter 1.

Measurement of Planar Features

To measure the orientation of cleavage cracks and twin lamellae in a crystal requires the use of only A1 and A4. Therefore A2, A3, and A5 are kept clamped in zero position.

Each feature to be measured should first be rotated parallel to the E–W cross-hair using A1. Rotation about A4 is then made until the cleavage planes or twin boundaries are seen as the sharpest and finest possible lines. With a pleochroic mineral such as biotite, it is sometimes advisable to rotate the polarizer of the microscope so as to lighten the color of the mineral.

In noting readings, be careful to note whether A4 was rotated away or towards you (e.g., by an arrow such as 30°↑ or 30°↓).

To plot a reading such as (A1) 41°, (A4) 29° ↓ , rotate the tracing paper until the index mark reads 41° on the perimeter of the net. The pole to the planar feature is plotted 29° along the N–S diameter of the net from the north side of the perimeter (for lower hemisphere projection). Figure 6.3 illustrates the position of this pole (*P*) once the tracing paper has been rotated back to its reference position (note that the perimeter of the net is graduated in degrees with 0° at its south end corresponding to the graduations on the A1 axis of a Leitz U-stage). The great circle projection of the plane (*C*) was found by rotating *P* to the E–W diameter of the net, and drawing the great circle 90° from P.

An alternative to the above is to clamp A3, A4, and A5, and place the planar features in a N–S and vertical orientation using A1 and A2. Poles then plot along the E–W diameter of the net.

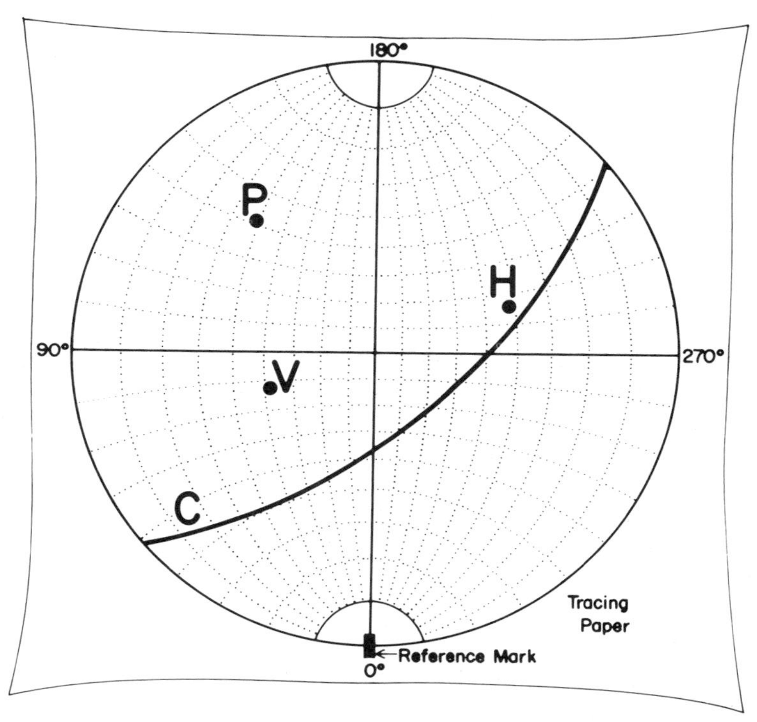

Figure 6.3 Examples of plotting a pole (*P*) to a planar feature (*C*), and two orientations of quartz optic axes (horizontal and vertical) determined with a universal stage, as discussed in text. Lower-hemisphere projection.

Determination of the Optical Indicatrix of a Mineral

Distinction of Uniaxial from Biaxial Minerals

The aim of standard U-stage procedures is to identify the symmetry axes of the optical indicatrix of a mineral. The number of axes differ for uniaxial and biaxial minerals.

Uniaxial minerals. The optic axis and all directions at right angles to the optic axis are symmetry axes of the indicatrix.

Biaxial minerals. The optical directions *X, Y,* and *Z* are the only symmetry axes of the indicatrix.

Distinction of uniaxial and biaxial minerals can normally be made as follows. If a uniaxial mineral is put into extinction by rotation about A1, in every case a symmetry axis will lie parallel to either A2 or A4, and the grain will remain extinguished when one of either A2 or A4 is rotated. If a biaxial mineral is similarly extinguished by rotation about A1, rotation about both the A2 and A4 axes will result in illumination except in the special cases where *X, Y,* or *Z* lies precisely parallel to A2 or A4.

The following sections outline the detailed procedure for determining the orientation of the indicatrix of uniaxial and biaxial minerals. Flow charts have been devised by Flinn (1973) as a practical alternative to the following written instructions.

Determination of the Optic Axis (c-axis) of a Uniaxial Mineral

The optic axis (*c*-axis) is the only direction that can be determined by examining the orientation of the uniaxial indicatrix with the U-stage. With a mineral such as quartz, which lacks visible crystallographic features such as cleavage, to obtain any information about the position of *a*-axes, it is necessary to use X-rays.

To determine the position of the optic axis (a symmetry axis of the indicatrix), we aim to set it parallel to A4, or if this is not possible, to set it vertical. Proceed as follows:

1. With all other axes at zero, rotate about A1 to an extinction position.
2. A trial tilt is made about A2. If extinction is maintained, the correct extinction position under (1) above was chosen (but see *notes* below). If the grain illuminates on tilting, A2 is restored to zero, and A1 rotated to the alternative extinction position.
3. A control tilt about A4 is made. The grain will normally show some illumination.
4. With the control tilt on A4 maintained (say 30°), extinction is restored by tilt about A2. The optic axis should now be either parallel to A4 or vertical, and extinction will be maintained on rotating A4.
5. Return A4 to zero and rotate the microscope stage A5. If the *c*-axis is parallel to A4 the grain will illuminate, if the *c*-axis is vertical the grain will remain extinguished.

Notes. A complication arises if the *c*-axis is initially horizontal in the section. Proceeding under (1) and (2) above, it will be found that such a grain in either extinction position about A1 will remain extinguished when tilting A2. The correct position is chosen by first tilting A4, and then tilting A2. If the grain remains extinguished the *c*-axis is N–S, and the alternative extinction position about A1 must be chosen.

With some experience, it is usually possible to anticipate whether the *c*-axis is to be set horizontal or vertical, and step (5) can be omitted. A grain with a near vertical *c*-axis has an interference color which is initially low, and which becomes progressively lower near the extinction position reached with the U-stage. On the other hand, a grain with a near horizontal *c*-axis has initially a higher interference color, and near the final extinction position reached with the U-stage, the interference color actually increases in order.

If the grain remains extinguished during rotation of A1, it must not be assumed that the *c*-axis is vertical. Errors of up to 20° are made in this way. The correct position of A1 under (1) above must be found by making trial tilts about A2 in various positions of A1 until extinction is maintained.

It is worth noting that the common strain shadows of quartz are nearly always subparallel to the *c*-axis. It will save time therefore if they are set E–W when finding the correct extinction position about A1.

Plotting the optic axis. Having set the optic axis horizontal or vertical, the A1 and A2 readings are taken. It is essential to note whether A2 has been tilted up to the left (L) or up to the right (R), and whether the optic axis is vertical. To plot a reading such as (A1) 20°, (A2) 40° (R)—horizontal, rotate the tracing paper until the index mark reads 20° on the perimeter of the net, and plot the optic axis 40° along the E–W diameter of the net from the east side (for the lower hemisphere projection). This is shown as point *H* on Figure 6.3. If the *c*-axis has been set vertical, say 40° (R), then it must be plotted 40° along the E–W diameter to the west from the center of the net. This is shown as point *V* on Figure 6.3.

Determination of the Orientation of the Indicatrix in Biaxial Minerals

The *X, Y,* and *Z* optical directions are all symmetry axes of the biaxial indicatrix. The aim of the following procedure is to set these axes in turn parallel to A4. Proceed as follows:

1. Rotate about A1 to an extinction position.
2. Tilt A4. The grain will normally illuminate. If so, restore extinction by tilting A2.
3. Retaining the tilt of A2, restore A4 to zero.
4. If the grain illuminates further, the procedure is repeated, first rotating A1 to extinction, tilting A4, restoring extinction by tilting A2, and returning A4 to zero, until extinction is maintained throughout the entire rotation of A4.
5. To determine whether *X, Y,* or *Z* has been set parallel to A4, rotate A5 to a 45° position.

   a. If it is *Y,* then the OAP is vertical with A4 as its normal. On rotating A4, a rapid fall of interference color will be seen with extinction if an optic axis can be brought into the vertical position. Sometimes both optic axes can be placed in a vertical position. For each optic axis, note the reading of A4 making sure to note the direction of rotation. A difficulty arises if the $2V$ is small and the acute bisectrix makes a small angle with the section. Distinction between Y and the obtuse bisectrix may then be difficult. To cope with such problems, it is best to examine first a grain of low partial birefringence, thus ensuring a determination of $2V$.

b. Except in those difficult cases mentioned above, if neither optic axis can be found, or if there is no noticeable fall in birefringence, then X or Z has been set parallel to A4. Keeping A5 in the 45° position, insert an accessory plate to determine whether the direction is fast (X) or slow (Z).

6. Find the orientation of the other optical directions by rotating to an alternative extinction position with A1 and following the above procedure. Sometimes the three directions can be directly determined, but if not, the third direction can be plotted normal to the other two.

The above procedure does not always work easily, particularly if an optic axis lies almost perpendicular to the microscope stage. In such cases it may be necessary to search for the correct tilt of the stage axes by trial and error; it helps if rotations are followed on the stereographic net remembering always that X, Y, and Z must lie at 90° to each other.

Plotting the results. The optical directions are plotted from readings of A1 and A2 in the same way as for the uniaxial optic axis. It should be checked that the three directions are at an angle of 90° from each other. The optic axes plot on the XZ plane. If only one optic axis can be directly determined, the other may be plotted at an equal angle along the OAP from X or Z. To plot a reading for an optic axis, the tracing paper should be rotated so that Y lies on the E–W diameter of the net. If the A4 rotation was away from you, the reading should be plotted on the great circle to which Y is the pole, towards the north from the E–W diameter (lower hemisphere projection). The axis is plotted to the south if A4 was rotated towards you.

Figure 6.4 shows how the following data can be plotted on the lower hemisphere projection: the optical direction Y was found at (A1) 10°, (A2) 30° (R); rotation about Y brought an optic axis into the vertical position at (A4) 30°↓; the fast optical direction X was found at (A1) 90°, (A2) 17° (L). Neither Z nor the other optic axis could be inspected directly. From this information the OAP was drawn as a great circle at 90° from Y. It passes through X and must contain Z which can be projected to lie 90° along the great circle from X. The second optic axis (which was too near horizontal to be viewed directly with the U-stage) must lie along the OAP equiangular from Z (in this case 50°). The $2V$ can be read off as $2V_z = 100°$ or $2V_x = 80°$. In practice, cleavage or twin-lamellae orientation for a particular crystal can be plotted on the same diagram allowing symmetry relations and angles such as $Z \wedge c$ to be determined.

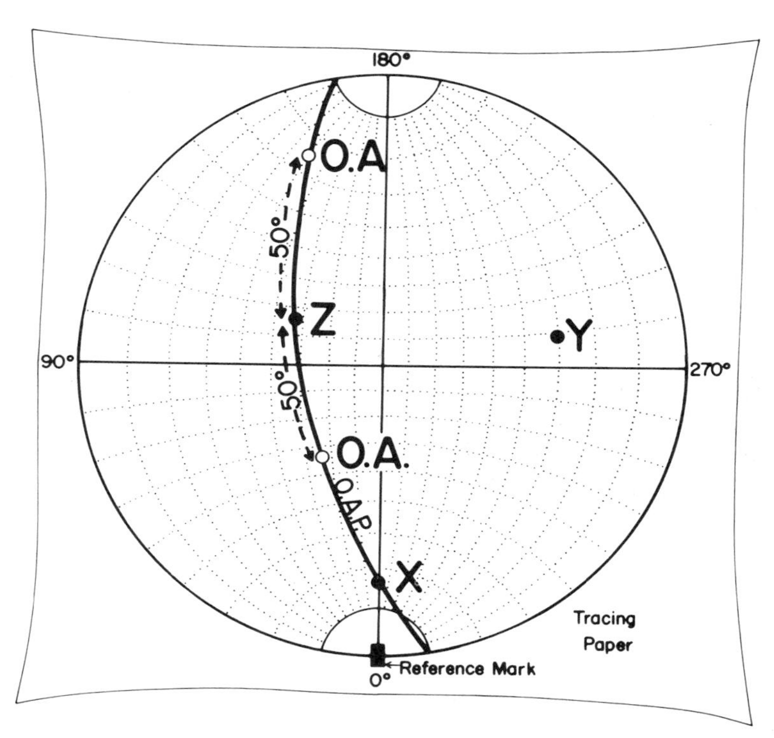

Figure 6.4 Lower-hemisphere projection of optical data for a biaxial mineral determined with a universal stage, as discussed in text.

Common Applications of the Universal Stage

Plagioclase Determination

The Michel–Lévy method for determination of anorthite content in plagioclase is described on page 257. The maximum angle of extinction between α' and (010) is determined with flat-stage techniques by a method of elimination involving examination of a number of grains. There is no guarantee that the optimum orientation has been found. This difficulty is easily overcome with the U-stage. With A3, A4, and A5 clamped in their zero positions, rotate A1 until the albite twins are N–S and rotate A2 until they are vertical (sharply bounded). With A1 and A2 set, unclamp A5 and A4 and measure the extinction angle between α' and (010) with the microscope stage (A5) for various trial positions of A4. The position of α' (the fast direction) must be checked constantly using a sensitive-tint plate. A4 is rotated until the maximum extinction angle about A5 is found. The anorthite content is read from Figure 9.67, using the high-temperature curve for volcanic plagioclase, the low-temperature curve for all others.

During rotation about A4, the section perpendicular to *a* (displaying two cleavages) may be encountered, and Figure 9.68 used; if Carlsbad and albite twinning are present, the changes in extinction angles can be plotted on Figure 9.70.

These methods are rapid, do not require a stereographic net for plotting results, and can be used with zoned crystals.

Olivine and Orthopyroxene Determination

These minerals form relatively simple, solid-solution series, and a determination of $2V$ (together with RI for orthopyroxenes) enables a good estimation of composition to be made (Figures 9.1 and 9.26). The angle of $2V$ is measured as described on page 114, and grains displaying a low partial birefringence should be chosen to ensure that at least one optic axis can be brought into a vertical position. The main problem likely to be encountered is in achieving sharp extinction positions when an optic axis is near vertical; the measured orientations of optical direction can be refined by checking that X, Y, and Z plot at 90° from each other.

Preferred Orientation Studies

The U-stage is the principal means of measuring the statistical preferred orientation of crystal axes for minerals in metamorphic rocks (less commonly igneous and sedimentary rocks). Results are always plotted on the *lower hemisphere* of an *equal-area Schmidt net* (Figure 1.8).

The orientation of a quartz *c*-axis (= optic axis) can be determined for any grain in a thin section, and typically as many as 300 or 400 measurements are made from the one section and plotted together on a stereographic net. To be of any use, the *c*-axis orientations need to be related to geographical coordinates and geological features such as bedding or schistosity. Therefore the section must be cut from the rock in a known orientation and kept in a fixed orientation on the U-stage. Special glass hemispheres with square edges and section guides (Figure 6.1) help to attain this. The basic methods of sampling, plotting, and contouring results are described in Turner and Weiss (1963). The experienced worker can measure and plot 300 *c*-axes in one day.

A similar method is often followed for calcite *c*-axes in marbles, though abundant twin lamellae may make observations difficult. The orientation of the twin lamellae themselves can provide information on stress and strain during metamorphism. The various procedures are discussed comprehensively by Turner and Weiss (1963) who point out that it may require 3 days' work to gain useful information from one specimen.

Measurement of the preferred orientations of biaxial minerals is similarly time-consuming in that a separate construction to determine X, Y, and Z directions is necessary for each crystal. The accumulated data are, of course, plotted together usually on three stereographic nets, one for each optical direction. Such information is most immediately useful for orthorhombic minerals such as olivine in which optical and crystallographic directions are parallel (for olivine: $X = b$, $Y = c$, and $Z = a$); consequently, there is a considerable literature on olivine orientations in peridotites (e.g., Nicolas and Poirier, 1976). For monoclinic or triclinic minerals such as feldspar, it is not usually possible to make a simple correlation between optics and crystallography; consequently, less work has been done even though strong feldspar-preferred orientations exist (Shelley, 1979). Several days' work may be involved to gain useful information from any one specimen.

Much easier to measure is the preferred orientation of (001) for sheet silicates. The cleavage planes are set either E–W or N–S and vertical, as described on page 110; poles to the planes are plotted together on one net. The principal difficulty encountered is that cleavage planes within about 45° of the thin-section plane cannot be directly observed. Details of the procedures involved are dealt with comprehensively by Turner and Weiss (1963).

Spindle-Stage Methods

The spindle stage is a remarkably simple and inexpensive piece of equipment that allows an individual mineral grain to be mounted on the end of a needle, immersed in a suitable liquid, and rotated 180° about a horizontal axis as well as 360° about the vertical microscope stage axis. These two rotation axes are capable of bringing all of the principal optical directions into a horizontal position parallel to the polarizer vibration direction, thus allowing all principal RIs to be measured in a single grain. The angle of $2V$ for biaxial minerals can also be computed with great accuracy.

Although the U-stage has been used in the past for such work, it is not really ideal because the A2 and A4 axes have limited rotations which do not necessarily permit all principal RIs to be measured. The U-stage is also very expensive.

It is not possible in this book to provide a completely comprehensive account of spindle-stage techniques. For that the reader is referred to the book by Bloss (1981), who has done much in recent years to promulgate and develop these techniques. However, the principles involved are straightforward enough, and I provide here an introduction to the subject.

The Equipment

For an illustrated account of a variety of spindle stages refer to Hartshorne and Stuart (1970). More recently, Bloss and Light (1973) have developed the *detent spindle stage,* this being an improvement on the earlier Wilcox spindle stage (Wilcox, 1959). The Bloss and Light stage (Figure 6.5) has a spindle, which can be made from a length of stainless steel tubing (from a hypodermic needle), bent into a right angle, and with the longer section fitted into a groove in the base plate of the stage; a clamp holds this spindle in position. The spindle can be rotated through 180° from its bent end, and the angles of rotation are read from a graduated semicircular protractor attached to the end of the plate; 10° click stops are provided. A needle is inserted into the end of the tubular spindle, and the mineral is glued to the end of the needle. A glass slide can be fitted into the recess of the plate so that the mineral grain rests in a small cell filled with an immersion liquid. The cell itself can be constructed from half a large staple over which is placed a small cover slip (Figure 6.5). Such equipment is sufficiently simple to be constructed in any laboratory, but the spindle stage illustrated is available commercially.

According to Bloss (1981), the optimum grain size for examination with the stage is between 0.074 and 0.149 mm, and the glue that he recommends for attaching the grain to the needle is a mixture of four parts water-soluble carpenter's glue plus one part crude "black strap" molasses; I have found modern rapid adhesives to be satisfactory too. Obviously the grain should only be touched with glue; otherwise any coating of the grain will prevent comparison of its RI with the immersion liquid.

The horizontal spindle-axis of rotation will be termed S and the microscope stage axis M in the following account. It will be assumed that the polarizer vibration direction is in the recently adopted international standard E–W position.

Figure 6.5 The detent spindle stage (after Bloss and Light, 1973) manufactured by Technical Enterprises, Blacksburg, Virginia.

Orthoscopic Extinction Angle Methods of Orienting a Grain

Having attached the spindle stage to the microscope (it can be held in position by a standard mechanical stage), the first step is to determine the reference position for M at which S (the spindle-stage axis) runs from the right-hand side of the stage exactly parallel to the polarizer vibration direction. This can be done by eye using a low-power objective and setting the spindle parallel to the E–W cross-hair.

A stereographic net is used to record extinction data for 10° intervals of rotation of S. The net is set so that the great circles intersect in S, and the great circles are labeled 0°–180° from top to bottom (Figures 6.6 and 6.7) corresponding to projection of data onto the upper hemisphere as S is rotated clockwise from 0° to 180°.

By using the microscope stage axis (M), the two extinction angles of the crystal are measured with S set first at 0°. These are plotted on the northern part of the perimeter (great circle 0°); examples are given in Figures 6.6 and 6.7. For a quartz crystal (Figure 6.6), the slow and fast vibration directions were brought in turn into an E–W orientation by 58°

Figure 6.6 Extinction curves for a quartz grain mounted on a spindle stage. Extinction data are given in Table 6.1. Upper-hemisphere projection. Full discussion in text.

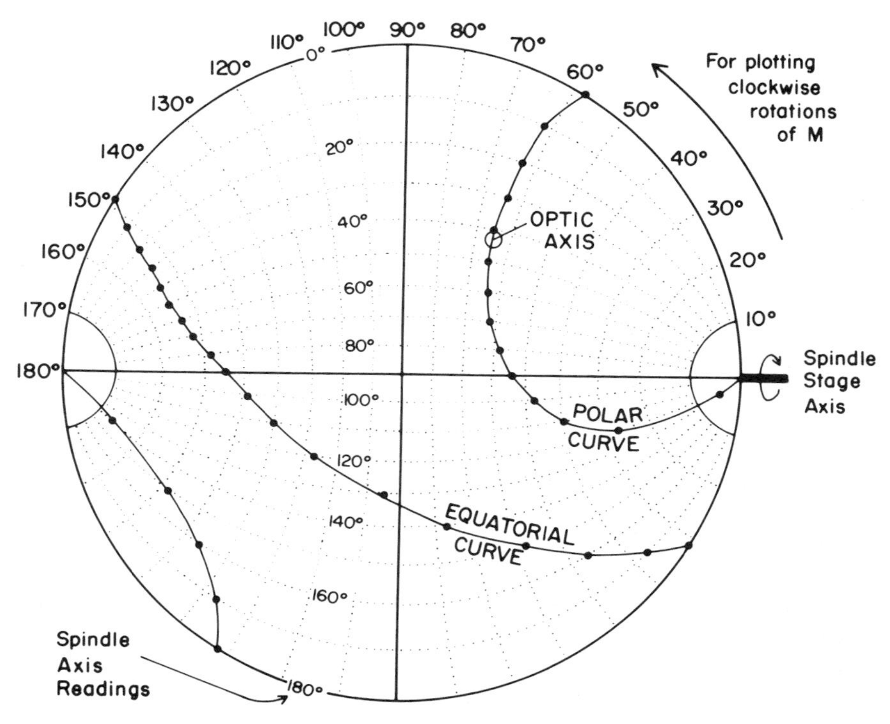

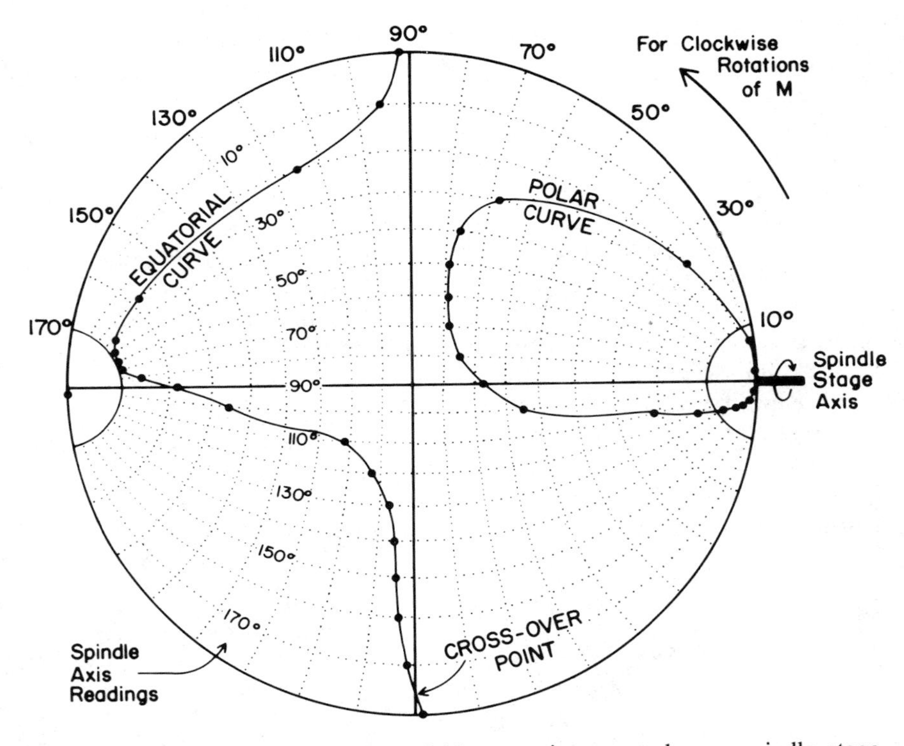

Figure 6.7 Extinction curves for a feldspar grain mounted on a spindle stage. Extinction data are given in Table 6.1. Upper-hemisphere projection. Full discussion in text.

and 148° clockwise rotations of *M* from the *M* reference position. For a feldspar crystal (Figure 6.7), 2° and 92° clockwise rotations were necessary to bring the fast and slow directions in turn into an E–W position. The spindle-stage axis *S* is then set at 10°, and the two new extinction angles about *M* measured and plotted along the 10° great circle using the small circles for angles about *M*. In our examples, *M* was rotated clockwise (from its reference position) 62° and 152° in order to place the quartz slow and fast directions E–W; for the feldspar, clockwise rotations of 6° and 96° were necessary to place fast and slow directions E–W. It is essential to realize that, for upper hemisphere projection, the two extinction positions must be measured as clockwise rotations of *M* from the reference position if they are to be plotted using the graduated scale of small circles as shown in Figures 6.6 and 6.7.

This process is continued with *S* set successively at 10° intervals until two extinction curves are plotted across the stereographic net. The data used for Figures 6.6 and 6.7 are given in Table 6.1. One of the curves always passes through *S* and is called a *polar extinction curve*; the other

Table 6.1 Spindle-Stage Extinction Data for a Quartz (Figure 6.6) and a Feldspar (Figure 6.7) Grain

	M			
	Quartz		Feldspar	
S	Slow	Fast	Slow	Fast
0	58	148	92	2
10	62	152	96	6
20	63	153	115	25
30	64	154	158	68
40	65	155	167	77
50	65	155	169	79
60	63	153	169	79
70	62	152	169	79
80	58	148	166	76
90	55	145	158	68
100	49	139	145	55
110	40	130	111	21
120	27	117	103	13
130	5	95	98	8
140	167	77	96	6
150	149	59	95	5
160	136	46	94	4
170	127	37	92	2
180	122	32	89	179

S, spindle-stage axis position; *M*, the clockwise angle of rotation from the reference position necessary to bring slow and fast directions to an E–W position.

always crosses over the N–S diameter of the net and is called an *equatorial extinction curve*. These two curves are, of course, precisely 90° from each other along any great circle.

In *uniaxial minerals* the equatorial extinction curve is always a great circle and represents ω (except for the special case when ω is exactly parallel to *S*). The polar extinction curve contains ε and its exact location can be identified as being the pole to the ω great circle (Figure 6.6).

In *biaxial minerals* the polar extinction curve always contains either *X* (α) or *Z* (γ) and can be ascertained by noting whether the extinction data were for a fast or slow vibration direction. The equatorial extinction curve (which is not a great circle for biaxial minerals) contains *Y* (β) and the other vibration direction. Because *X, Y,* and *Z* are always exactly 90° to each other, their positions on the extinction curves can be discovered by inspection (Figure 6.7). A *false solution* is also given in that the "crossover" point of the equatorial curve necessarily constitutes a pole to a great circle that intersects both extinction curves at a 90° interval. The "false" solution is only true when *X, Y,* or *Z* is exactly perpendicular to *S*. The positions of *X, Y,* and *Z* can be confirmed conoscopically as described below.

The method of Garaycochea and Wittke (1964) can be used to determine which of the two directions lying in the equatorial curve is Y. For the feldspar data of Figure 6.7, draw three great circles through the crossover point, one to the spindle axis (*GC1*), and one to each of the two vibration directions (*GC2* and *GC3*) on the equatorial curve (Figure 6.8). Note that the extinction curves lie entirely between *GC2* and *GC3* or *GC1* and *GC2*. *The area from which an extinction curve is absent* (in this case *GC1* and *GC3*) *is always bounded by the great circle passing through Y.*

Refractive Index Determinations

After having identified the positions of ω, ε, X, Y, and Z, as outlined above, it is an easy matter to rotate each of them in turn into an E–W horizontal position so that the principal RIs can be measured. For example, to measure γ of the feldspar represented by the data of Figure 6.7, rotate M 94° clockwise from the reference position, and rotate S to 153°; to measure ε of the quartz (Figure 6.6), rotate M 65° clockwise and S to 43°.

Figure 6.8 Position of optical directions of feldspar represented by extinction curves of Figure 6.7. Position of Y has been determined using the method of Garaycochea and Wittke (1964). Full discussion in text.

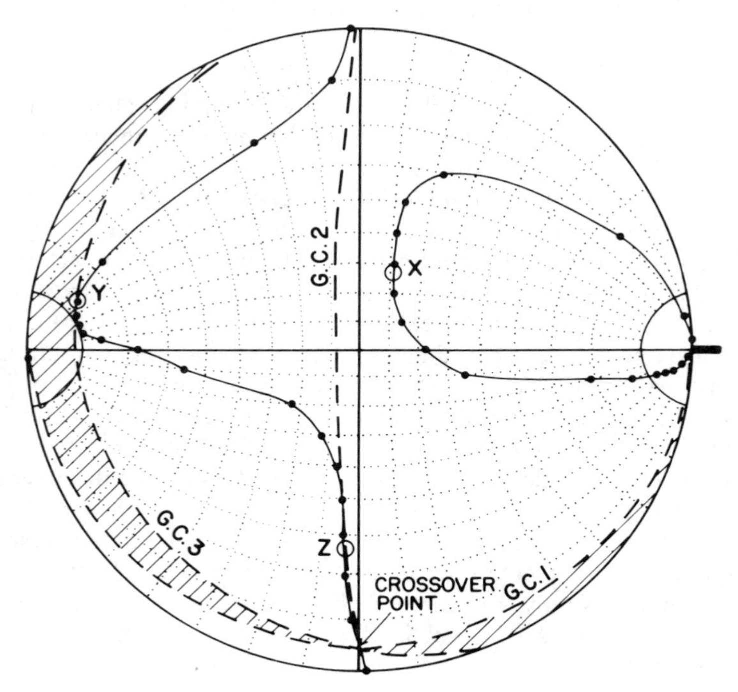

The immersion liquid cell is easily removed from the detent spindle stage, thus enabling a succession of liquids to be introduced. Of course, it is best if an approximate RI determination has already been made so that appropriate immersion liquids with RI close to those of the mineral are employed with the spindle stage. Louisnathan et al. (1978) show that, by using high-dispersion liquids, a variable wavelength monochromator, and a controlled room temperature, measurements can be made with an accuracy of ± 0.0003. There are a number of commercially available continuously variable monochromators that can be attached directly over the microscope illuminator. The principle of the method is illustrated by Figure 6.9, which shows how dispersion curves for α, β, and γ can be constructed by matching RIs with high-dispersion liquids at a variety of wavelengths. The precise form of the dispersion curves is discussed in detail by Louisnathan et al. (1978) and Bloss (1981), and, for very precise work, a certified set of Cargille immersion oils with RI intervals of 0.002 is desirable.

Some workers have advocated the so-called *double variation method* (Emmons, 1943) for which temperature is also varied systematically; this changes the RI of liquids more rapidly than solids, and allows a wider range of mineral RI to be measured without changing the immersion liquid. This method is discussed by Bloss (1981), who shows that it may introduce unnecessary complications and uncertainties.

Figure 6.9 Use of the intersecting liquid and mineral dispersion curves to determine the values of α, β, and γ for any wavelength of light.

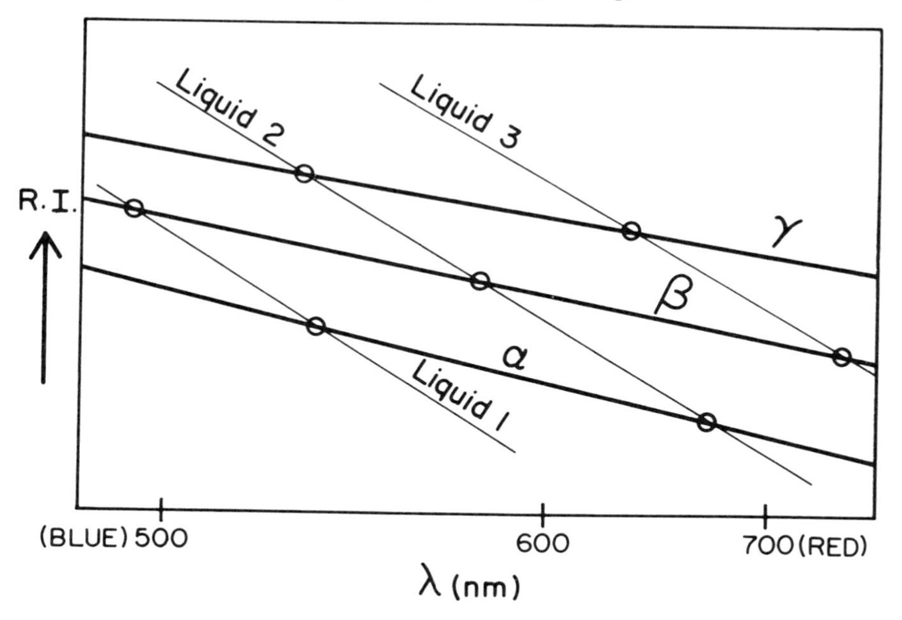

Determination of $2V$

If α, β, and γ have been measured accurately, $2V$ may be calculated using the formula given on page 43. Unfortunately, small errors in RI determination lead to large errors in $2V$, especially for minerals with a low birefringence.

The value of $2V$ may be computed to within a degree from the extinction curves, and details of this method are given in Bloss (1981).

More readily, a good estimation of $2V$ may be made from interference figures, as described below.

Conoscopic Methods Using the Spindle Stage

Before proceeding with conoscopic methods, it is advisable to check that the substage system and Bertrand lens are properly centered (refer to Chapter 2). The object of conoscopic methods is to use interference figures, which by manipulation of the spindle stage, are rotated into particular orientations with principal optical directions set parallel to the polarizer vibration direction. Immersion liquids can then be used to determine RIs, as described above. It may be possible to measure $2V$ either by direct observation or in conjunction with orthoscopically derived extinction curves. To obtain interference figures it may be necessary to use an objective with lower than normal NA (e.g., 0.65) because of the larger working distance required with the spindle stage.

It will be assumed that the reader is already familiar with the appearance of interference figures as described in Chapters 3 and 4.

For *uniaxial minerals,* with M in its reference position, rotate S until an isogyre lies along the E–W cross-hair. The optic axis now lies within the vertical E–W plane, and it can be placed in the horizontal plane of the microscope stage simply by rotating S exactly 90°. A centered flash figure will then be observed, and both ε and ω lie in the horizontal plane. These directions are exactly N–S and E–W when the mineral fragment extinguishes on rotation of M. To determine which is which, observe the moving pattern of isogyres in the figure. When the stage is rotated, ε lies in the quadrants out of which the isogyres move.

There are two special orientations in which the mineral fragment may by chance have been initially mounted. If ε is parallel to S, then centered flash figures will be obtained for all possible positions of S; ω and ε can then be measured for any position of S. If ω is parallel to S, a centered uniaxial cross will be viewed at some point of rotation of S; measurement of ω and ε is made after S has been rotated 90° from this position and a flash figure obtained.

For *biaxial minerals* it has already been shown how a spindle stage is capable of bringing X, Y, and Z into the E–W horizontal position. For all three such situations an interference figure will display an isogyre running

N–S along the cross-hair. The manipulation of the stage to find all three positions consists of a sequence of small rotations. Starting with S at 0°, rotate M until an isogyre crosses the center of the field of view. Rotate S by a small amount, but not so much that the isogyre leaves the field of view. Then rotate M until the isogyre once more crosses the center of the field of view. These latter steps are repeated until an isogyre is exactly N–S running along the cross-hair. Three such positions should be found, and for each a RI measurement is made. The distinction between X, Y, and Z is made from the relative values of RI.

$2V$ can be measured if either of the optic axes is brought into a vertical position by rotating S. Such an orientation is recognized if the isogyre passes through the cross-hair intersection, and if rotation of the microscope stage causes the isogyre to rotate about that intersection. The $2V$ may be estimated from the curvature of the isogyre in the 45° position (Figure 4.11), though ideally β should be known (Kamb, 1958). More accurately, the optic axis may be plotted on a stereographic net, and $2V$ measured along the great circle containing X and Z (already determined orthoscopically or conoscopically as described above).

If a suitable acute bisectrix figure is obtained, $2V$ may be measured using the Tobi chart given previously (Figure 4.15). Again, the refractive index β must be known, and a correction made to the value $2D/2R$ if an objective of NA other than 0.85 is used (as discussed on page 75).

Chapter 7

Routine Laboratory Procedures

Because the various theoretical and technical aspects of mineral identification are so interdependent, the student requires a good deal of work and experience in order to develop confidence in the subject. Many students will find some difficulty in charting a course through what might seem at first to be a morass of indigestible material; perhaps because of this, many will run the danger of relying too much on a casual instant visual recognition approach to the subject.

The following guide is an attempt to put the mass of detail presented in this book into some sort of perspective, so that the student can develop an efficient laboratory routine for ordinary work involving flat-stage techniques. The preceding chapters should be constantly referred to for detail. The more advanced techniques, especially those employing the universal and spindle stages, are probably best reserved until such time as the student has developed a sound routine along the lines suggested below.

Thin-Section Procedures

A

Many inexperienced students faced with a thin section, spend considerable time either wondering where to start, or examining the most insignificant mineral in the section. They eventually find that there is no time to examine the abundant and important minerals! To help overcome this, it is always advisable to hold the thin section against a sheet of white paper or a window before using the microscope. The position and percentage of minerals with differing color, opacity, alteration, and grain size can be noted, thus providing a proper perspective for what has to be done with the microscope. It must be realized, of course, that an area of the thin section that appears to be of the same material, may prove under the microscope to be composed of two or more minerals.

B

Color, relief, grain size, and shape enable an initial subdivision of the minerals present to be made. To do this, the thin section should be examined in plane-polarized light, usually with a low- or medium-power objective.

First, make a list of minerals grouped according to color. Second, partially close the lower diaphragm, and examine the relief of every group. This may result in further subdivision. Third, the list should again be subdivided if there are marked differences in grain size or shape within any group.

It is recommended that the student make this preliminary list of minerals by quickly scanning the entire thin section.

C

We now have some idea as to the *probable* number of minerals in the rock, and the next step is to examine each mineral in some detail. It is preferable to start with the most abundant mineral present.

If the mineral is opaque, refer immediately to Determinative Table X.

If the mineral is transparent, it is necessary to measure the following three properties before referring to the Determinative Tables.

(1) *Refractive index.* Use relief and Becke line to *estimate* this.

(2) *Birefringence (for anisotropic minerals).* This can be measured with reasonable accuracy, provided several grains of random orientation are present; look for the grain with the highest interference color.

(3) *Nature of the indicatrix.* Determine whether the mineral is isotropic, uniaxial +ve or −ve, biaxial +ve or −ve, and measure $2V$; look for the grain with the lowest interference color, and obtain an optic-axis figure.

The Determinative Tables can now be referred to. If the mineral is strongly colored, it is often best to start with Determinative Table IV. Otherwise start with Tables V–IX, in which minerals are grouped according to the nature of their indicatrices.

D

The above three properties are sometimes sufficient to identify a mineral. Even so, the mineral identification, or the short list of possibilities sorted from the Tables should be checked by reference to the orientation diagrams of Chapter 9. The presence or absence of cleavages and twinning, values of extinction angles, and the position of fast and slow directions are features that nearly always allow a final identification to be made. A knowledge of common mineral assemblages for a wide range of rock type is a considerable help, too (refer to sections on *occurrence* in the mineral descriptions of Chapter 9).

Instant Recognition of Minerals

With even a small amount of experience, it becomes inevitable that certain common minerals are "instantly recognized." Thus a brown, strongly pleochroic grain will instantly suggest biotite or hornblende, and a featureless, low-relief, colorless grain quartz. The procedure adopted in such cases is one that seeks confirmation, and shortcuts in identification may be taken. Such instant recognitions can hardly be avoided, and are to be encouraged, but, it is most important that confirmation be sought in a rigorous way.

For example, the following mistakes may result from the "instant recognition" of pyroxenes, hornblende, mica, and feldspar, or quartz.

(1) *Pyroxene*. The two cleavages approximately at 90° in sections at right angles to the *c*-axis are so familiar that a student will frequently not bother to examine the prismatic sections. The minerals olivine and epidote, which are common associates of pyroxene, may be overlooked in this way. Furthermore, in many igneous rocks, orthopyroxene occurs with augite, and because the properties of the two minerals overlap, the orthopyroxene may be overlooked when present in small quantities. A number of $2V$ determinations on suitable sections may be necessary to establish the presence of the orthopyroxene.

(2) *Hornblende*. The pleochroism and two cleavages at 120° are so familiar that many students misidentify the (100) sections of hornblende. Such sections are not pleochroic and show no cleavage at all, and are commonly identified as another mineral (usually tourmaline). The interference figure obtained from such sections readily distinguishes hornblende from tourmaline.

(3) *Micas*. As with hornblende, sections of micas do not always display cleavages (nor do all sections of the colored micas display pleochroism). Basal sections of biotite and muscovite frequently cause confusion, simply as a result of an unsystematic approach to mineral identification.

(4) *Feldspar and quartz*. It is incorrect to rely on the common alteration and twinning of feldspar to distinguish it from quartz. Characteristic

as these features are, they are nevertheless not always present or visible. The more essential properties are those of RI, birefringence, and the nature of the indicatrix. Quartz, for example, should always be confirmed by obtaining a uniaxial +ve interference figure.

It is important, therefore, always to check essential properties, and to make a very thorough preliminary examination of the thin section.

Loose- or Crushed-Grain Mount Procedures

These mounts, described in Chapter 5, are usually of two types: (1) a sedimentary grain assemblage in which the grain sizes, shapes, and degree of roundness are to be observed as well as mineral types; and (2) a crushed grain of a single mineral, scraped from a rock to enable a quick optical determination to be made, obviating the need for a thin section. Refractive index may be approximated very quickly in temporary mounts using immersion liquids. The following recommended methods and procedures for routine work differ slightly from those for thin sections.

A

A quick scan of the mount in plane-polarized light is important for sedimentary grain assemblages, where more than one mineral may be present. As for thin sections, a list of the probable number of minerals present should be made on the basis of color, relief, grain size, and shape.

B

An estimate of RI, birefringence, and the nature of the indicatrix must then be made for each mineral.

(1) *Birefringence*. Because there is no standard thickness of loose grains, it may be difficult to estimate birefringence from interference colors. Furthermore, sedimentary grains are sometimes so thick that the interference colors cannot be easily determined. Some guide to thickness may be made from a measurement of the smallest dimension of a grain, as viewed down the microscope (Figure 5.3). Grains with a prominent crystallographic shape or cleavage may not provide the necessary range of orientations for birefringence measurement. Despite these problems it is worthwhile to make as good an *estimate* of birefringence as is possible.

(2) *Refractive index*. RI can be measured more accurately with loose grains than in thin section, because we may immerse the grains in any number of liquids of known RI. We may also separately measure ω and ε, or α, β, and γ by employing the methods described on page 94, although this may not be considered necessary in routine work.

(3) *Nature of the indicatrix.* Sometimes the nature of the indicatrix is easily determined. In other cases it may be difficult to obtain accurate information. For example, an euhedral prismatic mineral will always lie on the slide with a preferred orientation; in many cases (e.g., all prisms of uniaxial minerals), only flash figures will be obtained. Even in these cases, some intelligent guesses can be made. For example, consistently straight extinction suggests the mineral is either uniaxial, or biaxial and orthorhombic. If flash figures are always obtained on the prismatic sections, the mineral is positive if length slow, and negative if length fast.

Once the above three properties have been determined, the Determinative Tables can be referred to. More emphasis should be placed on the accurate determination of RI made in grain mounts. As was the procedure for thin sections, a final identification is made by reference to the systematic descriptions of Chapter 9 where details of minerals sorted from the Tables should be checked.

Chapter 8

Determinative Tables

In addition to the Michel–Lévy interference color chart (at back of book), the following determinative tables are provided to assist in the identification of minerals:

Table I.	List of Minerals in Order According to Lowest Refractive Index
Table II.	Minerals that Commonly Display Anomalous Interference Colors
Table III.	List of Minerals in Order According to Birefringence
Table IV.	Colored Minerals in Thin Section
Table V.	Isotropic Minerals in Order According to Refractive Index
Table VI.	Uniaxial Positive Minerals in Order According to Refractive Index
Table VII.	Uniaxial Negative Minerals in Order According to Refractive Index
Table VIII.	Biaxial Positive Minerals in Order According to Refractive Index
Table IX.	Biaxial Negative Minerals in Order According to Refractive Index
Table X.	Opaque Minerals

Determinative Table I List of Minerals in Order According to Lowest Refractive Index

Lower RI		Higher RI		Mineral	δ^a	Indicatrix
1.41	(n)	1.47	(n)	opal (71)	0	isotropic
1.433	(n)	1.44	(n)	fluorite (111)	0	isotropic
1.461	(n)	1.509	(n)	sodalite group (77)	0	isotropic
1.466	(α)	1.494	(γ)	carnallite (114)	M	$2V_z = 66°$
1.468–1.479	(α)	1.473–1.483	(γ)	tridymite (69)	VL–L	$2V_z = 30°–90°$
1.468	(n)	1.512	(n)	allophane (66)	0	isotropic
1.470–1.500	(ε)	1.472–1.505	(ω)	chabazite group (87)	VL–L	uniax −ve, +ve
1.472–1.483	(α)	1.477–1.487	(γ)	mordenite (78)	VL	$2V_x = 76°–104°$
1.473–1.483	(α)	1.485–1.496	(γ)	natrolite (79)	L	$2V_z = 58°–64°$
1.476–1.505	(α)	1.479–1.512	(γ)	heulandite (83)	VL–L	$2V_z = 0°–74°$
1.479	(n)	1.524	(n)	analcime (85)	0	isotropic
1.480–1.600	(α)	1.500–1.640	(γ)	smectite group (62)	M–H	$2V_x = 0°$ – moderate
1.482–1.500	(α)	1.496–1.513	(γ)	stilbite (84)	L	$2V_x = 28°–49°$
1.483–1.505	(α)	1.486–1.514	(γ)	phillipsite (89)	VL–L	$2V_z = 60°–80°$
1.484	(ε)	1.487	(ω)	cristobalite (70)	VL	uniax −ve
1.485	(n)	1.620	(n)	volcanic glass (91)	0	isotropic
1.486	(ε)	1.658	(ω)	calcite—pure (124)	VH	uniax −ve
1.488–1.503	(ε)	1.490–1.528	(ω)	cancrinite (76)	VL–M	uniax −ve
1.490	(n)	1.490	(n)	sylvite (113)	0	isotropic
1.490–1.522	(α)	1.505–1.555	(γ)	palygorskite–sepiolite (65)	L–H	$2V_x = 0°–70°$
1.494	(α)	1.516	(γ)	kainite (116)	M	$2V_x = 85°$
1.497–1.530	(α)	1.518–1.544	(γ)	thomsonite (82)	L–M	$2V_z = 38°–75°$
1.498	(α)	1.502	(γ)	wairakite (86)	VL	$2V_z = 70°–105°$
1.500	(ε)	1.679	(ω)	dolomite—pure (124)	VH	uniax −ve
1.502–1.514	(α)	1.514–1.526	(γ)	laumontite (88)	L	$2V_x = 26°–47°$
1.504	(β)	1.508	(β)	mesolite (80)	VL	$2V_z =$ ca. 80°
1.507–1.513	(α)	1.517–1.521	(γ)	scolecite (81)	L	$2V_x = 36°–56°$
1.508–1.511	(ω)	1.509–1.511	(ε)	leucite (74)	0–VL	isotropic or uniax +ve
1.509	(ε)	1.700	(ω)	magnesite—pure (124)	VH	uniax −ve

1.510–1.548 (ε)	1.690–1.750 (ω)	ankerite (124)	VH	uniax −ve
1.518–1.527 (α)	1.524–1.539 (γ)	alkali-feldspar (72)	L	$2V_x = 0°–98°$
1.519–1.521 (α)	1.529–1.531 (γ)	gypsum (121)	L	$2V_z = 58°$
1.520 (α)	1.584 (γ)	kieserite (117)	H	$2V_z = 55°$
1.52 (ε)	1.63 (ω)	dahllite–francolite (125)	VL–M	uniax −ve
1.522–1.560 (α)	1.527–1.578 (γ)	cordierite (27)	L–M	$2V_x = 42°–104°$
1.525–1.548 (α)	1.551–1.587 (γ)	lepidolite (49)	M–H	$2V_x = 0°–58°$
1.525–1.560 (α)	1.545–1.585 (γ)	vermiculite (64)	M	$2V_x$ small
1.526–1.542 (ε)	1.529–1.547 (ω)	nepheline (75)	VL–L	uniax −ve
1.526–1.544 (ω)	1.532–1.553 (ε)	chalcedony (68)	L	uniax +ve
1.526–1.563 (α)	1.556–1.571 (γ)	kaolinite group (61)	VL–L	$2V_x = 24°–50°$, $2V_z = 52°–80°$
1.526–1.575 (α)	1.536–1.588 (γ)	plagioclase-feldspar (73)	L	$2V_z = 73°–135°$
1.530–1.531 (α)	1.685–1.686 (γ)	aragonite (123)	VH	$2V_x = 18°$
1.530–1.564 (α)	1.596–1.601 (γ)	pyrophyllite (54)	H	$2V_x = 46°–62°$
1.530–1.625 (α)	1.558–1.696 (γ)	biotite–phlogopite (50)	M–VH	$2V_x = 0°–25°$
1.532–1.570 (α)	1.545–1.584 (γ)	serpentine (53)	VL–M	$2V_x = 20°–60°$
1.533–1.544 (ω)	1.535–1.544 (ε)	apophyllite (60)	0–VL	uniax +ve, −ve
1.534 (n)	1.534 (n)	langbeinite (115)	0	isotropic
1.539–1.550 (α)	1.584–1.596 (γ)	talc (55)	H	$2V_x = 0°–30°$
1.540–1.571 (ε)	1.546–1.600 (ω)	scapolite (90)	L–H	uniax −ve
1.543–1.634 (α)	1.576–1.745 (γ)	stilpnomelane (56)	H–VH	$2V_x = 0°–40°$
1.544 (n)	1.544 (n)	halite (112)	0	isotropic
1.544 (ω)	1.553 (ε)	quartz (67)	L	uniax +ve
1.545–1.630 (α)	1.570–1.670 (γ)	illite (63)	M–H	$2V_x$ small
1.547 (α)	1.567 (γ)	polyhalite (118)	M	$2V_x = 64°$
1.550–1.671 (α)	1.550–1.690 (γ)	chlorite (52)	0–L	$2V_z = 0°–70°$, $2V_x = 0°–60°$
1.552–1.578 (α)	1.587–1.617 (γ)	muscovite (47)	H	$2V_x = 0°–47°$
1.559–1.590 (ω)	1.579–1.600 (ε)	brucite (104)	L–M	uniax +ve
1.564–1.580 (α)	1.600–1.609 (γ)	paragonite (48)	M–H	$2V_x = 0°–46°$
1.564–1.600 (ε)	1.568–1.608 (ω)	beryl (26)	VL–L	uniax −ve

[a] Birefringence is abbreviated: 0 = zero; VL = (very low) 0.00–0.005; L = (low) 0.005–0.015; M = (moderate) 0.015–0.030; H = (high) 0.030–0.065; VH = (very high) >0.065.

(continued)

Determinative Table I List of Minerals in Order According to Lowest Refractive Index *(continued)*

Lower RI	Higher RI	Mineral	δ^a	Indicatrix
1.565–1.571 (α)	1.580–1.595 (γ)	gibbsite (105)	L–M	$2V_z < 20°$
1.567–1.660 (ω)	1.572–1.655 (ε)	eudialyte (eucolite) (30)	0–L	uniax +ve, −ve
1.569–1.574 (α)	1.609–1.618 (γ)	anhydrite (120)	H	$2V_z = 42°–44°$
1.572 (ω)	1.592 (ε)	alunite (122)	M	uniax +ve
1.585–1.616 (α)	1.600–1.644 (γ)	glauconite (51)	L–H	$2V_x = 0°–20°$
1.592–1.643 (α)	1.621–1.674 (γ)	humite group (3)	M–H	$2V_z = 65°–85°$
1.595–1.610 (α)	1.632–1.645 (γ)	pectolite (38)	H	$2V_z = 50°–63°$
1.596–1.695 (α)	1.615–1.722 (γ)	anthophyllite–gedrite (42)	L–M	$2V_x = 69°–120°$
1.597 (ε)	1.816 (ω)	rhodochrosite—pure (124)	VH	uniax −ve
1.599–1.688 (α)	1.622–1.705 (γ)	tremolite–Fe-actinolite (44)	M	$2V_x = 64°–86°$
1.606–1.634 (α)	1.616–1.644 (γ)	topaz (10)	L	$2V_z = 46°–70°$
1.606–1.702 (α)	1.627–1.718 (γ)	glaucophane–riebeckite (46)	L–M	$2V_x = 0°–89°$
1.610–1.650 (ε)	1.635–1.675 (ω)	tourmaline (28)	M–H	uniax −ve
1.611–1.632 (α)	1.632–1.665 (γ)	prehnite (59)	M–H	$2V_z = 65°–69°$
1.612–1.700 (α)	1.630–1.710 (γ)	eckermannite–arfvedsonite (46)	L–M	$2V_x = 0°–80°$
1.613–1.705 (α)	1.632–1.731 (γ)	hornblende (44)	L–M	$2V_x = 10°–120°$
1.616–1.690 (ε)	1.624–1.70 (ω)	melilite (25)	0–L	uniax +ve, −ve
1.616–1.640 (α)	1.631–1.653 (γ)	wollastonite (37)	L	$2V_x = 38°–60°$
1.622 (α)	1.631 (γ)	celestine (119)	L	$2V_z = 51°$
1.628–1.665 (ε)	1.632–1.668 (ω)	apatite (125)	VL–L	uniax −ve
1.629–1.642 (α)	1.638–1.653 (γ)	andalusite (5)	L	$2V_x = 73°–86°$
1.630–1.638 (α)	1.644–1.650 (γ)	margarite (57)	L	$2V_x = 40°–67°$
1.635 (ε)	1.875 (ω)	siderite—pure (124)	VH	uniax −ve
1.635–1.696 (α)	1.655–1.729 (γ)	cummingtonite–grunerite (43)	M–H	$2V_x = 84°–115°$
1.635–1.732 (α)	1.670–1.775 (γ)	Mg-olivine (1)	H	$2V_x = 73°–98°$
1.636 (α)	1.648 (γ)	baryte (119)	L	$2V_z = 37°$
1.638–1.640 (α)	1.678–1.680 (γ)	spurrite (15)	H	$2V_x = 40°$
1.639–1.654 (α)	1.650–1.674 (γ)	monticellite (2)	L–M	$2V_x = 69°–82°$
1.640–1.649 (α)	1.660–1.668 (γ)	boehmite (107)	M	$2V_z = 74°–86°$

1.640–1.670 (α)	1.651–1.690 (γ)	mullite (8)	L–M	$2V_z = 45°–61°$
1.640–1.681 (α)	1.652–1.692 (γ)	jadeite (35)	L	$2V_z = 67°–90°$
1.640–1.682 (α)	1.660–1.690 (γ)	katophorite (45)	L–M	$2V_x = 0°–50°$
1.643–1.649 (α)	1.655–1.663 (γ)	clintonite (58)	L	$2V_x = 2°–40°$
1.648–1.663 (α)	1.662–1.679 (γ)	spodumene (36)	L–M	$2V_z = 55°–70°$
1.650–1.702 (α)	1.689–1.796 (γ)	oxyhornblende (44)	M–VH	$2V_x = 56°–88°$
1.651–1.769 (α)	1.658–1.788 (γ)	orthopyroxene (31)	L–M	$2V_x = 45°–128°$
1.654–1.661 (α)	1.673–1.683 (γ)	sillimanite (7)	M	$2V_z = 20°–30°$
1.655–1.686 (α)	1.685–1.723 (γ)	dumortierite (9)	L–H	$2V_x = 13°–63°$
1.656–1.665 (α)	1.680–1.685 (γ)	lawsonite (23)	M	$2V_z = 76°–87°$
1.656–1.694 (α)	1.668–1.705 (γ)	axinite (29)	L	$2V_x = 63°–109°$
1.659–1.743 (α)	1.688–1.772 (γ)	augite group (32)	M–H	$2V_z = 25°–83°$
1.665–1.728 (α)	1.683–1.754 (γ)	pumpellyite (24)	L–M	$2V_z = 10°–92°$, $2V_x = 0°–80°$
1.670–1.715 (α)	1.690–1.734 (γ)	clinozoisite (20)	L	$2V_z = 14°–90°$
1.675 (n)	1.734 (n)	hydrogrossular (4)	0–VL	isotropic
1.682–1.722 (α)	1.705–1.751 (γ)	pigeonite (33)	M	$2V_z = 0°–30°$
1.682–1.730 (α)	1.694–1.734 (γ)	chloritoid (12)	L–M	$2V_z = 40°–125°$
1.685–1.705 (α)	1.697–1.725 (γ)	zoisite (19)	VL–M	$2V_z = 0°–70°$
1.685–1.706 (α)	1.730–1.752 (γ)	diaspore (106)	H	$2V_z = 84°–86°$
1.690–1.813 (α)	1.706–1.891 (γ)	allanite (22)	L–VH	$2V_x = 0°–123°$
1.697–1.750 (ε)	1.703–1.762 (ω)	vesuvianite (18)	VL–L	uniax −ve
1.700–1.750 (α)	1.730–1.800 (γ)	aegirine-augite (34)	H	$2V_x = 70°–110°$
1.701–1.731 (α)	1.705–1.745 (γ)	sapphirine (41)	VL–L	$2V_x = 40°–114°$
1.702–1.710 (α)	1.718–1.726 (γ)	merwinite (17)	L–M	$2V_z = 52°–76°$
1.706–1.718 (α)	1.719–1.734 (γ)	kyanite (6)	L–M	$2V_x = 77°–82°$
1.707 (α)	1.730 (γ)	larnite (16)	M	$2V_z$ moderate
1.711–1.738 (α)	1.724–1.751 (γ)	rhodonite (39)	L–M	$2V_z = 61°–87°$
1.714–1.751 (α)	1.730–1.797 (γ)	epidote (20)	M–H	$2V_x = 64°–90°$
1.715 (n)	1.98 (n)	spinel (100)	0	isotropic
1.720–1.724 (ω)	1.810–1.828 (ε)	xenotime (127)	VH	uniax +ve

[a] Birefringence is abbreviated: 0 = zero; VL = (very low) 0.00–0.005; L = (low) 0.005–0.015; M = (moderate) 0.015–0.030; H = (high) 0.030–0.065; VH = (very high) >0.065.

(*continued*)

Determinative Table I List of Minerals in Order According to Lowest Refractive Index *(continued)*

Lower RI		Higher RI		Mineral	δ^a	Indicatrix
1.720	(n)	2.00	(n)	garnet (4)	0	isotropic
1.732–1.794	(α)	1.762–1.829	(γ)	piemontite (21)	M–VH	$2V_z = 64°–99°$
1.732–1.827	(α)	1.775–1.879	(γ)	Fe-olivine (1)	H	$2V_x = 46°–73°$
1.736	(n)	1.745	(n)	periclase (93)	0	isotropic
1.736–1.747	(α)	1.749–1.762	(γ)	staurolite (11)	L	$2V_z = 80°–92°$
1.750–1.776	(α)	1.800–1.836	(γ)	aegirine (34)	H	$2V_x = 60°–70°$
1.759–1.763	(ε)	1.765–1.772	(ω)	corundum (94)	L	uniax −ve
1.770–1.800	(α)	1.825–1.850	(γ)	monazite (126)	H–VH	$2V_z = 3°–19°$
1.790–1.81	(α)	1.87 –1.9	(γ)	aenigmatite (40)	VH	$2V_z = 27°–55°$
1.840–1.950	(α)	1.943–2.110	(γ)	titanite (sphene) (13)	VH	$2V_z = 20°–56°$
1.920	(ω)	1.937	(ε)	scheelite (128)	M	uniax +ve
1.92 –1.96	(ω)	1.96 –2.02	(ε)	zircon (14)	H	uniax +ve
1.94	(α)	2.51	(γ)	lepidocrocite (103)	VH	$2V_x = 83°$
1.990–2.010	(ω)	2.093–2.100	(ε)	cassiterite (95)	VH	uniax +ve
2.0	(n)	2.0	(n)	limonite (103)	0	isotropic
2.00	(n)	2.16	(n)	chromite (100)	0	isotropic
2.217–2.275	(α)	2.356–2.415	(γ)	goethite (103)	VH	$2V_x = 0°–27°$
2.30	(n)	2.38	(n)	perovskite (99)	0–VL	isotropic or $2V = 90°$
2.488	(ε)	2.561	(ω)	anatase (97)	VH	uniax −ve
2.583	(α)	2.700–2.741	(γ)	brookite (98)	VH	$2V_z = 0°–30°$
2.605–2.616	(ω)	2.899–2.903	(ε)	rutile (96)	VH	uniax +ve
2.87 –2.94	(ε)	3.15 –3.22	(ω)	hematite (102)	VH	uniax −ve

[a] Birefringence is abbreviated: 0 = zero; VL = (very low) 0.00–0.005; L = (low) 0.005–0.015; M = (moderate) 0.015–0.030; H = (high) 0.030–0.065; VH = (very high) >0.065.

Determinative Table II Minerals that Commonly Display Anomalous Interference Colors

Allanite (22)	Melilite (25)
Apophyllite (60)	Na-rich amphiboles
Brucite (104)	Pumpellyite (24)
Chlorite (52)	Vesuvianite (18)
Chloritoid (12)	Zoisite (19)
Clinozoisite–epidote (20)	

Determinative Table III List of Minerals in Order According to Birefringence

δ	Mineral	RI range	Indicatrix
0.00 –0.002	analcime (85)	1.479–1.524	(isotropic)
0.00 –0.001	leucite (74)	1.508–1.511	uniax +ve
0.00 –0.002	apophyllite (60)	1.533–1.544	uniax +ve, −ve
0.00 –0.002	Ca-garnet (4)	1.675–2.0	(isotropic)
0.00 –0.002	perovskite (99)	2.30 –2.38	$2V$ = ca. 90° or isotropic
0.00 –0.008	eudialyte (eucolite) (30)	1.567–1.660	uniax +ve, −ve
0.00 –0.013	melilite (25)	1.616–1.70	uniax +ve, −ve
0.00 –0.015	chlorite (52)	1.550–1.690	$2V_z = 0°–70°$, $2V_x = 0°–60°$
0.00 –0.016	dahllite–francolite (125)	1.52 –1.63	uniax −ve
0.001	mesolite (80)	1.504–1.508	$2V_z$ = ca. 80°
0.001–0.009	kaolinite group (61)	1.526–1.571	$2V_x = 24°–50°$, $2V_z = 52°–80°$
0.001–0.012	vesuvianite (18)	1.697–1.762	uniax −ve
0.002–0.005	mordenite (78)	1.472–1.487	$2V_x = 76°–104°$
0.002–0.007	tridymite (69)	1.468–1.483	$2V_z = 30°–90°$
0.002–0.008	heulandite (83)	1.476–1.512	$2V_z = 0°–74°$
0.002–0.008	apatite (125)	1.628–1.668	uniax −ve
0.002–0.015	chabazite group (87)	1.470–1.505	uniax −ve, +ve
0.002–0.025	cancrinite (76)	1.488–1.528	uniax −ve
0.003 ca.	cristobalite (70)	1.484–1.487	uniax −ve
0.003–0.007	nepheline (75)	1.526–1.547	uniax −ve
0.003–0.010	phillipsite (89)	1.483–1.514	$2V_z = 60°–80°$
0.004	wairakite (86)	1.498–1.502	$2V_z = 70°–105°$
0.004–0.009	beryl (26)	1.564–1.608	uniax −ve
0.004–0.015	sapphirine (41)	1.701–1.745	$2V_x = 40°–114°$
0.004–0.017	serpentine (53)	1.532–1.584	$2V_x = 20°–60°$
0.004–0.022	zoisite (19)	1.685–1.725	$2V_z = 0°–70°$
0.005–0.009	chalcedony (68)	1.526–1.553	uniax +ve
0.005–0.009	corundum (94)	1.759–1.772	uniax −ve
0.005–0.010	alkali-feldspar (72)	1.518–1.539	$2V_x = 0°–98°$
0.005–0.015	clinozoisite (20)	1.670–1.734	$2V_z = 14°–90°$
0.005–0.018	cordierite (27)	1.522–1.578	$2V_x = 42°–104°$

0.006–0.022	chloritoid (12)	1.682–1.740	$2V_z = 40°–125°$
0.006–0.022	glaucophane–riebeckite (46)	1.606–1.718	$2V_x = 0°–89°$
0.006–0.036	scapolite (90)	1.540–1.600	uniax −ve
0.007–0.010	scolecite (81)	1.507–1.521	$2V_x = 36°–56°$
0.007–0.011	topaz (10)	1.606–1.644	$2V_z = 46°–70°$
0.007–0.013	plagioclase-feldspar (73)	1.526–1.588	$2V_z = 73°–135°$
0.007–0.014	axinite (29)	1.656–1.705	$2V_x = 63°–109°$
0.007–0.019	orthopyroxene (31)	1.651–1.788	$2V_x = 45°–128°$
0.007–0.021	katophorite (45)	1.640–1.690	$2V_x = 0°–50°$
0.008–0.014	stilbite (84)	1.482–1.513	$2V_x = 28°–49°$
0.008–0.015	jadeite (35)	1.640–1.692	$2V_z = 67°–90°$
0.008–0.023	merwinite (17)	1.702–1.726	$2V_z = 52°–76°$
0.009	quartz (67)	1.544–1.553	uniax +ve
0.009	celestine (119)	1.622–1.631	$2V_z = 51°$
0.009–0.011	andalusite (5)	1.629–1.653	$2V_x = 73°–86°$
0.009–0.035	palygorskite–sepiolite (65)	1.490–1.555	$2V_x = 0°–70°$
0.010	gypsum (121)	1.519–1.531	$2V_z = 58°$
0.010–0.015	laumontite (88)	1.502–1.526	$2V_x = 26°–47°$
0.010–0.015	staurolite (11)	1.736–1.762	$2V_z = 80°–92°$
0.010–0.021	brucite (104)	1.559–1.600	uniax +ve
0.010–0.028	pumpellyite (24)	1.665–1.754	$2V_z = 10°–92°$, $2V_x = 0°–80°$
0.010–0.037	dumortierite (9)	1.655–1.723	$2V_x = 13°–63°$
0.011–0.017	rhodonite (39)	1.711–1.751	$2V_z = 61°–87°$
0.011–0.020	monticellite (2)	1.639–1.674	$2V_x = 69°–82°$
0.012	baryte (119)	1.636–1.648	$2V_z = 37°$
0.012 ca.	natrolite (79)	1.473–1.496	$2V_z = 58°–64°$
0.012–0.014	clintonite (58)	1.643–1.663	$2V_x = 2°–40°$
0.012–0.014	margarite (57)	1.630–1.650	$2V_x = 40°–67°$
0.012–0.016	kyanite (6)	1.706–1.734	$2V_x = 77°–82°$
0.012–0.028	mullite (8)	1.640–1.690	$2V_z = 45°–61°$
0.013–0.015	wollastonite (37)	1.616–1.653	$2V_x = 38°–60°$
0.013–0.028	anthophyllite–gedrite (42)	1.596–1.722	$2V_x = 69°–120°$
0.013–0.078	allanite (22)	1.690–1.891	$2V_x = 0°–123°$

(continued)

Determinative Table III List of Minerals in Order According to Birefringence (*continued*)

δ	Mineral	RI range	Indicatrix
0.014–0.027	spodumene (36)	1.648–1.679	$2V_z = 55°–70°$
0.014–0.029	hornblende (44)	1.612–1.731	$2V_x = 10°–120°$
0.014–0.030	gibbsite (105)	1.565–1.595	$2V_z < 20°$
0.014–0.032	glauconite (51)	1.585–1.644	$2V_x = 0°–20°$
0.015–0.020	boehmite (107)	1.640–1.668	$2V_z = 74°–86°$
0.015–0.049	epidote (20)	1.714–1.797	$2V_x = 64°–90°$
0.017	scheelite (128)	1.920–1.937	uniax +ve
0.017–0.027	tremolite–Fe-actinolite (44)	1.599–1.705	$2V_x = 64°–86°$
0.017–0.035	tourmaline (28)	1.610–1.675	uniax −ve
0.018–0.033	augite group (32)	1.659–1.772	$2V_z = 25°–83°$
0.018–0.039	lepidolite (49)	1.525–1.587	$2V_x = 0°–58°$
0.019–0.024	sillimanite (7)	1.654–1.683	$2V_z = 20°–30°$
0.020	polyhalite (118)	1.547–1.567	$2V_x = 64°$
0.020	alunite (122)	1.572–1.592	uniax +ve
0.020–0.024	lawsonite (23)	1.656–1.685	$2V_z = 76°–87°$
0.020–0.030	vermiculite (64)	1.525–1.585	$2V_x$ small
0.020–0.040	smectite group (62)	1.480–1.640	$2V_x = 0°$–moderate
0.020–0.045	cummingtonite–grunerite (43)	1.635–1.729	$2V_x = 84°–115°$
0.020–0.094	oxyhornblende (44)	1.650–1.796	$2V_x = 56°–88°$
0.021–0.035	prehnite (59)	1.611–1.665	$2V_z = 65°–69°$
0.022	kainite (116)	1.494–1.516	$2V_x = 85°$
0.022–0.041	humite group (3)	1.592–1.674	$2V_z = 65°–85°$
0.022–0.055	illite (63)	1.545–1.670	$2V_x$ small
0.023	larnite (16)	1.707–1.730	$2V_z$ moderate
0.023–0.029	pigeonite (33)	1.682–1.751	$2V_z = 0°–30°$
0.025–0.088	piemontite (21)	1.732–1.829	$2V_z = 64°–99°$
0.028	carnallite (114)	1.466–1.494	$2V_z = 66°$
0.028–0.038	paragonite (48)	1.564–1.609	$2V_x = 0°–46°$
0.028–0.078	biotite–phlogopite (50)	1.530–1.696	$2V_x = 0°–25°$
0.030–0.038	pectolite (38)	1.595–1.645	$2V_z = 50°–63°$
0.030–0.050	aegirine-augite (34)	1.700–1.800	$2V_x = 70°–110°$

0.033–0.111	stilpnomelane (56)	1.543–1.745	$2V_x = 0°–40°$
0.035–0.052	olivine (1)	1.635–1.879	$2V_x = 46°–98°$
0.036–0.049	muscovite (47)	1.552–1.617	$2V_x = 0°–47°$
0.039–0.050	talc (55)	1.539–1.596	$2V_x = 0°–30°$
0.040	spurrite (15)	1.638–1.680	$2V_x = 40°$
0.040–0.047	anhydrite (120)	1.569–1.618	$2V_z = 42°–44°$
0.040–0.050	diaspore (106)	1.685–1.752	$2V_z = 84°–86°$
0.040–0.060	zircon (14)	1.92 –2.02	uniax +ve
0.045–0.068	pyrophyllite (54)	1.530–1.601	$2V_x = 46°–62°$
0.045–0.075	monazite (126)	1.770–1.850	$2V_z = 3°–19°$
0.050–0.060	aegirine (34)	1.750–1.836	$2V_x = 60°–70°$
0.064	kieserite (117)	1.520–1.584	$2V_z = 55°$
0.07 –0.08	aenigmatite (40)	1.79 –1.90	$2V_z = 27°–55°$
0.073	anatase (97)	2.488–2.561	uniax −ve
0.086–0.107	xenotime (127)	1.720–1.828	uniax +ve
0.090–0.103	cassiterite (95)	1.990–2.100	uniax +ve
0.100–0.192	titanite (sphene) (13)	1.840–2.110	$2V_z = 20°–56°$
0.117–0.158	brookite (98)	2.583–2.741	$2V_z = 0°–30°$
0.139–0.140	goethite (103)	2.217–2.415	$2V_x = 0°–27°$
0.155	aragonite (123)	1.530–1.686	$2V_x = 18°$
0.172	calcite (124)	1.486–1.658	uniax −ve
0.179	dolomite (124)	1.500–1.679	uniax −ve
0.182–0.202	ankerite (124)	1.510–1.750	uniax −ve
0.191	magnesite (124)	1.509–1.700	uniax −ve
0.219	rhodochrosite (124)	1.597–1.816	uniax −ve
0.242	siderite (124)	1.635–1.875	uniax −ve
0.28	hematite (102)	2.87 –3.22	uniax −ve
0.286–0.294	rutile (96)	2.605–2.903	uniax +ve
0.57	lepidocrocite (103)	1.94 –2.51	$2V_x = 83°$

Determinative Table IV Colored Minerals in Thin Section[a]

	Pink	Red	Violet-purple	Blue	Green
Isotropic	garnet (4) sodalite group (77) spinel (100) volcanic glass (91)	volcanic glass (91)	fluorite (111)	sodalite group (77) spinel (100) fluorite (111)	garnet (4) [chlorite (52)] perovskite (99) spinel (100)
Uniaxial +ve	eudialyte (30)	cassiterite (95) rutile (96)	[Cr-chlorite (52)]		vesuvianite (18) [chlorite (52)]
Uniaxial −ve	tourmaline (28) corundum (94) [Cr-chlorite (52)] eucolite (30)	hematite (102) [biotite (50)]	[Cr-chlorite (52)]	tourmaline (28) corundum (94) anatase (97)	vesuvianite (18) tourmaline (28) [biotite–phlogopite (50)] [chlorite (52)] corundum (94)
Baxial +ve	titanite (sphene) (13) sapphirine (41) zoisite (19) piemontite (21) rhodonite (39) diaspore (106) Cr-chlorite (52) *Pyroxenes:* orthopyroxene (31) augite (32) pigeonite (33)	allanite (22) iddingsite (1) aenigmatite (40)	piemontite (21) augite (32) Cr-chlorite (52) axinite (29)	chloritoid (12) sapphirine (41) lawsonite (23) axinite (29)	chloritoid (12) sapphirine (41) pumpellyite (24) chlorite (52) perovskite (99) *Amphiboles:* anthophyllite–gedri (42) cummingtonite (43) hornblende (44)
Biaxial −ve	andalusite (5) dumortierite (9) sapphirine (41) Cr-chlorite (52) orthopyroxene (31) piemontite (21)	biotite (50) allanite (22) iddingsite (1) *Amphiboles:* oxyhornblende (44) katophorite (45)	dumortierite (9) axinite (29) Cr-chlorite (52) piemontite (21)	dumortierite (9) chloritoid (12) sapphirine (41) axinite (29) glaucophane–riebeckite (46)	bowlingite (1) andalusite (5) chloritoid (12) sapphirine (41) epidote (20) pumpellyite (24) *Sheet silicates:* fuchsite (47) biotite–phlogopite (50) glauconite (51) chlorite (52) serpentine (53) stilpnomelane (56) clintonite (58) nontronite (62) vermiculite (64)

[a] Examples in brackets may appear to have the indicatrix indicated.

Determinative Table IV (*continued*)

Yellow		Brown		Grey
	garnet (4) [melilite (25)] limonite (103) collophane (125)	garnet (4) opal (71) perovskite (99) chromite (100) limonite (103) collophane (125) volcanic glass (91)		sodalite group (77) perovskite (99) spinel (100) collophane (125) volcanic glass (91)
	melilite (25) cassiterite (95) rutile (96) xenotime (127) eudialyte (30) scheelite (128)	zircon (14) vesuvianite (18) xenotime (127) cassiterite (95) rutile (96) scheelite (128)		
	melilite (25) tourmaline (28) [biotite–phlogopite (50)] [chlorite (52)] corundum (94) anatase (97) dahllite–francolite (125) eucolite (30)	vesuvianite (18) tourmaline (28) [biotite–phlogopite (50)] anatase (97) dahllite–francolite (125)		tourmaline (28) anatase (97) dahllite–francolite (125)
Pyroxenes: orthopyroxene (31) hedenbergite (32) augite (32) omphacite (32) fassaite (32) pigeonite (33) aegirine-augite (34)	humite group (3) staurolite (11) chloritoid (12) sapphirine (41) zoisite (19) clinozoisite (20) lawsonite (23) orthopyroxene (31) brookite (98) monazite (126)	titanite (sphene) (13) allanite (22) pumpellyite (24) perovskite (99) brookite (98) gibbsite (105) diaspore (106) aenigmatite (40) *Pyroxenes:* hedenbergite (32) augite (32)	*Amphiboles:* anthophyllite–gedrite (42) cummingtonite (43) hornblende (44)	perovskite (99)
Pyroxenes: orthopyroxene (31) aegirine (34) aegirine-augite (34) *Amphiboles:* anthophyllite (42) actinolite (44) grunerite (43) hornblende (44) oxyhornblende (44) glaucopane (46) katophorite (45)	Fe-olivine (1) andalusite (5) staurolite (11) chloritoid (12) sapphirine (41) epidote (20) axinite (29) *Sheet silicates:* biotite–phlogopite (50) chlorite (52) stilpnomelane (56) nontronite (62) palygorskite–sepiolite orthopyroxene (31) goethite (103) lepidocrocite (103)	dumortierite (9) allanite (22) pumpellyite (24) goethite (103) lepidocrocite (103) *Sheet silicates:* biotite–phlogopite (50) stilpnomelane (56) clintonite (58) nontronite (62) vermiculite (64) *Pyroxene:* acmite (34)	*Amphiboles:* anthophyllite (42) grunerite (43) actinolite (44) hornblende (44) oxyhornblende (44) glaucophane (46) eckermannite–arfvedsonite (46) katophorite (45)	

Determinative Table V Isotropic Minerals in Order According to Refractive Index[a]

RI	Mineral	Remarks
1.41 –1.47	opal (71)	organic; secondary
1.433–1.44	fluorite (111)	often blue or violet
1.461–1.509	sodalite group (77)	feldspathoid; six-sided crystals
1.468–1.512	allophane (66)	clay mineral
1.479–1.524	analcime (85)	zeolite
1.485–1.62	volcanic glass (91)	
1.490	sylvite (113)	water soluble, salty
1.508–1.511	leucite (74)	feldspathoid; often weak δ and twins
1.52 –1.63	collophane (125)	phosphate; yellow, brown
1.534	langbeinite (115)	water soluble
1.544	halite (112)	water soluble, salty
1.675–1.734	hydrogrossular (4)	garnet; often weak δ
1.715–1.98	spinel (100)	octahedral crystals
1.720–1.770	pyrope (4)	garnet in ultramafics, eclogites
1.735–1.770	grossular (4)	garnet; often weak δ
1.736–1.745	periclase (93)	cubic cleavage; alters to brucite
1.770–1.820	almandine (4)	common metamorphic garnet
1.790–1.810	spessartine (4)	garnet; usually in pegmatites
1.850–2.0	andradite (4)	garnet; Ti-rich varieties dark-brown in section; often weak δ
1.86	uvarovite (4)	Cr-garnet; green
2.0	limonite (103)	yellow-brown; secondary
2.00 –2.16	chromite (100)	dark brown; in ultramafics
2.30 –2.38	perovskite (99)	often brown and weak δ

[a] Refer to Determinative Table III for minerals that are pseudo-istropic with a zero or very weak birefringence.

Determinative Table VI Uniaxial Positive Minerals in Order According to Refractive Index[a]

Mineral	ω	ε	δ
Chabazite group (87)	ca. 1.47 –1.50	ca. 1.47 –1.50	0.002–0.015
Leucite (74)	1.508–1.511	1.509–1.511	0.00 –0.001
Chalcedony (68)	1.526–1.544	1.532–1.553	0.005–0.009
Apophyllite (60)	1.533–1.544	1.535–1.544	0.00 –0.002
Quartz (67)	1.544	1.553	0.009
Brucite (104)	1.559–1.590	1.579–1.600	0.010–0.021
Eudialyte (30)	1.567–1.660	1.572–1.655	0.00 –0.008
Alunite (122)	1.572	1.592	0.020
Melilite (25)	1.630–1.650	1.637–1.650	0.00 –0.008
Xenotime (127)	1.720–1.724	1.810–1.828	0.086–0.107
Scheelite (128)	1.920	1.937	0.017
Zircon (14)	1.92 –1.96	1.96 –2.02	0.04 –0.06
Cassiterite (95)	1.990–2.010	2.093–2.100	0.090–0.103
Rutile (96)	2.605–2.616	2.899–2.903	0.286–0.294

[a] Refer to Determinative Table VIII for pseudo-uniaxial minerals with $2V = 0°$ or small.

Determinative Table VII Uniaxial Negative Minerals in Order According to Refractive Index[a]

Mineral	ω	ε	δ
Chabazite group (87)	1.472–1.505	1.470–1.500	0.002–0.015
Cristobalite (70)	1.487	1.484	0.003
Cancrinite (76)	1.490–1.528	1.488–1.503	0.002–0.025
Dahllite–francolite (125)	(1.52–1.63)		0.00 –0.016
Nepheline (75)	1.529–1.547	1.526–1.542	0.003–0.007
Apophyllite (60)	1.544–1.5445	1.544	<0.001
Scapolite (90)	1.546–1.600	1.540–1.571	0.006–0.036
Eucolite (30)	1.567–1.660	1.572–1.655	0.00 –0.008
Beryl (26)	1.568–1.608	1.564–1.600	0.004–0.009
Melilite (25)	1.624–1.700	1.616–1.690	0.00 –0.013
Apatite (125)	1.632–1.668	1.628–1.665	0.002–0.008
Tourmaline (28)	1.635–1.675	1.610–1.650	0.017–0.035
Calcite—pure (124)	1.658	1.486	0.172
Dolomite—pure (124)	1.679	1.500	0.179
Ankerite (124)	ca. 1.69 –1.75	1.510–1.548	0.182–0.202
Magnesite—pure (124)	1.700	1.509	0.191
Vesuvianite (18)	1.703–1.762	1.697–1.750	0.001–0.012
Corundum (94)	1.765–1.772	1.759–1.763	0.005–0.009
Rhodochrosite—pure (124)	1.816	1.597	0.219
Siderite—pure (124)	1.875	1.635	0.242
Anatase (97)	2.561	2.488	0.073
Hematite (102)	3.15 –3.22	2.87 –2.94	0.28

[a] Refer to Determinative Table IX for pseudo-uniaxial minerals with $2V = 0°$ or small.

Determinative Table VIII Biaxial Positive Minerals in Order According to Refractive Index

Mineral	α	β	γ	δ	$2V$	Remarks
Carnallite (114)	1.466	1.475	1.494	0.028	66°	water soluble
Tridymite (69)	1.468–1.479	1.469–1.480	1.473–1.483	0.002–0.007	30°–90°	SiO_2; platy crystals with twinning
Mordenite (78)	1.472–1.483	1.475–1.485	1.477–1.487	0.002–0.005	76°–90°	a zeolite
Natrolite (79)	1.473–1.483	1.476–1.486	1.485–1.496	ca. 0.012	58°–64°	a zeolite
Heulandite (83)	1.476–1.505	1.477–1.508	1.479–1.512	0.002–0.008	0°–74°	a zeolite
Phillipsite (89)	1.483–1.505	1.484–1.510	1.486–1.514	0.003–0.010	60°–80°	a zeolite
Thomsonite (82)	1.497–1.530	1.513–1.533	1.518–1.544	0.006–0.016	38°–75°	a zeolite
Wairakite (86)	1.498		1.502	0.004	70°–90°	a zeolite
Mesolite (80)		1.504–1.508		0.001	ca. 80°	a zeolite
Gypsum (121)	1.519–1.521	1.522–1.526	1.529–1.531	0.010	58°	evaporite or secondary mineral
Kieserite (117)	1.520	1.533	1.584	0.064	55°	water soluble
Cordierite (27)	1.522–1.547	1.524–1.552	1.527–1.557	0.005–0.016	76°–90°	sector twins, if present, are distinctive
Plagioclase-feldspar (73)	1.526–1.575	1.532–1.584	1.536–1.588	0.007–0.013	73°–90°	refer to Figure 9.65 for $2V$ variation
Alkali-feldspar (72)	1.527–1.529	1.531–1.533	1.536–1.539	0.009–0.010	82°–90°	
Chlorite (52)	1.550–1.670	1.550–1.670	1.550–1.670	0.00 –0.018	0°–70°	commonly green with anomalous brown interference colors
Dickite (61B)	1.560–1.562		1.566–1.571	0.006–0.009	52°–80°	a clay mineral
Gibbsite (105)	1.565–1.571	ca. = α	1.580–1.595	0.014–0.030	<20°	in bauxites
Anhydrite (120)	1.569–1.574	1.574–1.579	1.609–1.618	0.040–0.047	42°–44°	evaporite or secondary mineral
Humite group (3)	1.592–1.643	1.602–1.653	1.621–1.674	0.022–0.041	65°–85°	usually yellow
Pectolite (38)	1.595–1.610	1.605–1.615	1.632–1.645	0.030–0.038	50°–63°	usually in veins and cavities
Topaz (10)	1.606–1.634	1.609–1.637	1.616–1.644	0.007–0.011	46°–70°	
Prehnite (59)	1.611–1.634	1.615–1.642	1.632–1.665	0.021–0.035	65°–69°	may resemble muscovite
Hornblende (44)	1.613–1.650	1.618–1.660	1.635–1.670	0.018–0.024	60°–90°	an amphibole
Anthophyllite–gedrite (42)	1.62 –1.69	1.63 –1.705	1.645–1.715	0.013–0.025	60°–90°	an amphibole in metamorphics; straight extinction
Celestine (119)	1.622	1.624	1.631	0.009	51°	in evaporites and veins
Cummingtonite (43)	1.635–1.663	1.644–1.680	1.655–1.696	0.020–0.035	65°–90°	a metamorphic amphibole; usually elongate
Olivine (1)	1.635–1.665	1.651–1.684	1.670–1.702	0.035–0.037	82°–90°	no good cleavage
Baryte (119)	1.636	1.637	1.648	0.012	37°	in veins and some sediments
Boehmite (107)	1.640–1.649		1.660–1.668	0.015–0.020	74°–86°	in bauxites
Mullite (8)	1.640–1.670	1.642–1.675	1.651–1.690	0.012–0.028	45°–61°	in very high-temperature metamorphics

Jadeite (35)	1.640–1.681	1.645–1.684	1.652–1.692	0.008–0.015	67°–90°	in very high-pressure metamorphics
Spodumene (36)	1.648–1.663	1.655–1.669	1.662–1.679	0.014–0.027	55°–70°	Li-pyroxene in pegmatites
Enstatite (31)	1.651–1.665	1.653–?	1.658–1.675	0.007–0.010	52°–90°	orthopyroxene
Sillimanite (7)	1.654–1.661	1.658–1.662	1.673–1.683	0.019–0.024	20°–30°	often fibrous; in high-grade metamorphics
Axinite (29)	1.656–?	1.660–?	1.668–?	0.012	71°–90°	wedge-shaped crystals common; usually −ve
Lawsonite (23)	1.656–1.665	1.667–1.674	1.680–1.685	0.020–0.024	76°–87°	in high-pressure metamorphics
Augite group (32)	1.659–1.743	1.670–1.750	1.688–1.772	0.018–0.033	25°–83°	clinopyroxenes
Pumpellyite (24)	1.665–1.710	1.670–1.730	1.683–1.730	0.010–0.020	10°–90°	often bluish-green or brown; similar to epidote
Clinozoisite (20)	1.670–1.715	1.674–1.725	1.690–1.734	0.005–0.015	14°–90°	anomalous interference colors
Pigeonite (33)	1.682–1.722	1.684–1.722	1.705–1.751	0.023–0.029	0°–30°	clinopyroxene in volcanics
Chloritoid (12)	1.682–1.730	1.688–1.734	1.694–1.740	0.006–0.022	40°–90°	platy crystals; similar to chlorite
Zoisite (19)	1.685–1.705	1.688–1.710	1.697–1.725	0.004–0.022	0°–70°	anomalous interference colors
Diaspore (106)	1.685–1.706	1.705–1.725	1.730–1.752	0.040–0.050	84°–86°	in bauxites
Allanite (22)	1.690–1.813	1.700–1.857	1.706–1.891	0.013–0.078	57°–90°	dark brown or red-brown
Aegirine-augite (34)	1.700–1.725	1.710–1.742	1.730–1.760	0.030–0.035	70°–90°	a green clinopyroxene; alkaline igneous rocks
Sapphirine (41)	1.701–1.731	1.703–1.741	1.705–1.745	0.004–0.015	66°–90°	yellow, pink, or blue; in high-grade metamorphics
Merwinite (17)	1.702–1.710	1.710–1.718	1.718–1.726	0.008–0.023	52°–76°	high-temperature Ca-silicate
Larnite (16)	1.707	1.715	1.730	0.023	moderate	high-temperature Ca-silicate
Rhodonite (39)	1.711–1.738	1.716–1.741	1.724–1.751	0.011–0.017	61°–87°	in Mn-rich deposits
Piemontite (21)	1.732–1.794	1.750–1.807	1.762–1.829	0.025–0.088	64°–90°	pink or purple; similar to epidote
Staurolite (11)	1.736–1.747	1.741–1.754	1.749–1.762	0.010–0.015	80°–90°	yellow; in medium-grade metamorphics
Orthoferrosilite (31)	1.755–1.769	?–1.771	1.772–1.788	0.017–0.019	60°–90°	orthopyroxene
Monazite (126)	1.770–1.800	1.777–1.801	1.825–1.850	0.045–0.075	3°–19°	yellow, brown, or red; an accessory mineral
Aenigmatite (40)	1.790–1.81	1.805–1.826	1.87 –1.89	0.07 –0.08	27°–55°	dark red, brown, opaque; in alkaline igneous rocks
Titanite (Sphene) (13)	1.840–1.950	1.870–2.034	1.943–2.110	0.100–0.192	20°–56°	wedge-shaped or droplike (sugary) brown crystals
Perovskite (99)		2.30 –2.38		0.00 –0.002	ca. 90°	pseudocubic; often brown
Brookite (98)	2.583	2.584–2.586	2.700–2.741	0.117–0.158	0°–30°	yellow or brown; incomplete extinction

Determinative Table IX Biaxial Negative Minerals in Order According to Refractive Index

Mineral	α	β	γ	δ	$2V$	Remarks
Mordenite (78)	1.472–1.483	1.475–1.485	1.477–1.487	0.002–0.005	76°–90°	a zeolite
Smectite group (62)	1.48 –1.60		1.50 –1.64	0.020–0.040	0°–moderate	clay minerals
Stilbite (84)	1.482–1.500	1.489–1.507	1.496–1.513	0.008–0.014	28°–49°	a zeolite
Palygorskite–sepiolite (65)	1.490–1.522		1.505–1.555	0.009–0.035	0°–70°	fibrous clay; possibly yellow
Kainite (116)	1.494	1.505	1.516	0.022	85°	water soluble
Wairakite (86)	1.498		1.502	0.004	75°–90°	a zeolite
Laumontite (88)	1.502–1.514	1.512–1.525	1.514–1.526	0.010–0.015	26°–47°	a zeolite
Scolecite (81)	1.507–1.513	1.516–1.520	1.517–1.521	0.007–0.010	36°–56°	a zeolite
Alkali-feldspar (72)	1.518–1.527	1.522–1.533	1.524–1.536	0.005–0.009	0°–90°	
Cordierite (27)	1.522–1.560	1.524–1.574	1.527–1.578	0.005–0.018	42°–90°	sector twins, if present, are distinctive
Lepidolite (49)	1.525–1.548	1.548–1.585	1.551–1.587	0.018–0.039	0°–58°	Li-mica
Vermiculite (64)	1.525–1.560		1.545–1.585	0.020–0.030	small	a clay mineral; often pseudomorphing biotite
Kaolinite group (61)	1.526–1.563		1.556–1.570	0.001–0.007	24°–50°	clay minerals
Plagioclase-feldspar (73)	1.526–1.575	1.532–1.584	1.536–1.588	0.007–0.013	45°–90°	refer to Figure 9.65 for $2V$ variation
Aragonite (123)	1.530–1.531	1.680–1.682	1.685–1.686	0.155	18°	$CaCO_3$; effervesces in dilute HCl
Pyrophyllite (54)	1.530–1.564	1.586–1.592	1.596–1.601	0.045–0.068	46°–62°	similar to colorless micas
Biotite–phlogopite (50)	1.530–1.625	1.557–1.696	1.558–1.696	0.028–0.078	0°–25°	a mica; biotite is brown or green pleochroic
Serpentine (53)	1.532–1.570		1.545–1.584	0.004–0.017	20°–60°	fine grained, flaky or fibrous
Talc (55)	1.539–1.550	1.584–1.594	1.584–1.596	0.039–0.050	0°–30°	soft, soapy feel; similar to colorless micas
Stilpnomelane (56)	1.543–1.634	ca. = γ	1.576–1.745	0.033–0.111	0°–40°	similar to biotite; usually fine grained
Illite (63)	1.545–1.630		1.570–1.670	0.022–0.055	small	a clay mineral
Chlorite (52)	1.550–?	1.550–1.70	1.550–?	0.00 –0.020	0°–60°	commonly green with anomalous blue interference colors

Muscovite (47)	1.552–1.578	1.582–1.615	1.587–1.617	0.036–0.049	0°–47°	mica, usually colorless
Paragonite (48)	1.564–1.580	1.594–1.609	1.600–1.609	0.028–0.038	0°–46°	very similar to muscovite
Glauconite (51)	1.585–1.616	ca. = γ	1.600–1.644	0.014–0.032	0°–20°	green pellets in marine sediments
Anthophyllite (42)	1.596–1.62	1.605–1.63	1.615–1.645	0.013–0.025	69°–90°	a metamorphic amphibole; straight extinction
Tremolite–Fe-actinolite (44)	1.599–1.688	1.612–1.697	1.622–1.705	0.017–0.027	64°–86°	elongate amphibole in lower grade metamorphics
Glaucophane–riebeckite (46)	1.606–1.702	1.622–1.712	1.627–1.718	0.006–0.022	0°–89°	blue amphiboles; $2V$ of glaucophane $<50°$
Eckermannite–arfvedsonite (46)	1.612–1.700	1.625–1.709	1.630–1.710	0.005–0.020	0°–80°	amphibole in plutonic alkaline igneous rocks
Wollastonite (37)	1.616–1.640	1.627–1.650	1.631–1.653	0.013–0.015	38°–60°	usually in metamorphosed carbonate rocks
Hornblende (44)	1.618–1.705	1.628–1.729	1.636–1.731	0.014–0.029	10°–90°	common amphibole
Andalusite (5)	1.629–1.642	1.633–1.646	1.638–1.653	0.009–0.011	73°–86°	in low-pressure metamorphics
Margarite (57)	1.630–1.638	1.642–1.648	1.644–1.650	0.012–0.014	40°–67°	a colorless, brittle mica
Spurrite (15)	1.638–1.640	1.672–1.674	1.678–1.680	0.040	40°	high-temperature Ca-silicate
Monticellite (2)	1.639–1.654	1.646–1.664	1.650–1.674	0.011–0.020	69°–82°	similar to olivine
Katophorite (45)	1.640–1.682	1,658–1.688	1.660–1.690	0.007–0.021	0°–50°	an amphibole in alkaline igneous rocks
Clintonite (58)	1.643–1.649	1.655–1.662	1.655–1.663	0.012–0.014	2°–40°	a brittle mica; often green or brown
Oxyhornblende (44)	1.650–1.702	1.682–1.769	1.689–1.796	0.020–0.094	56°–88°	red-brown amphibole; phenocrysts in volcanics
Dumortierite (9)	1.655–1.686	1.667–1.722	1.685–1.723	0.010–0.037	13°–63°	usually blue, pleochroic; straight extinction
Axinite (29)	1.656–1.694	1.660–1.701	1.668–1.705	0.007–0.014	63°–90°	wedge-shaped crystals common
Grunerite (43)	1.663–1.696	1.680–1.708	1.696–1.729	0.030–0.045	84°–90°	elongate metamorphic amphibole
Orthopyroxene (31)	1.665–1.755		1.675–1.772	0.010–0.017	45°–90°	
Olivine (1)	1.665–1.827	1.684–1.869	1.702–1.879	0.037–0.052	46°–90°	no good cleavage
Chloritoid (12)	1.682–1.730	1.688–1.734	1.694–1.740	0.006–0.022	55°–90°	platy crystals; similar to chlorite

(continued)

Determinative Table IX Biaxial Negative Minerals in Order According to Refractive Index (*continued*)

Mineral	α	β	γ	δ	$2V$	Remarks
Fe-gedrite (42)	1.690–1.695	1.705–1.710	1.715–1.722	0.025–0.028	82°–90°	elongate metamorphic amphibole
Allanite (22)	1.690–1.813	1.700–1.857	1.706–1.891	0.013–0.078	0°–90°	dark brown or red-brown
Sapphirine (41)	1.701–1.731	1.703–1.741	1.705–1.745	0.004–0.015	40°–90°	yellow, pink, or blue; in high-grade metamorphics
Kyanite (6)	1.706–1.718	1.714–1.723	1.719–1.734	0.012–0.016	77°–82°	prominent cleavages in some sections
Pumpellyite (24)	1.710–1.728	1.730–1.748	1.730–1.754	0.020–0.028	0°–80° 88°–90°	often bluish-green or brown; similar to epidote
Epidote (20)	1.714–1.751	1.721–1.784	1.730–1.797	0.015–0.049	64°–90°	often a distinctive green; lack of first-order white
Aegirine-augite (34)	1.725–1.750	1.742–1.780	1.760–1.800	0.035–0.050	70°–90°	a green clinopyroxene; in alkaline rocks
Piemontite (21)	1.732–1.794	1.750–1.807	1.762–1.829	0.025–0.088	81°–90°	pink or purple; similar to epidote
Staurolite (11)	1.736–1.747	1.741–1.754	1.749–1.762	0.010–0.015	88°–90°	yellow; in medium-grade metamorphics
Aegirine (34)	1.750–1.776	1.780–1.820	1.800–1.836	0.050–0.060	60°–70°	a green clinopyroxene; in alkaline rocks
Lepidocrocite (103)	1.94	2.20	2.51	0.57	83°	FeO·OH; yellow-brown, pleochroic
Goethite (103)	2.217–2.275	2.346–2.409	2.356–2.415	0.139–0.140	0°–27°	FeO·OH; yellow or brown

Determinative Table X Opaque Minerals

Color in reflected light	Mineral	Remarks
Black, metallic lustre	graphite (92)	very soft, fine-grained or platy crystals
Black, metallic lustre	magnetite (100)	magnetic; octahedral crystals
Black, metallic lustre	hematite (102)	translucent red at thin edges; red streak
Black, metallic lustre	ilmenite (101)	nonmagnetic; platy crystals; alters to leucoxene
White, like cotton-wool	leucoxene (101)	alteration product of Ti-rich minerals
Brownish-black, metallic or submetallic lustre	chromite (100)	translucent brown at thin edges
Brassy-yellow, metallic lustre	pyrite, marcasite } (108)	
Bronze, metallic lustre	pyrrhotite (109)	magnetic
Golden-yellow, metallic lustre	chalcopyrite (110)	

Chapter 9

Mineral Descriptions

The following grouping of minerals according to structure and chemistry is adopted in this book:

A. Nesosilicates and sorosilicates
B. Cyclosilicates
C. Inosilicates
D. Phyllosilicates
E. Tectosilicates
F. Volcanic glass (not a mineral)
G. Nonsilicates

For a straightforward discussion of the classes of silicate structures the reader is referred to Hurlbut and Klein (1977).

A grouping of minerals according to their optical properties is not followed because: (1) the range of properties such as RI and δ overlaps from mineral to mineral, (2) some do not lend themselves to a classification on the basis of optics—for example, a mineral may be +ve or −ve, (3) it is better to retain the natural grouping in mineral families such as the feldspathoids, members of which would be separated on the basis of their optical properties. Groupings according to optical properties are presented in the Determinative Tables (Chapter 8).

Each mineral is given a number in order of description. Generally, every mineral is described separately, although the descriptions of some related types have been coordinated. Petrologically important and abundant minerals such as the feldspars, amphiboles, and pyroxenes are described in greater detail than those of lesser importance.

For the majority of minerals, the description is laid out as follows: name and composition; crystallography; color; (other) optical properties; orientation diagrams; occurrence; and distinguishing features. In addition to the numerical data provided, birefringence is described as very low (<0.005), low (0.005–0.015), moderate (0.015–0.030), high (0.030–0.065), and very high (>0.065), and relief (in Canada balsam) as low (1.49–1.57), moderate (<1.49 and 1.57–1.68), high (1.68–1.78), very high (1.78–1.90) and extreme (>1.90). Where appropriate and relevant to detrital grain studies, properties such as color and the orientation of cleavage fragments are noted.

All the important rock-forming minerals are described. The worker is referred to A. Winchell (1939, 1951), H. Winchell (1965), and Phillips and Griffen (1981) for more comprehensive listings and tables.

A. Nesosilicates and Sorosilicates

The nesosilicates contain isolated [SiO_4] tetrahedral groups, and the sorosilicates paired tetrahedral groups [Si_2O_7] (Hurlbut and Klein, 1977). Neso- and sorosilicates described here are: olivine; monticellite; the humite group; garnet; andalusite; kyanite; sillimanite; mullite; dumortierite; topaz; staurolite; chloritoid; titanite (sphene); zircon; spurrite; larnite; merwinite; vesuvianite (idocrase); zoisite; clinozoisite–epidote; piemontite; allanite; lawsonite; pumpellyite; and melilite.

No. 1. OLIVINE (+ve and −ve) $(Mg,Fe)_2[SiO_4]$

A series from *forsterite* Mg_2SiO_4 to *fayalite* Fe_2SiO_4
See Plates 1 and 2.
Orthorhombic. Imperfect {010} and poor {100} cleavages. Euhedral stout crystals with {010} {110} {021} and {001} forms prominent. Also subhedral or granular.

Color in thin section. Usually colorless, but Fe-rich members may be pale yellow and pleochroic with $Y > X = Z$.

Color in detrital grains. Usually a pale olive-green.

Optical properties. Biaxial +ve or −ve. $2V_x = 98°–46°$. $r > v$ (weak) about X.

$$\alpha = 1.635–1.827,\ \beta = 1.651–1.869,\ \gamma = 1.670–1.879.$$

$$\delta = 0.035–0.052.\ X = b,\ Y = c,\ Z = a.$$

RI and δ increase, and $2V_x$ decreases with Fe content (Figure 9.1). A useful estimate of composition can be made in thin section by estimating $2V$, and this estimate of composition may be improved by using the U-stage (Chapter 6, page 116). Zoning may be evident from changes of birefringence across the crystal, especially in (100) sections.

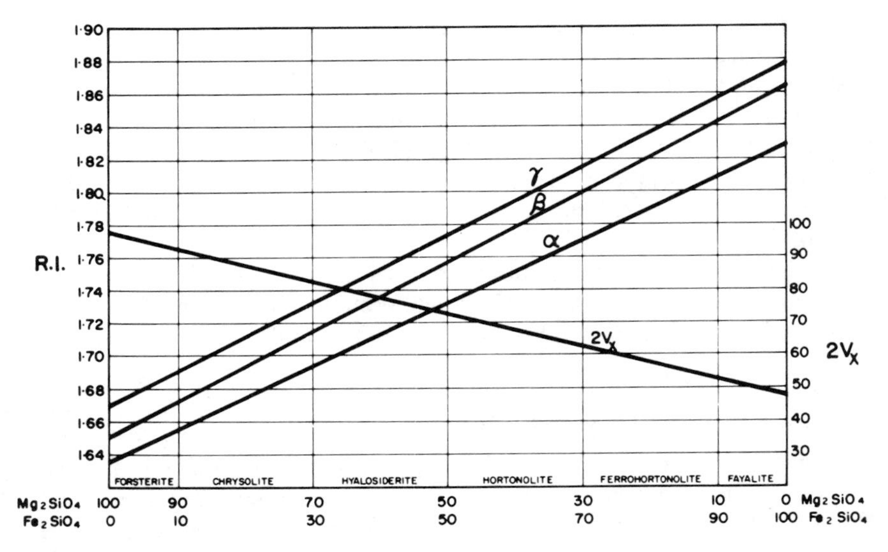

Figure 9.1 Variation of RI and $2V_x$ with composition in olivine. Based on Poldervaart (1950, Figure 2).

Orientation diagrams. *(100) section* (Figure 9.2a): acute (pure forsterite) or obtuse bisectrix figure; δ′ (moderate to high) = 0.016–0.042; slow along poorly developed cleavage; straight extinction; OAP across cleavage.
(010) section (Figure 9.2b): acute or obtuse (pure forsterite) bisectrix figure; δ′ (low to moderate) = 0.010–0.019; fast along; no cleavage; OAP across length of euhedral crystals.
(001) section (Figure 9.2c): flash figure; δ (high) = 0.035–0.052; poor cleavage may be developed parallel to slow direction.

Alteration products. Olivine is very susceptible to deuteric and hydrothermal alteration. Because it lacks a good cleavage, alteration characteristically proceeds along external surfaces and irregular cracks leaving "islands" of unaltered olivine. If total replacement occurs, pseudomorphs of the original olivine may be recognizable from their shape. There are four common alteration products:
Serpentine (page 229) is the common alteration product of olivine in dunites and peridotites. Often a meshlike structure of fibrous serpentine encloses almost isotropic "pools" of serpentine (Figure 9.48). Wicks and Whittaker (1977) discuss these textures in detail. Shearing of massive serpentinite is common, and pseudomorphic forms may then be destroyed.
Iddingsite frequently replaces olivine phenocrysts in basic volcanics. It is a blood-red substance which is sometimes fibrous, sometimes uniform in appearance. RI is high (ca. 1.76–1.89), birefringence moderate

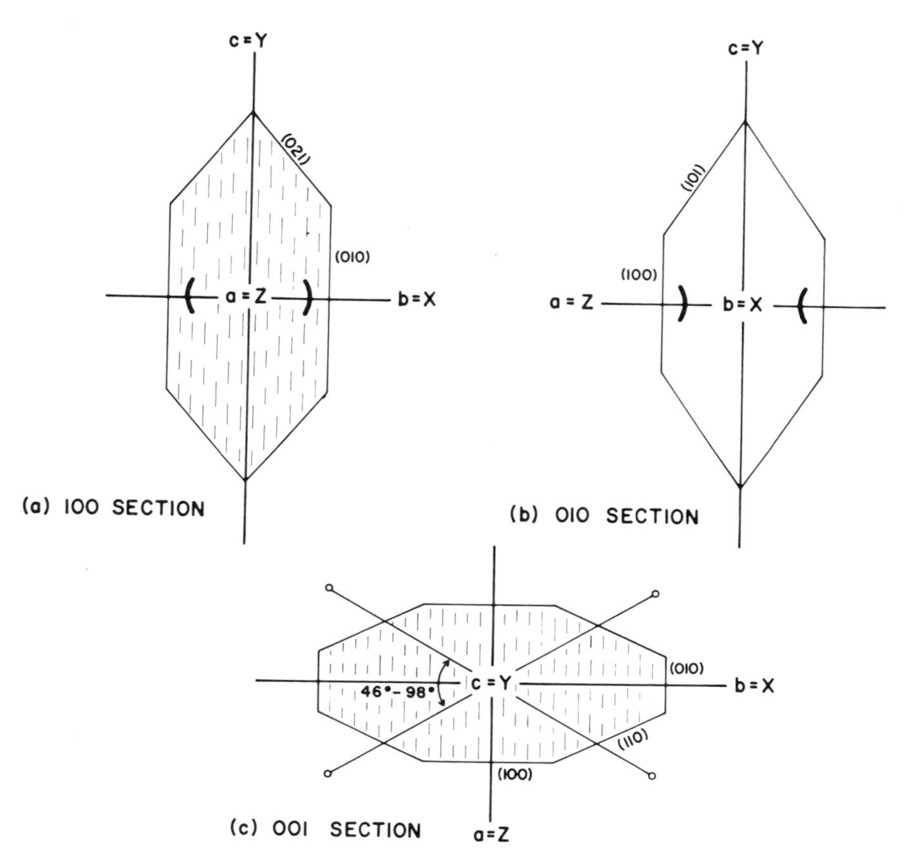

Figure 9.2 Orientation diagrams for olivine.

to high, optic sign +ve or −ve, and it may be pleochroic and superficially similar to biotite. It consists of an intimate mixture of smectite, chlorite, and goethite or hematite, sometimes with quartz and calcite, which inherits the oxygen framework of the original olivine.
Chlorophaeite is isotropic, orange to green, and RI = 1.50–1.62. Like iddingsite, it is an intimate mixture of minerals, probably saponite, chlorite, goethite, and calcite.
Bowlingite is a green, moderately birefringent alteration product with RI between 1.48 and 1.62. It too is an intimate mixture of minerals, but in a less oxidized state than in iddingsite.

A comprehensive review of olivine alteration is given by Deer, Howie, and Zussman (1982).

Occurrence. In a wide variety of igneous rocks, but most common in ultrabasic and basic types. Nearly pure forsterite occurs in dunite and peridotite, and moderately Mg-rich olivine in gabbro and basalt. Euhedral phenocrysts often display evidence of corrosion. Syenites and granites (and volcanic equivalents) may contain Fe-rich olivine. Almost pure forsterite occurs in thermally metamorphosed impure carbonate rocks. Fe-rich olivines are found in metamorphosed Fe-rich sediments. Olivine in metamorphosed gabbros may be surrounded by reaction rims (corona structures) usually composed of orthopyroxene and amphibole or garnet. Olivine is not a common detrital mineral because it is very susceptible to weathering and hydrothermal alteration (see above).

Deformation mechanisms. Olivine, as recrystallized mosaics in peridotites, commonly possesses a preferred orientation. Deformation mechanisms are discussed in detail by Nicolas and Poirier (1976). The most important slip systems are (with increasing temperature) (100)[001], {110}[001], {0k1}[100], and (010)[100]. The latter two systems are responsible for the common deformation bands parallel to (100), which represent dislocations concentrated in arrays following recovery.

Distinguishing features. Characterized by lack of color (except for Fe-rich members which are pale-yellow), poor cleavage, high RI and δ, alteration products, and occurrence. Irregular cracking is characteristic (Plates 1 and 2). May be confused with diopside, augite, epidote, monticellite, and humite. Distinguished from diopside and augite by lack of good cleavage, higher $2V$ and δ. Epidote has inclined extinction and anomalous interference colors. Monticellite and humite generally have a lower RI and/or δ.

No. 2. MONTICELLITE (−ve) $CaMg[SiO_4]$

Fe may substitute for Mg, but most natural monticellites are very Mg-rich.
Orthorhombic. Imperfect {010} cleavage. Sometimes euhedral but often granular. Twinning with twin plane (031).

Color in thin section. Colorless.

Optical properties. Biaxial −ve. $2V_x = 69°–82°$. $r > v$ (weak).

$$\alpha = 1.639–1.654,\ \beta = 1.646–1.664,\ \gamma = 1.650–1.674.$$

$$\delta = 0.011–0.020.\ X = b,\ Y = c,\ Z = a.$$

Fe-rich varieties may have a higher RI and δ, and a smaller $2V$.

Orientation diagrams. Very similar to those for olivine, but always −ve and with a smaller δ.

Occurrence. In metamorphosed impure carbonate rocks. Also in some alkaline basic and ultrabasic rocks, often as parallel overgrowths on olivine.

Distinguishing features. Similar to olivine. Distinguished from olivine by its lower δ and from diopside and augite by its poor cleavage and −ve character.

No. 3. HUMITE GROUP (+ve)

Chondrodite	$2Mg_2[SiO_4] \cdot Mg(OH,F)_2$
Humite	$3Mg_2[SiO_4] \cdot Mg(OH,F)_2$
Clinohumite	$4Mg_2[SiO_4] \cdot Mg(OH,F)_2$

Humite is orthorhombic. Chondrodite ($\beta = 109°$) and clinohumite ($\beta = 100°$) are monoclinic. Imperfect {100} cleavage. Subhedral or rounded anhedral crystals. Chondrodite, humite, and clinohumite may coexist in parallel intergrowths. Simple and lamellar twinning on {001} common in chondrodite and clinohumite.

Color in thin section. Pleochroic, colorless to pale yellow or pale to dark golden-yellow with $X > Z > Y$.

Optical properties. Biaxial +ve. $2V_z = 65°–85°$. $r > v$ (weak to strong).

$$\alpha = 1.592–1.643,\ \beta = 1.602–1.653,\ \gamma = 1.621–1.674.$$

$$\delta = 0.022–0.041.$$

RI of clinohumite generally > humite > chondrodite, but ranges overlap. RI increases with Fe substitution for Mg. Ti may raise RI in excess of 1.7.

Orientation diagrams. *(010) sections* (Figure 9.3): acute bisectrix figure; δ' (low) = 0.010; humite extinguishes parallel to the poor {100} cleavage, and the OAP is across the cleavage; chondrodite and clinohumite have inclined extinction and the OAP at an acute angle to the cleavage.
(001) sections: trace of the poor cleavage and straight extinction in all species; (001) section of humite displays the maximum δ with Z parallel to the cleavage.
(100) sections: no cleavage; (100) sections of chondrodite and clinohumite display the maximum δ.

Occurrence. In thermally metamorphosed and metasomatized carbonate rocks.

Distinguishing features. Pale-yellow color, lack of good cleavage, moderate relief, and moderate to high δ characteristic. The colorless species may be confused with olivine, but they usually have a smaller +ve $2V$.

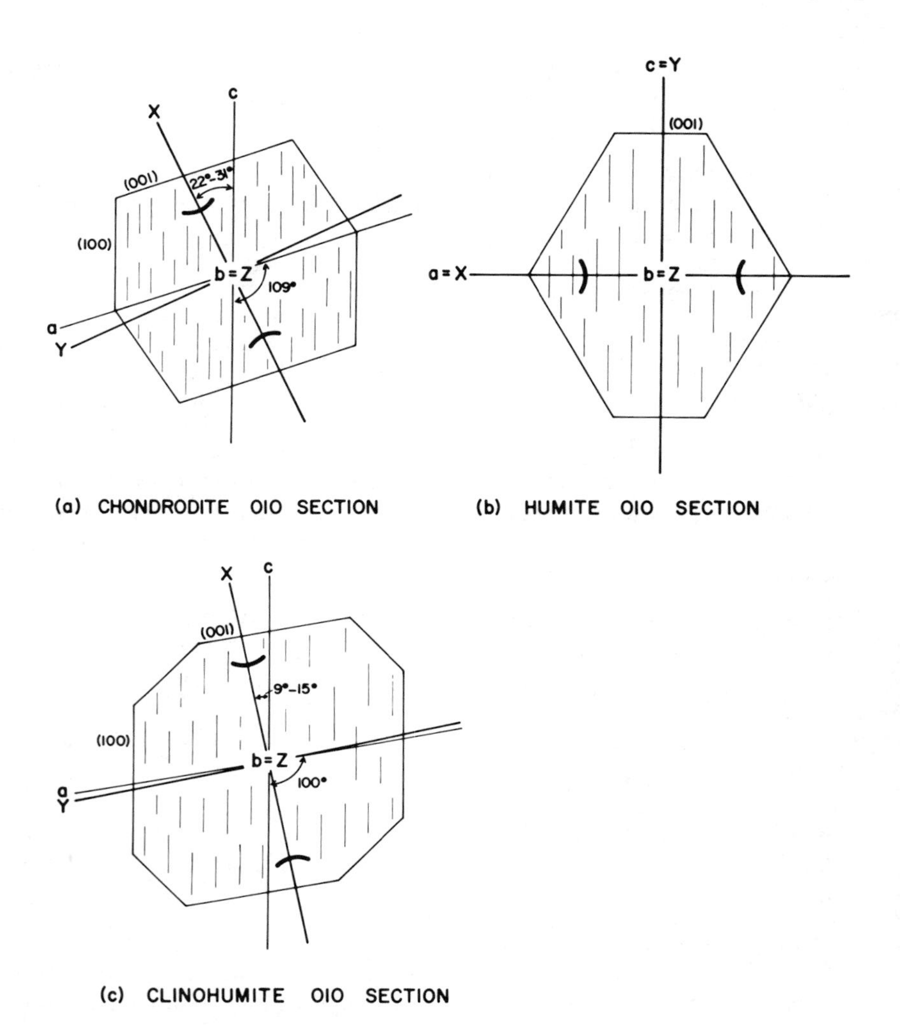

Figure 9.3 Orientation diagrams for the humite group.

The yellow species may be confused with yellow tourmaline or staurolite. However, tourmaline is uniaxial, and staurolite has a different pleochroic scheme with $Z > X$, a higher RI and lower δ.

No. 4. GARNET Group (isotropic)

See Plates 13, 14, 25, and 26.

Pyralspite	pyrope	$Mg_3Al_2Si_3O_{12}$
	almandine	$Fe_3^{2+}Al_2Si_3O_{12}$
	spessartine	$Mn_3Al_2Si_3O_{12}$

Ugrandite	uvarovite	$Ca_3Cr_2Si_3O_{12}$
	grossular	$Ca_3Al_2Si_3O_{12}$
	andradite	$Ca_3Fe^{3+}_2Si_3O_{12}$
Hydrogrossular		$Ca_3Al_2Si_2O_8(SiO_4)_{1-m}(OH)_{4m}$

Melanite and *schorlomite* are Ti-rich andradites (Plates 13 and 14).

There is continuous variation in composition within the two series pyralspite and ugrandite, but limited mixing between the two. The name given to any particular garnet is that of the dominant molecule present; pure end members are rare.

Cubic. No cleavage. Euhedral dodecahedra (six-sided sections) and trapezohedra (eight-sided sections) common; also anhedral. Zoning is often present but may be visible only in the strongly colored varieties. Complex twin patterns in some birefringent varieties.

Color in thin section. Common varieties are colorless or pink; melanite and schorlomite are dark brown; uvarovite is green.

Color in detrital grains. As above but deeper colors, often red; hydrogrossular is usually pale green or white.

Optical properties. Isotropic, though members of the ugrandite series, hydrogrossular, and spessartine may be weakly birefringent.

RI values for natural varieties with the following end members, dominant are: pyrope = 1.720–1.770; almandine = 1.770–1.820; spessartine = 1.790–1.810; uvarovite = ca. 1.86; grossular = 1.735–1.770; andradite = 1.850–1.890 (Ti-rich andradite = ca. 1.86–2.0); hydrogrossular = 1.675–1.734.

Occurrence. Detrital grains of garnet are common in sediments. *Pyrope* occurs in some ultrabasic rocks, especially kimberlite. Eclogites and amphibolites contain a mixed pyrope–almandine garnet. *Almandine* is the typical garnet of regional metamorphism in pelitic and semi-pelitic schists, forming first in the upper part of the greenschist facies; earlier formed garnets usually contain significant spessartine. Less commonly found in thermally metamorphosed rocks and rare in igneous rocks. *Spessartine* is most common in granite-pegmatites. *Uvarovite* is very rare; found in some serpentinites and skarns. *Grossular* is most common in metamorphosed carbonate rocks, and in metasomatic rocks, often with diopside. *Andradite* occurs in thermally metamorphosed impure carbonate rocks and skarns. *Melanite* and *schorlomite* are found in alkaline igneous rocks and skarns. *Hydrogrossular* occurs in metamorphosed and metasomatized carbonate rocks, in altered gabbros, and in rodingites with diopside.

Distinguishing features and identification of particular garnet species. The high relief, crystal shape, and isotropic character are distinctive.

Garnet may be confused with spinel, but spinel crystallizes as octahedra, and is commonly grey or green in thin section.

Individual species of garnet are not easy to determine. Various diagrams have been constructed to estimate the composition of a garnet from the three properties, RI, density, and cell size (Winchell, 1958), or from the two properties, RI and cell size (Sriramadas, 1957). Unfortunately, density measurements are not easy to make due to the common presence of inclusions in garnets; cell-size determination requires the use of X-rays. Neither of these methods nor chemical analysis provides satisfactory results for zoned garnets, and the best method at present is analysis using the X-ray microprobe. In some cases, RI by itself provides an estimate of composition. For example, hydrogrossular has the lowest RI, andradite the highest. In addition, occurrence is a useful guide. The color in thin section or hand specimen may be distinctive; although most garnets are red in hand specimen and colorless or pale pink in thin section, melanite and schorlomite are dark brown in thin section, uvarovite is green, and hydrogrossular in hand specimen is often white or pale green. Members of the ugrandite series, hydrogrossular, and spessartine may exhibit a weak birefringence.

No. 5. ANDALUSITE (−ve) Al_2SiO_5

See Figure 9.5a.

Small amounts of Fe or Mn may be present.

Orthorhombic. Distinct {110} and poor {100} cleavages. Usually euhedral elongate prismatic crystals with well-developed {110} faces giving a square cross section; also anhedral. The variety *chiastolite* contains graphitic inclusions, which, when viewed in cross section, appear in the form of a cross running from the edges of the prism faces through the center of the crystal (Figure 9.5a).

Color in thin section. Colorless or pleochroic with X = pink, Y and Z = pale green or yellow.

Color in detrital grains. More strongly colored and pleochroic as above; may be color-zoned.

Optical properties. Biaxial −ve. $2V_x = 73°–86°$. $r < v$ (weak).

$$\alpha = 1.629–1.642,\ \beta = 1.633–1.646,\ \gamma = 1.638–1.653.$$

$$\delta = 0.009–0.011.\ X = c,\ Y = b,\ Z = a.$$

RI, δ, and dispersion increase with Fe and Mn. Some Mn-rich andalusites are +ve and have a higher RI than the range given above.

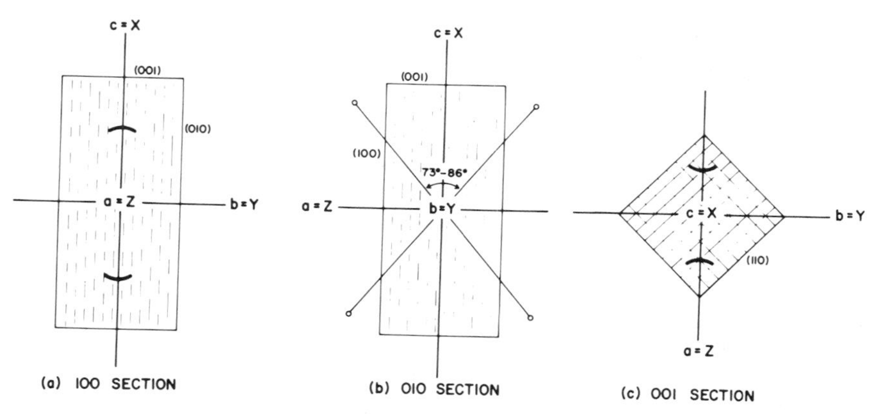

Figure 9.4 Orientation diagrams for andalusite.

Orientation diagrams. *(100) section* (Figure 9.4a): obtuse bisectrix figure; δ′ (very low) = 0.004; fast along and straight extinction; cleavage is oblique to the section and may not be visible.
(010) section (Figure 9.4b): flash figure; δ (low) up to 0.011; fast along and straight extinction; cleavage is oblique to the section and may not be visible.
(001) section (Figure 9.4c): acute bisectrix figure; δ′ (low) up to 0.006; symmetrical extinction, two cleavages almost at right angles to each other; chiastolite displays inclusion cross parallel to *Y* and *Z*.

Occurrence. Typically developed in thermally metamorphosed pelitic rocks, but also in pelitic rocks regionally metamorphosed under high geothermal gradients. Andalusite is the low-pressure aluminium-silicate polymorph (Figure 9.6). Commonly associated with cordierite. Andalusite changes to sillimanite at higher grades of metamorphism. Rarely occurs in granites and pegmatites. Frequently alters to sericite.

Distinguishing features. The shape of the crystals, the inclusion cross, the moderate relief, low δ, and occurrence are characteristic. The two cleavages and pleochroic scheme may cause confusion with orthopyroxene, but orthopyroxene is length slow.

No. 6. KYANITE (−ve) Al_2SiO_5

See Figure 9.5b.
Triclinic, $\alpha = 90°5'$, $\beta = 101°$, $\gamma = 106°$. Perfect {100} and less perfect {010} cleavages. Also {001} parting. Subhedral {100} tablets elongate parallel to *c* (blade-shaped crystals). Simple twins with {100} composition plane common; also multiple twins with {001} composition plane.

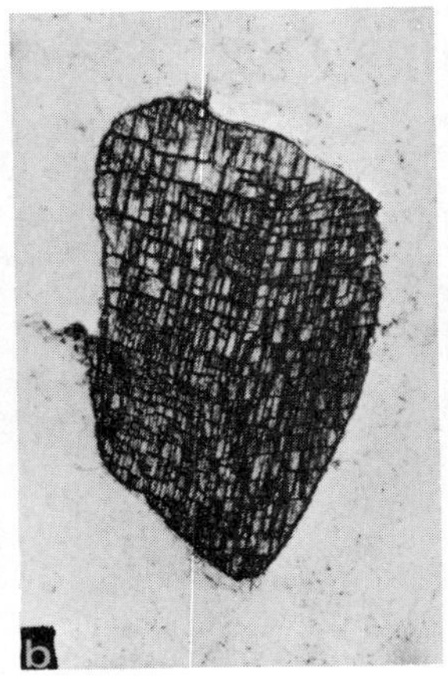
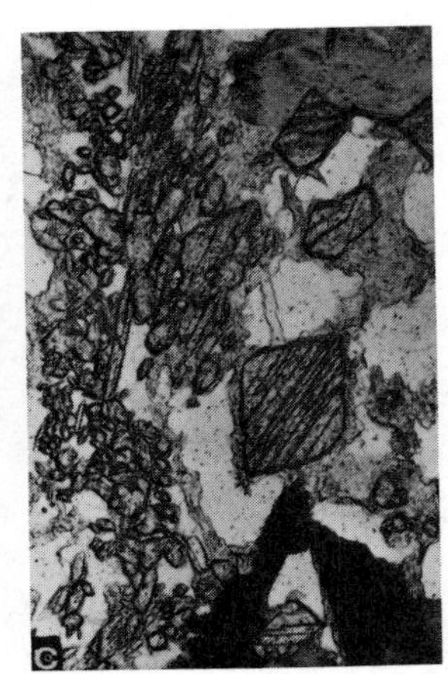

Figure 9.5 The aluminum silicate polymorphs. **(a)** Andalusite (var. chiastolite) in slate, Berridale, N.S.W., Australia (view measures 3.2 × 2.0 mm). **(b)** Kyanite, in granulite (view measures 0.8 × 0.5 mm). **(c)** Sillimanite, in gneiss, Broken Hill, Australia (view measures 0.8 × 0.5 mm). Plane-polarized light.

Color in thin section. Colorless.

Color in detrital grains. Blue, pleochroic with $Z > Y > X$.

Optical properties. Biaxial −ve. $2V_x = 77°–82°$. $r > v$ (weak).

$\alpha = 1.706–1.718$, $\beta = 1.714–1.723$, $\gamma = 1.719–1.734$.

$\delta = 0.012–0.016$. $X \wedge a =$ ca. 0°–5°, $Y \wedge b =$ ca. 30°, $Z \wedge c =$ ca. 30°.

Rare Cr-rich kyanites have been reported with γ up to 1.776.

Orientation diagrams. *(100) section (and cleavage fragments)* (Figure 9.7a): acute bisectrix figure; δ' (low) = 0.006–0.011; Z onto good cleavage = ca. 30°; also (001) cross-parting visible.
(001) section (Figure 9.7b): off-centered flash figure; δ' (low) = ca. 0.013; two well-developed cleavages at 74° to each other and a small extinction angle onto both; X is usually across the length of the crystals.
Optic-axis figures obtained from sections displaying no cleavage.
Maximum interference colors seen in sections showing (100) cleavage and nearly straight extinction (slow along).

Occurrence. Typically in high-grade regionally metamorphosed pelitic rocks. Also in granulites. More rarely in thermally metamorphosed rocks. Kyanite is the high-pressure aluminum–silicate polymorph (Figure 9.6) and may change to sillimanite at higher grades of metamor-

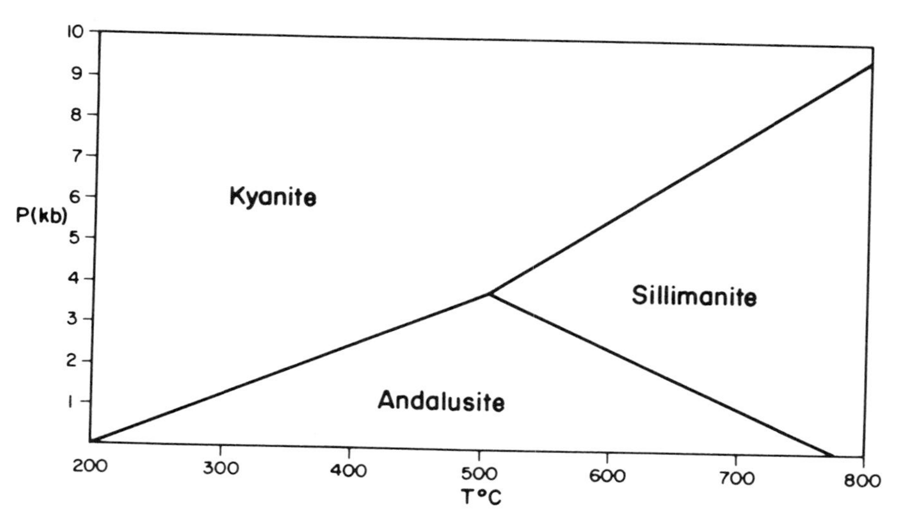

Figure 9.6 Phase diagram for the aluminum silicate polymorphs. Based on Holdaway (1971).

phism. Also occurs in quartz segregation veins. Often conspicuous as detrital grains. Alters to sericite.

Distinguishing features. The high-relief, low δ, very well-developed cleavages (Figure 9.5b), and inclined extinction (especially on (100) sections) are distinctive. Not easily confused with other minerals. Detrital grains are characterized by their high relief, conspicuous cleavages, and pale-blue color.

No. 7. SILLIMANITE (+ve) Al_2SiO_5

See Plates 37 and 38 and Figure 9.5c.
Orthorhombic. Perfect {010} cleavage. Usually euhedral: stout prisms with a nearly square cross-section or commonly as minute fibers (variety *fibrolite*).

Color in thin section. Colorless.

Color in detrital grains. Colorless or pleochroic with X = pale brown, Y = brown or green, Z = dark brown or blue.

Optical properties. Biaxial +ve. $2V_z = 20°–30°$. $r > v$ (strong).

$$\alpha = 1.654–1.661,\ \beta = 1.658–1.662,\ \gamma = 1.673–1.683.$$

$$\delta = 0.019–0.024.\ X = a,\ Y = b,\ Z = c.$$

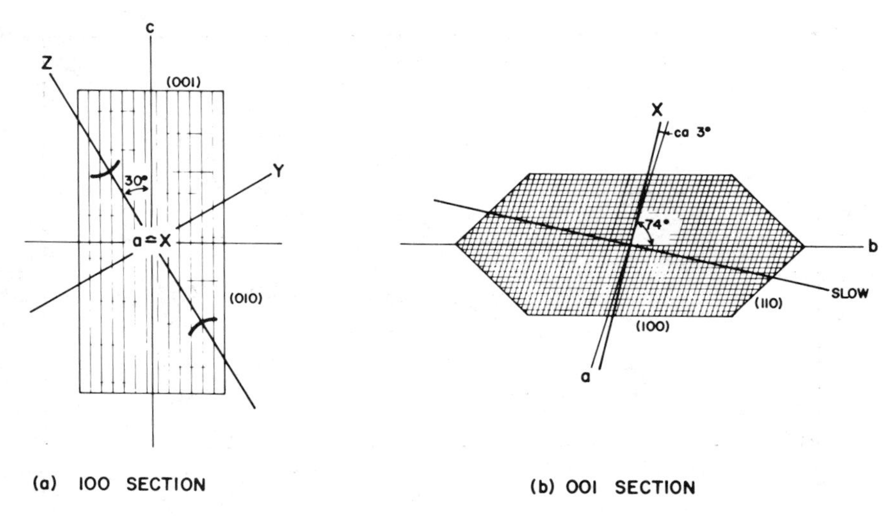

Figure 9.7 Orientation diagrams for kyanite.

Orientation diagrams. *(100) section* (Figure 9.8a): obtuse bisectrix figure; δ′ (moderate) up to 0.021; slow along and straight extinction; cleavage rarely seen in small fibers.
(010) section (and cleavage fragments) (Figure 9.8b): flash figure; δ (moderate) up to 0.024; slow along and straight extinction.
(001) section (Figure 9.8c): acute bisectrix figure; δ′ (low) up to 0.009; extinction positions often indistinct due to proximity of optic axes.
Optic-axis figures obtained in sections slightly oblique to (001).

Occurrence. A metamorphic mineral characteristic of high-grade schists, gneisses, migmatites, and hornfelses. Sillimanite is the high-temperature aluminium-silicate polymorph (Figure 9.6). Fibrolite masses are often embedded in micas and quartz.

Distinguishing features. The single cleavage in stout prisms or fibrous nature (Figure 9.5c and Plates 37 and 38), the straight extinction, length-slow character, moderate δ, and occurrence are characteristic. Small fibers may be confused with anthophyllite, apatite, wollastonite, or mullite. Apatite is fast along and has hexagonal cross sections, although both these features are difficult to observe with small fibers embedded in other minerals. Anthophyllite has a more restricted occurrence, and usually a slightly lower RI. It may be necessary to use X-rays to distinguish sillimanite from mullite. Wollastonite has slightly inclined extinction and some fibers are length fast.

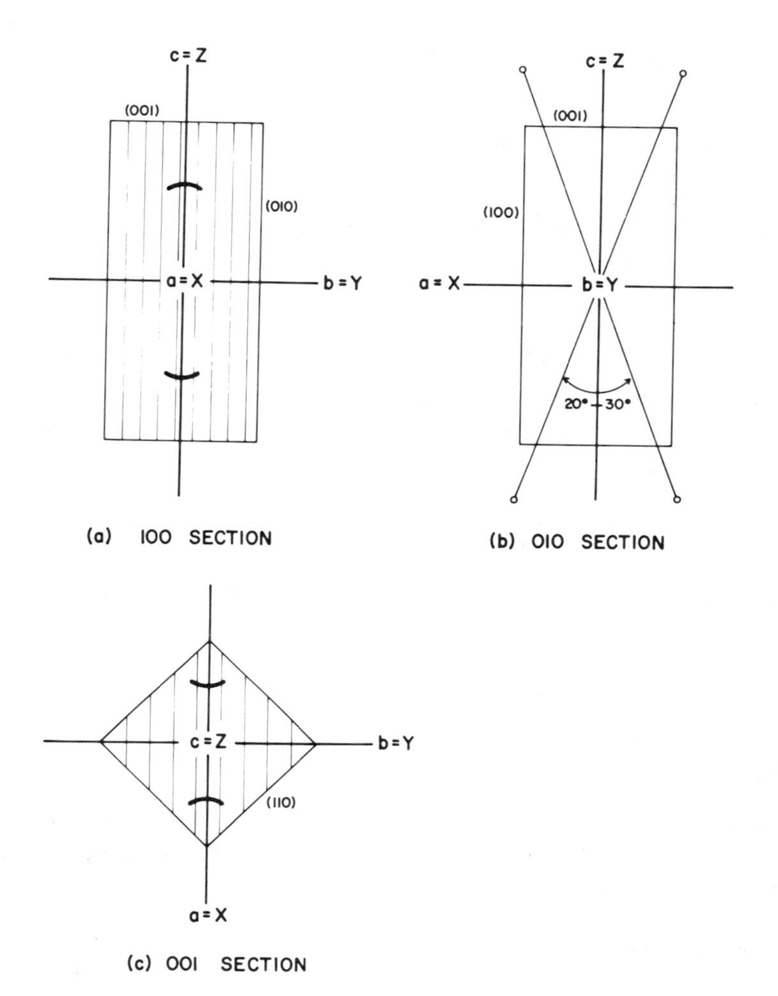

Figure 9.8 Orientation diagrams for sillimanite.

No. 8. MULLITE (+ve) $3Al_2O_3 \cdot 2SiO_2$

Orthorhombic. Perfect {010} cleavage. Long prisms with nearly square cross sections, or fibrous.

Color in thin section. Colorless.

Optical properties. Biaxial +ve. $2V_z = 45°–61°$. $r > v$ (weak to distinct).

$$\alpha = 1.640–1.670,\ \beta = 1.642–1.675,\ \gamma = 1.651–1.690.$$

$$\delta = 0.012–0.028.\ X = a,\ Y = b,\ Z = c.$$

Orientation diagrams. Almost identical to those for sillimanite.

Occurrence. Restricted to pelitic rocks thermally metamorphosed or fused (buchites) at very high temperatures; usually in xenoliths. A common refractory product.

Distinguishing features. Almost indistinguishable from sillimanite, but its occurrence is very restricted. X-ray tests may, with difficulty, distinguish the two.

No. 9. DUMORTIERITE (−ve) $(Al,Fe)_7O_3(BO_3)[SiO_4]_3$

Orthorhombic. Distinct {100} and poor {110} cleavages. Usually fibrous or bladed crystals elongate parallel to *c*. Sector twins with {110} composition planes.

Color in thin section. Strongly pleochroic, usually blue or violet, also brown or pink, with $X > Y > Z$.

Optical properties. Biaxial −ve. $2V_x = 13°–63°$. $r < v$ (strong).

$$\alpha = 1.655–1.686, \beta = 1.667–1.722, \gamma = 1.685–1.723.$$

$$\delta = 0.010–0.037. X = c, Y = b, Z = a.$$

Crystals have straight extinction, are length fast, and have the maximum absorption parallel to their length.

Occurrence. Usually associated with metasomatic or hydrothermal activity, and found in granite-pegmatites, gneisses, quartz veins, quartzites, and altered igneous rocks.

Distinguishing features. The strong pleochroism and blue color are distinctive, together with the straight extinction and length-fast character. May be confused with tourmaline, but the maximum absorption of tourmaline is across the length of the crystals. Pale-colored fine crystals of dumortierite may be confused with sillimanite, but sillimanite is length slow.

No. 10. TOPAZ (+ve) $Al_2[SiO_4](OH,F)_2$

Orthorhombic. Perfect {001} cleavage. Sometimes euhedral prismatic crystals. Often anhedral.

Color in thin section. Colorless.

Color in detrital grains. X and Y = yellow, Z = pink.

Optical properties. Biaxial +ve. $2V_z = 46°–70°$. $r > v$ (distinct).

$\alpha = 1.606–1.634$, $\beta = 1.609–1.637$, $\gamma = 1.616–1.644$.

$\delta = 0.007–0.011$. $X = a$, $Y = b$, $Z = c$.

$2V$ increases and RI decreases with increase in F.

Orientation diagrams. *(100) section* (Figure 9.9a): obtuse bisectrix figure; δ' (low) = ca. 0.007; OAP across cleavage; slow along length of prisms; straight extinction.
(010) section: similar appearance to (100) section but flash figure and maximum interference colors seen.
(001) section and cleavage fragments (Figure 9.9b): acute bisectrix figure; δ' (very low) = ca. 0.003; no cleavage visible.
Optic-axis figures obtained in sections slightly oblique to (001).

Occurrence. Present in some granites and rhyolites, but most common in granite pegmatites and greisen associated with pneumatolytic activity. Alters to sericite and fluorite.

Distinguishing features. The moderate relief, low δ, and biaxial +ve character are distinctive. Apatite resembles topaz but is uniaxial −ve.

No. 11. STAUROLITE (+ve, rarely −ve) $Fe_2^{2+}Al_9Si_4O_{22}(O,OH)_2$

See Plates 35 and 36.

There may be minor substitutions of Mg for Fe^{2+}, and Fe^{3+} for Al. Monoclinic, but pseudo-orthorhombic with β = ca. 90°. Imperfect {010}

Figure 9.9 Orientation diagrams for topaz.

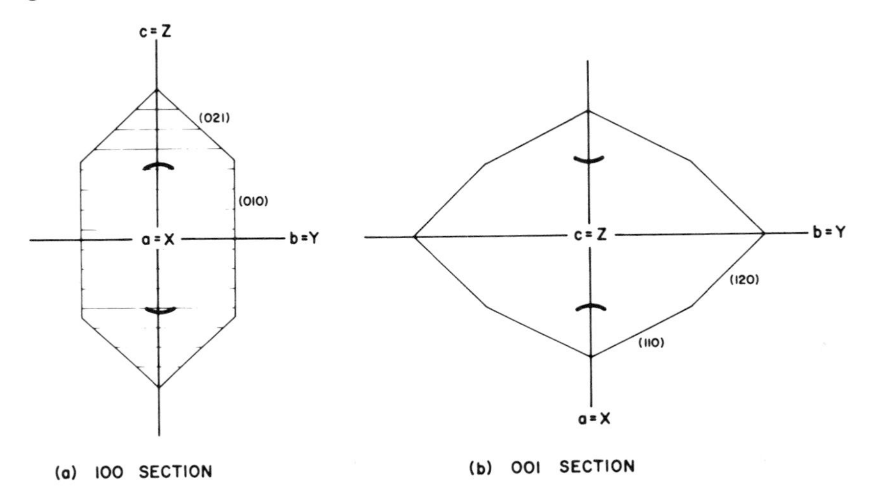

cleavage. Euhedral or subhedral prismatic crystals with {110} and {010} dominating. Penetration twins, not usually seen in thin section.

Color in thin section. Pleochroic from colorless to yellow with $Z > Y > X$.

Color in detrital grains. As above but deeper colors.

Optical properties. Biaxial +ve or −ve. $2V_z$ = 80°–92°. $r > v$ (weak to distinct).

$$\alpha = 1.736\text{–}1.747,\ \beta = 1.741\text{–}1.754,\ \gamma = 1.749\text{–}1.762.$$

$$\delta = 0.010\text{–}0.015.\ X = b,\ Y = a,\ Z = c.$$

RI and δ increase and $2V_z$ decreases with iron content (Griffen and Ribbe, 1973). A zincian staurolite has been described with RI 0.016 less than reported above.

Orientation diagrams. *(100) section* (Figure 9.10a): flash figure; δ (low) up to 0.015; slow parallel to weak cleavage; straight extinction.
(010) section (Figure 9.10b): obtuse bisectrix figure; δ′ (low) = ca. 0.007; straight extinction to prismatic edges; no cleavage.
(001) section (Figure 9.10c): acute bisectrix figure; δ′ (low) = ca. 0.006; OAP across weak cleavage; straight extinction to cleavage.

Occurrence. Often as poikiloblasts in medium-grade regionally metamorphosed pelites. Often associated with chloritoid or kyanite. Kyanite sometimes replaces staurolite. May alter to sericite and chlorite, but is resistant to weathering and common as a detrital mineral.

Distinguishing features. High relief, low δ, and yellow color very distinctive. May be confused with yellow tourmaline, which, however, is uniaxial and fast along.

No. 12. CHLORITOID (+ve, sometimes −ve)

$(Fe^{2+},Mg,Mn)_2(Al,Fe^{3+})Al_3O_2[SiO_4]_2(OH)_4$

See Plates 29 and 30.

Ottrelite is Mn-rich chloritoid.

Monoclinic or triclinic (β = 102°, γ = ca. 90°). Perfect {001} and poor {110} ? cleavages. Subhedral tablets parallel to {001} with pseudohexagonal outlines. Also anhedral. Lamellar twinning with {001} composition plane common. Hour-glass structure common.

Color in thin section. Colorless to green or blue-green pleochroic usually with X = green, Y = blue, Z = colorless or pale yellow.

Color in detrital grains. As above but deeper colors.

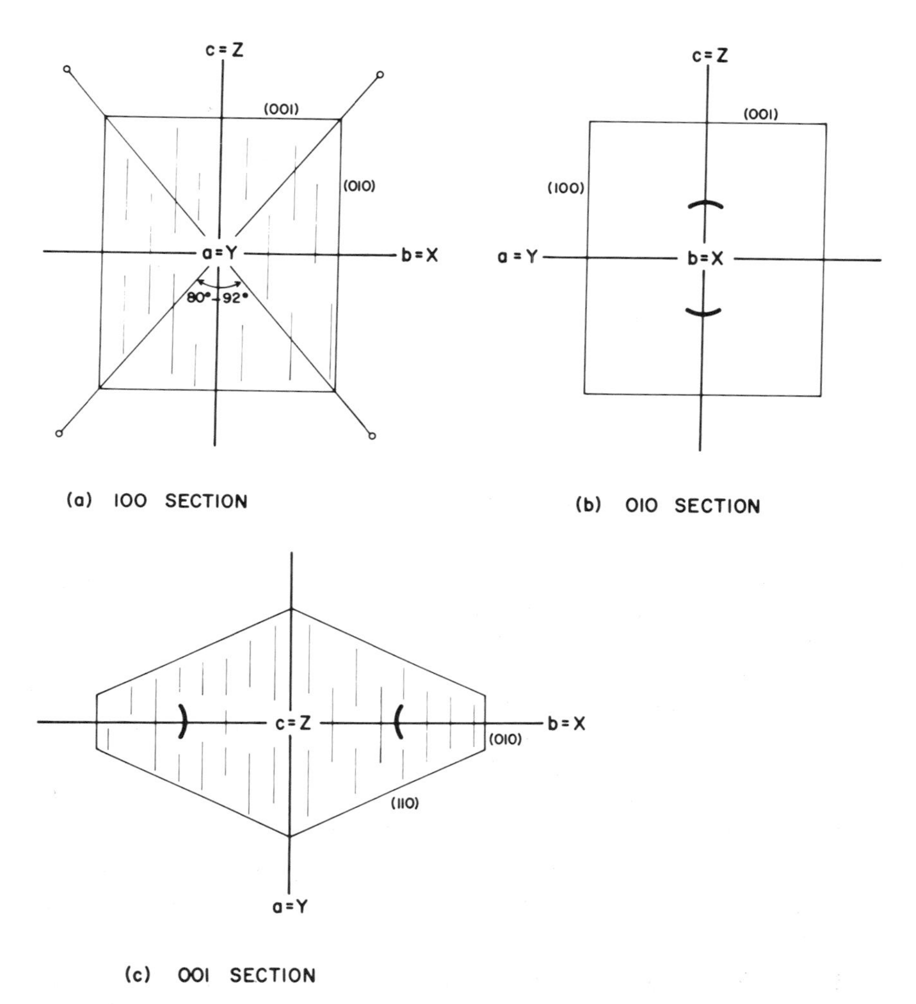

Figure 9.10 Orientation diagrams for staurolite.

Optical properties. Biaxial +ve or −ve. $2V_z = 40°–125°$ (usually <90°). $r > v$ (strong).

$\alpha = 1.682–1.730$, $\beta = 1.688–1.734$, $\gamma = 1.694–1.740$.

$\delta = 0.006–0.022$. $X = b$ (rarely $Y = b$), $Y \wedge a = 2°–30°$, $Z \wedge c = 14°–42°$ (Z onto $\perp$ 001 $= 2°–30°$).

Interference colors may be anomalous.
RI increases generally with total Fe and Mn content.

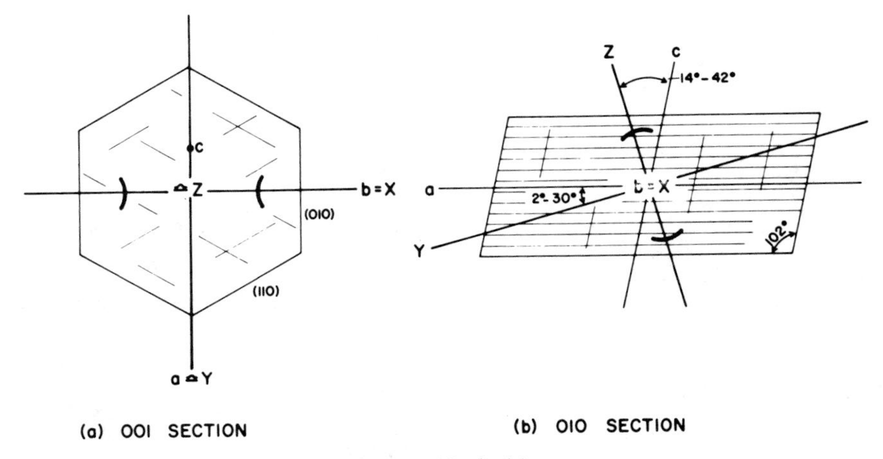

Figure 9.11 Orientation diagrams for chloritoid.

Orientation diagrams. *(001) section and cleavage fragments* (Figure 9.11a): acute bisectrix figure (slightly off-centered); δ′ (low) = ca. 0.006; extinction positions often indistinct due to proximity of optic axes; poor {110} cleavages rarely visible; very weak pleochroism.
(010) section (Fig. 9.11b): obtuse bisectrix figure; δ′ (low) = ca. 0.005; inclined extinction, fast onto perfect cleavage; twinning commonly observed; poor cleavages // *c* rarely visible.
(100) section: similar to (010) section, but flash figure, straight extinction, and displays maximum interference colors.

Occurrence. In low-grade regionally metamorphosed pelitic rocks, often as small posttectonic porphyroblasts (usually monoclinic). Rarely in quartz veins (usually triclinic).

Distinguishing features. High relief, low δ, green color, and platy crystals with twinning distinctive. May be confused with chlorite, but chlorite has a lower RI, and usually a smaller $2V$ and extinction angle. The brittle micas are optically −ve. Chloritoid does not display the mottled extinction many other platy minerals do.

No. 13. TITANITE (SPHENE) (+ve) $CaTiSiO_5$

See Plates 9 and 10.

Some substitution of Sr, Ba, Th, or the rare earths for Ca, Fe^{3+}, and Al for Ti, and OH and F for O is possible.

Monoclinic, $\beta = 120°$. Distinct {110} cleavages. Euhedral crystals with rhombic cross sections common. Also anhedral as lensoid or droplike grains. Twinning (lamellar) with {221} composition planes is not uncommon.

Color in thin section. Colorless or pleochroic sugary-brown with $Z > Y > X$.

Color in detrital grains. Colorless or sugary to dark brown.

Optical properties. Biaxial +ve. $2V_z = 20°–56°$. $r > v$ (strong).

$\alpha = 1.840–1.950$, $\beta = 1.870–2.034$, $\gamma = 1.943–2.110$.

$\delta = 0.100–0.192$. $X \wedge a =$ ca. 21°, $Y = b$, $Z \wedge c =$ ca. 51°.

Orientation diagrams. *Section perpendicular to c* (Figure 9.12a): off-centered figure which may be close to an optic-axis figure; δ' usually high; symmetrical extinction with respect to two cleavages; traces of twin lamellae may be visible parallel to cleavages.
(010) section (Figure 9.12b): flash figure; δ (very high) = 0.1–0.192; a cleavage trace may be visible with $Z \wedge c = 51°$; twin lamellae may be visible oblique to cleavage trace.
Optic-axis figures obtained in sections slightly oblique to (100) and approximately parallel to (001).

Occurrence. Very common as an accessory mineral in intermediate and acid plutonic igneous rocks, particularly syenites. Also in a wide range of metamorphic rocks, and as detrital grains.

Distinguishing features. The very high relief and δ and crystal shapes are distinctive. Superficially similar to the carbonates in thin section, but carbonates have a lower RI and often "twinkle." May be confused with monazite which has a lower δ, and with detrital grains of xenotime, cassiterite or rutile, all of which are uniaxial. Incomplete extinction in low δ' grains (because of high dispersion) is characteristic, especially in grain mounts.

Figure 9.12 Orientation diagrams for titanite (sphene).

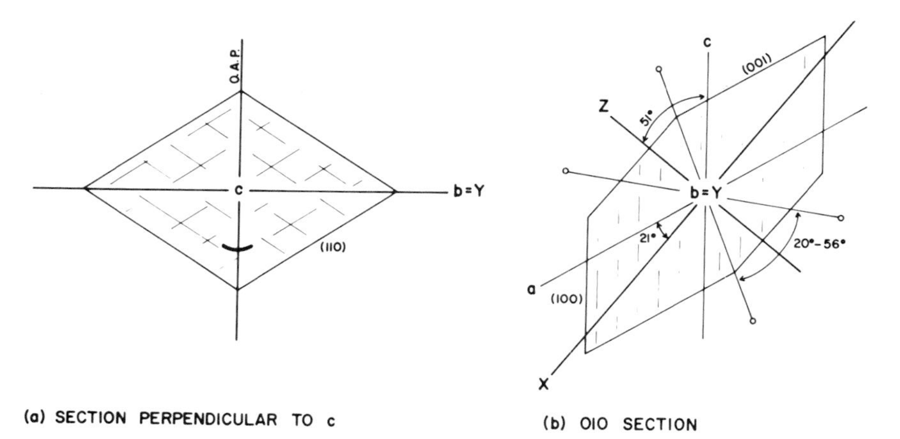

No. 14. ZIRCON (+ve) $ZrSiO_4$

See Figure 9.13.

Usually contains Hf, and often Y, Th, U, and Fe^{3+}.

Tetragonal. Poor {110} cleavages. Usually as small euhedral or subhedral prismatic crystals with {110} dominating.

Color in thin section. Colorless or pale brown.

Color in detrital grains. Colorless, yellow, brown, pink, or purple; pleochroism $\varepsilon > \omega$.

Optical properties. Uniaxial +ve.

$$\omega = 1.92\text{–}1.96, \ \varepsilon = 1.96\text{–}2.02.$$

$$\delta = 0.04\text{–}0.06.$$

Orientation diagrams. *Section parallel to c, and elongate detrital grains* (Figure 9.13): flash figure; δ (high) = 0.04–0.06; straight extinction, slow along.

Uniaxial cross figures obtained in sections cut ⊥ *c*, with equidimensional cross sections (often square) and low interference colors. Usu-

Figure 9.13 Detrital grains of zircon. Note the prisms with pyramidal terminations well preserved. Pleistocene sands, Westport, New Zealand (view measures 1.7 × 1.3 mm). Plane-polarized light.

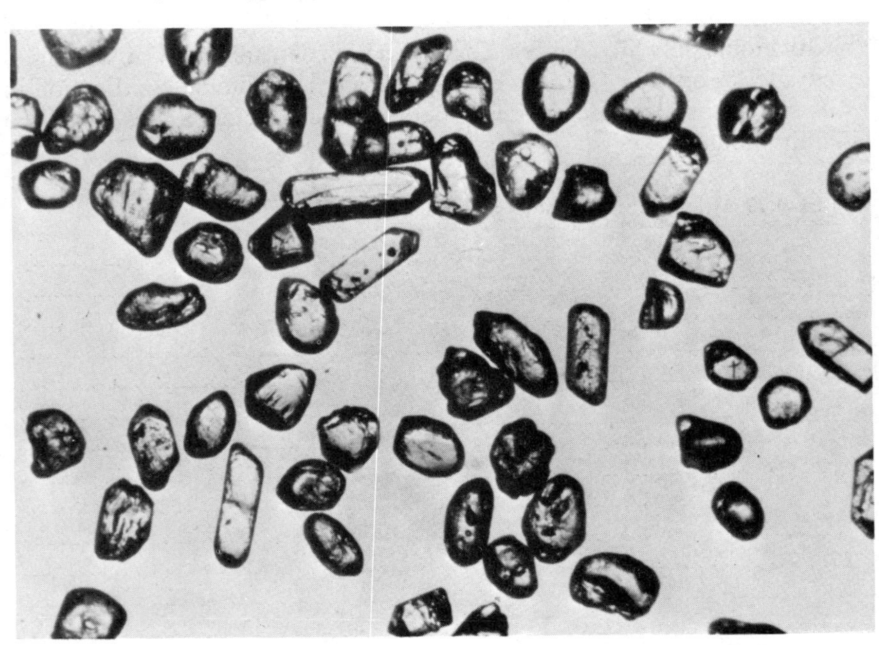

ally not easy to obtain in thin sections because of small size of crystals, or in detrital grains because of grain shape.

Occurrence. A common accessory mineral in plutonic igneous rocks, particularly granites and syenites. May form large crystals in pegmatites. Also as an accessory mineral in schists and gneisses. It is very resistant to weathering, and is abundant as detrital grains in many sediments, and may be preserved through several cycles of deposition. Detrital grains, sometimes with overgrowths, may also be recognized in metamorphic rocks and granites. Zircon produces pleochroic haloes in biotite. Radioactive elements in zircon may destroy the atomic structure producing "metamict" isotropic pseudomorphs.

Distinguishing features. Extreme relief, high δ, crystal shapes and straight extinction are distinctive. May be confused with cassiterite and rutile, but these minerals are usually deeper brown, and have a higher δ. Xenotime has a lower RI and higher δ.

High-Temperature Calcium Silicate Minerals

In addition to the more common minerals wollastonite, tremolite, vesuvianite, grossular, scapolite, epidote, diopside, monticellite, and the humite group, a variety of rare calcium silicate minerals develop at high temperatures as a result of the metamorphism and metasomatism of carbonate rocks. Three of the most common such minerals, spurrite, larnite, and merwinite, are described below briefly.

No. 15. SPURRITE (−ve) $2Ca_2[SiO_4] \cdot CaCO_3$

Monoclinic, $\beta = 123°$. Distinct {001} and imperfect {100} cleavages. Usually anhedral, granular. Lamellar twinning with {001} composition planes.

Color in thin section. Colorless.

Optical properties. Biaxial −ve. $2V_x = 40°$. $r > v$ (weak).

$\alpha = 1.638–1.640$, $\beta = 1.672–1.674$, $\gamma = 1.678–1.680$.

$\delta = 0.040$. $X = b$, $Y \wedge c = 33°$, Z approximately parallel to a.

Occurrence. Restricted to high-grade thermally metamorphosed and metasomatized carbonate rocks.

Distinguishing features. Occurrence is distinctive. Distinguished from the associated minerals larnite and merwinite by its −ve character and higher δ.

No. 16. LARNITE (+ve) $Ca_2[SiO_4]$

Monoclinic, $\beta = 95°$. Distinct {100} cleavage. Usually anhedral, granular. Lamellar twinning with {100} composition planes.

Color in thin section. Colorless.

Optical properties. Biaxial +ve. $2V_z$ moderate.

$$\alpha = 1.707,\ \beta = 1.715,\ \gamma = 1.730.$$

$$\delta = 0.023.\ X \wedge c = 13°,\ Y \wedge a = 8°,\ Z = b.$$

Occurrence. Restricted to high-grade thermally metamorphosed and metasomatized carbonate rocks.

Distinguishing features. Occurrence is distinctive. Often associated with spurrite and merwinite. Distinguished from spurrite by its +ve character, and from merwinite by its different cleavage, twinning, and usually higher δ.

No. 17. MERWINITE (+ve) $Ca_3Mg[Si_2O_8]$

Monoclinic, $\beta = 92°$. Poor {100} cleavage. Usually anhedral, granular. Intersecting lamellar twins with {100}, (611) and $(6\bar{1}1)$ composition planes.

Color in thin section. Colorless.

Optical properties. Biaxial +ve. $2V_z = 52°–76°$. $r > v$ (weak).

$$\alpha = 1.702–1.710,\ \beta = 1.710–1.718,\ \gamma = 1.718–1.726.$$

$$\delta = 0.008–0.023.\ X \wedge c = 13°,\ Y = b.$$

Occurrence. Restricted to high-grade thermally metamorphosed and metasomatized carbonate rocks.

Distinguishing features. Occurrence is distinctive. Often associated with spurrite and larnite. Distinguished from spurrite by its +ve character, and from larnite by its twinning, and usually lower δ.

No. 18. VESUVIANITE (IDOCRASE) (−ve)

$Ca_{19}(Mg,Fe,Al)_5Al_8[Si_2O_7]_4[SiO_4]_{10}(OH,F)_{10}$

Tetragonal. Imperfect {110} and {100} cleavages. Euhedral prismatic to anhedral, granular crystals.

Color in thin section. Colorless, plate brown or green; tends to become opaque if rich in rare earths.

Optical properties. Uniaxial −ve (rarely +ve or biaxial).

$$\omega = 1.703\text{–}1.762,\ \varepsilon = 1.697\text{–}1.750.$$

$$\delta = 0.001\text{–}0.012.$$

Anomalous deep-blue, brown, or purple interference colors common.

Occurrence. Typically occurs in thermally metamorphosed or metasomatized carbonate rocks. More rarely occurs in nepheline-syenites, and in veins cutting ultramafic rocks such as serpentine.

Distinguishing features. High relief, very low δ, anomalous interference colors, poor cleavage, and occurrence distinctive. May be confused with melilite, zoisite or clinozoisite, but zoisite and clinozoisite are +ve, and melilite has a better cleavage and lower RI. Vesuvianite that does not display anomalous interference colors is similar to apatite, but has a higher RI.

The Epidote Group

The five minerals—zoisite, clinozoisite, epidote, piemontite, and allanite—have in common a structure of independent $[SiO_4]$ and $[Si_2O_7]$ groups which link together chains of AlO_6 and $AlO_4(OH)_2$. Clinozoisite and epidote form a continuous series, and are described together.

No. 19. ZOISITE (+ve) $Ca_2Al_3O[Si_2O_7][SiO_4](OH)$

Fe^{3+} may substitute for up to about 5% of the Al in ferrian or α-zoisite (Myer, 1966). Some Mn substitutes for Al in the variety *thulite*. Orthorhombic. Perfect {100} cleavage (N.B. this is reported as {010} in many texts, but here, following Deer et al., 1962, the previous *a*, *b*, and *c* are changed to *c*, *a*, and *b*). Often subhedral aggregates of bladed or fibrous crystals elongate parallel to *b*.

Color in thin section. Colorless; thulite is pleochroic with X = pink, Y = colorless, Z = yellow.

Optical properties. Biaxial +ve. $2V_z = 0°\text{–}70°$ (usually >30°); a large $2V_x$ has also been reported. $r > $ or $< v$ (strong).

$\alpha = 1.685\text{–}1.705,\ \beta = 1.688\text{–}1.710,\ \gamma = 1.697\text{–}1.725.$

$\delta = 0.004\text{–}0.022$ (usually <0.010; the higher values are normally due to a high Fe or Mn content).

$X = b,\ Y = a,\ Z = c$ for Fe-rich ferrian or α-zoisite.

$X = a,\ Y = b,\ Z = c$ for β-zoisite.

Zoisite may display anomalous blue interference colors.

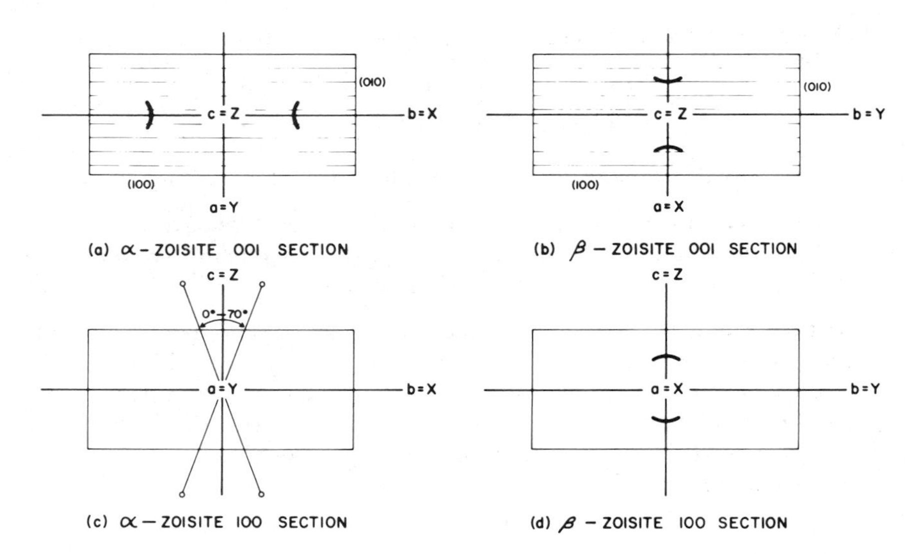

Figure 9.14 Orientation diagrams for zoisite.

Orientation diagrams. *(001) sections* (Figure 9.14a, b): acute bisectrix figure; OAP parallel to perfect cleavage in α form, and perpendicular to cleavage in β form; δ' (very low) < 0.005; may be fast (α-zoisite) or slow (β-zoisite) along, and straight extinction.
(100) sections and cleavage fragments (Figure 9.14c, d): obtuse bisectrix (β-zoisite) or flash figure (α-zoisite); δ' is very low to moderate; α-zoisite (100) sections display the highest interference colors; fast along, and straight extinction in euhedral grains.
Optic-axis figures obtained in sections showing good cleavage in α-zoisite, but in sections oblique to cleavage in β-zoisite.

Occurrence. Not as common as clinozoisite–epidote, but widespread, occurring in regionally and thermally metamorphosed (usually impure carbonate) rocks. Zoisite may be a component of saussurite—an alteration product of plagioclase.

Distinguishing features. High relief, low to moderate δ, single cleavage, and straight extinction characteristic. Most easily confused with clinozoisite which, however, has inclined extinction. May sometimes be confused with vesuvianite, apatite, and melilite, but these minerals are uniaxial and usually −ve.

No. 20. CLINOZOISITE (+ve) − EPIDOTE (−ve)

$Ca_2(Al,Fe^{3+})Al_2O[Si_2O_7][SiO_4](OH)$

See Plates 27 and 28.

There is increasing substitution of Fe for Al from clinozoisite to epidote.

Monoclinic, $\beta = 115°$. Perfect {001} and imperfect {100} cleavages. Often subhedral bladed or fibrous crystals elongate parallel to b. Also anhedral. Twinning with {100} composition plane not common. Zoning common.

Color in thin section. Clinozoisite, colorless; epidote, colorless or yellow-green pleochroic with $Z > Y > X$, or $Y > Z > X$; color-zoning common.

Color in detrital grains. Clinozoisite, colorless or pale yellow; epidote, yellowish green.

Optical properties. *Clinozoisite:* biaxial +ve. $2V_z = 14°–90°$. $r < v$ (distinct).

$\alpha = 1.670–1.715$, $\beta = 1.674–1.725$, $\gamma = 1.690–1.734$.

$\delta = 0.005–0.015$. $X \wedge c = 0°–97°$ (in obtuse angle β – decreases with increase in Fe), $Y = b$, $Z \wedge a = 0°–72°$.

Epidote: biaxial –ve. $2V_x = 64°–90°$. $r > v$ (strong).

$\alpha = 1.714–1.751$, $\beta = 1.721–1.784$, $\gamma = 1.730–1.797$.

$\delta = 0.015–0.049$. $X \wedge c = 0°–15°$ (in acute angle β – increases with increase in Fe), $Y = b$, $Z \wedge a = 25°–40°$.

First-order white interference color is absent in both clinozoisite and epidote, and replaced by anomalous blue-grey or yellow, especially in low-birefringent grains. RI and δ increase generally with increase in Fe. Compositional zoning may be marked by distinct changes in interference color.

Orientation diagrams. *(010) section* (Figure 9.15a): flash figure; δ varies from low for clinozoisite to moderate or high for epidote; inclined extinction $Z \wedge a$ varies from clinozoisite to epidote; a second cleavage parallel to (100) may sometimes be developed.
(100) section (Figure 9.15b): obtuse bisectrix figure for epidote, but positions of X and Z vary widely in clinozoisite; OAP across cleavage; δ' varies from low for clinozoisite to moderate or high for epidote; straight extinction; epidote is slow along.
(001) section and cleavage fragments (Figure 9.15c): close to optic-axis figure for epidote and some clinozoisite; no cleavage visible; δ' low, often anomalous interference colors; may be slow or fast along the length of euhedral grains.

Occurrence. Common in metamorphic rocks, particularly greenschists and metamorphosed and metasomatized carbonate rocks. Also in veins, as an alteration product in igneous rocks, and in vesicles. May be a component of saussuritized plagioclase. Common as a detrital mineral.

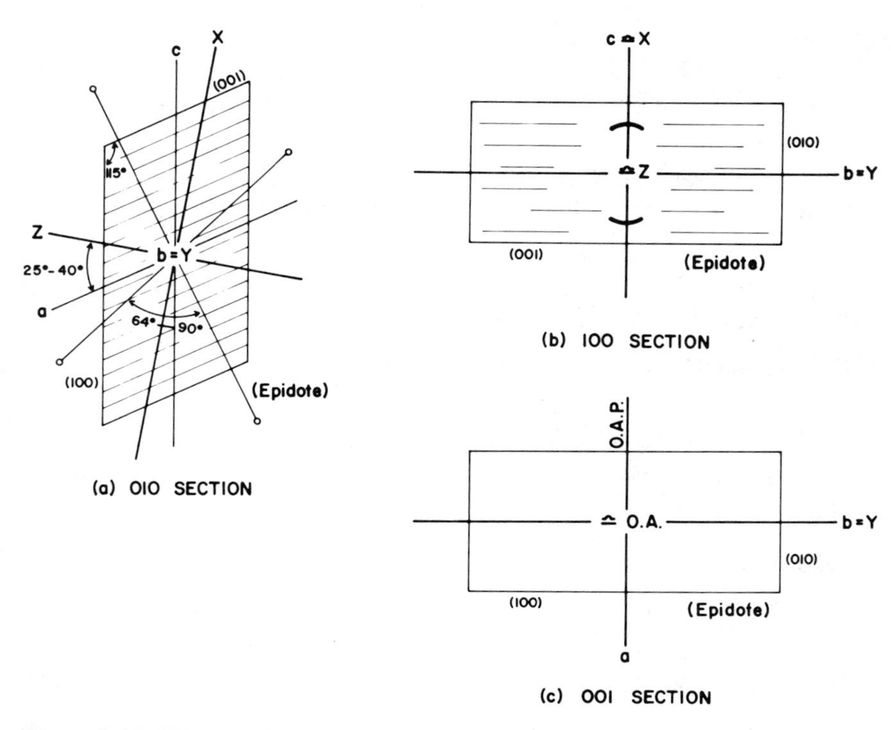

Figure 9.15 Orientation diagrams for epidote.

Distinguishing features. High relief and, for epidote, moderate to high δ, single cleavage, and inclined extinction in (010) sections are characteristic. The characteristic color of epidote, and the lack of first-order white interference color (especially at the edge of low-birefringent grains) are very distinctive. Clinozoisite is distinguished from epidote by its lower δ and +ve character. Compositional variations are marked by rapid changes in δ′. Olivine, pyroxene, amphibole, and zoisite may be confused with clinozoisite–epidote. It is distinguished from pyroxene and amphibole by its single cleavage with the OAP across it, from olivine and zoisite by its inclined extinction in some sections, and from olivine, pyroxene, and amphibole by its anomalous interference colors and characteristic yellow-green color (epidote). Pumpellyite is normally +ve and also has a different color (bluish-green) from epidote.

No. 21. PIEMONTITE (+ve, rarely −ve)

$Ca_2(Mn,Fe^{3+},Al)_2AlO[Si_2O_7][SiO_4](OH)$

Monoclinic, $\beta = 115°$. Perfect {001} cleavage. Subhedral crystals elongate parallel to *b*.

Color in thin section. Pink or purple, pleochroic with $Z > Y > X$, or $Y > X > Z$.

Optical properties. Biaxial +ve. $2V_z = 64°–99°$. $r > v$ (strong).

$$\alpha = 1.732–1.794,\ \beta = 1.750–1.807,\ \gamma = 1.762–1.829.$$

$$\delta = 0.025–0.088.\ X \wedge c = 2°–9° \text{ (in acute angle } \beta),\ Y = b,\ Z \wedge a = 27°–34°.$$

Orientation diagrams. Similar to epidote, but generally +ve.

Occurrence. In low-grade schists, altered igneous rocks, and metamorphic and hydrothermal manganese deposits.

Distinguishing features. Similar to clinozoisite–epidote, but its color and pleochroism are distinctive.

No. 22. ALLANITE (−ve, rarely +ve)

$(Ca,Ce)_2(Fe^{2+},Fe^{3+})Al_2O[Si_2O_7][SiO_4](OH)$

Also known as *orthite*.

Monoclinic, $\beta = 116°$. Imperfect {001} and {100} cleavages. Subhedral (100) tablets or bladed crystals elongate parallel to b; also anhedral. Twins with {100} composition planes, not common.

Color in thin section. Dark brown or red-brown, pleochroic usually with $Z > Y > X$, or $Y > Z > X$; color-zoning common.

Optical properties. Biaxial −ve (rarely +ve). $2V_x = 0°–123°$. $r >$ or $< v$ (strong).

$$\alpha = 1.690–1.813,\ \beta = 1.700–1.857,\ \gamma = 1.706–1.891.$$

$$\delta = 0.013–0.078.\ X \wedge c = 1°–42°,\ Y = b,\ Z \wedge a = 27°–68°.$$

Allanite may display anomalous blue interference colors.

Orientation diagrams. Similar to epidote, but lacks a well-developed cleavage.

Occurrence. An accessory mineral in granites, diorites, alkaline igneous rocks, carbonatites, and pegmatites. Also in metamorphosed carbonate rocks. Radioactive elements in allanite may destroy the atomic structure, reducing the mineral to an isotropic "metamict" state. Allanite inclusions in biotite produce pleochroic haloes.

Distinguishing features. The high or very high relief, moderate to very high δ, and brown color are distinctive. Orientation diagrams, and lack of good cleavage distinguish allanite from brown amphibole.

No. 23. LAWSONITE (+ve) $CaAl_2(OH)_2[Si_2O_7]H_2O$

See Plates 21 and 22.

The structure of lawsonite is similar to that of the epidote group, and contains chains of Al surrounded by O and OH. As for epidote, the direction of the chains is designated *b*, but in some other texts this direction is designated *c*.

Orthorhombic. Distinct {010} and {100}, and poor {101} cleavages. Euhedral crystals may be prismatic and elongate parallel to *b*, or (010) plates. Intersecting sets of twins with {101} composition planes common.

Color in thin section. Usually colorless, but may be pleochroic with X = blue, Y = yellow, and Z = colorless.

Optical properties. Biaxial +ve. $2V_z = 76°–87°$. $r > v$ (strong).

$$\alpha = 1.656–1.665,\ \beta = 1.667–1.674,\ \gamma = 1.680–1.685.$$

$$\delta = 0.020–0.024.\ X = c,\ Y = a,\ Z = b.$$

Orientation diagrams. *(010) section* (Figure 9.16a): acute bisectrix figure; δ′ (low) = 0.009–0.011; good {100} and poor {101} cleavages visible; fast direction parallel to good cleavage.

(100) section (Figure 9.16b): flash figure; δ (moderate) = 0.020–0.024; sections may be elongate parallel to *b*, or parallel to the {010} cleavage; trace of poor {101} cleavage may be visible; fast parallel to good cleavage; straight extinction.

Figure 9.16 Orientation diagrams for lawsonite.

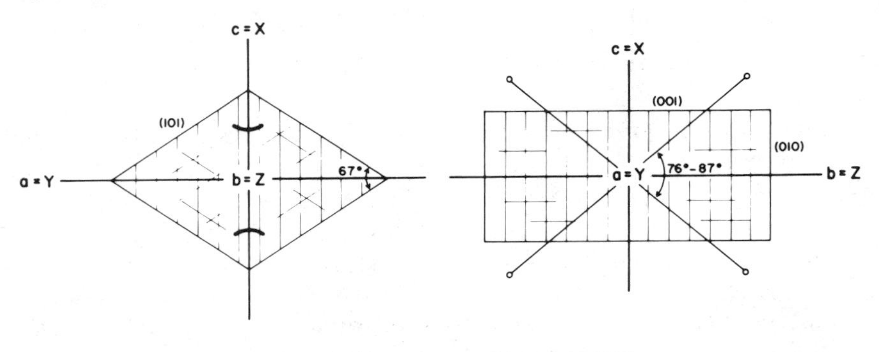

(001) section: extinction parallel to two good cleavages at right angles to each other; obtuse bisectrix figure.

Occurrence. Lawsonite is a comparatively rare mineral, but it is important in petrology and diagnostic of the low-temperature high-pressure lawsonite–albite facies of regional metamorphism. It is often found in glaucophane schists; also frequently present in metamorphic rocks that have retained their pre-metamorphic fabric and texture.

Distinguishing features. Moderate to high relief, moderate δ, cleavages, and occurrence are characteristic. May be confused with zoisite and prehnite, but zoisite has anomalous interference colors, and prehnite has a higher δ and lower RI. The length-fast character of the common (010) plates serves to pick out small amounts of lawsonite amongst micas in fine-grained slaty rocks.

No. 24. PUMPELLYITE (+ve, also −ve)

$Ca_4(Mg,Fe^{2+})(Al,Fe^{3+})_5[Si_2O_7]_2[SiO_4]_2O(OH)_3 2H_2O$

See Plate 16.

Monoclinic, $\beta = 97°$. Distinct {001} and imperfect {100} cleavages. Crystals elongate parallel to *b*. Twinning with {001} and {100} composition planes.

Color in thin section. Pleochroic, colorless to bluish-green or less commonly brown, with $Y > Z > X$; may be color-zoned.

Optical properties. Biaxial +ve (less commonly −ve). $2V_z = 10°–92°$; brown Fe-rich varieties may have $2V_x = 0°–80°$. $r < v$ (strong) about Z.

$\alpha = 1.665–1.728$, $\beta = 1.670–1.748$, $\gamma = 1.683–1.754$.

$\delta = 0.010–0.028$. $X \wedge a = 5°–37°$, $Y = b$, $Z \wedge c = 0°–30°$. Brown Fe-rich varieties may also have the OAP $\perp$ (010).

RI, $2V_z$, extinction angles and color increase with Fe; $\beta > 1.73$ for −ve varieties. Colored varieties often display anomalous interference colors.

Orientation diagrams. *(010) section, +ve varieties* (Figure 9.17): flash figure; δ (low to moderate) = 0.010–0.020; inclined extinction with fast onto the better cleavage 5°–37°.

Sections parallel to b: straight extinction; the trace of a cleavage parallel to *b* may be visible; slow or fast along.

Occurrence. Usually as small crystals in a wide range of low-grade metamorphic rocks that in many cases have retained their original fabric and texture. Also occurs as a secondary mineral in volcanic rocks and as an alteration product of biotite.

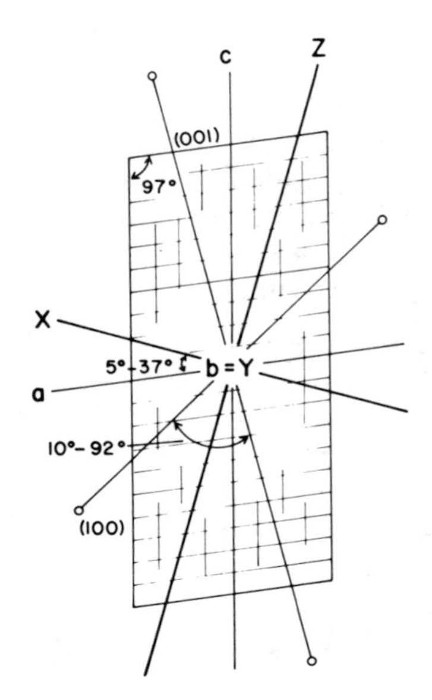

Figure 9.17 Orientation diagram for pumpellyite.

Distinguishing features. May be confused with the epidote-group minerals. Colorless (Fe-poor) pumpellyite is distinguished from zoisite and clinozoisite by its lower RI, and from zoisite by its inclined extinction in (010) sections. Green pumpellyite has a lower RI than epidote and is usually +ve; the bluish-green color of pumpellyite is quite different from the yellow-green of epidote. The angle between the {001} and {100} cleavages is 83° for pumpellyite and 65° for epidote.

No. 25. MELILITE (−ve and +ve)

A series from *gehlenite* $Ca_2Al_2SiO_7$ to *åkermanite* $Ca_2MgSi_2O_7$. Na may substitute for Ca, and Fe for Al.
Tetragonal. Imperfect {001} cleavage. Euhedral–subhedral tabular crystals parallel to (001) with square cross sections common. Rodlike inclusions aligned perpendicular to (001) are common (so-called *peg structure*).

Color in thin section. Usually colorless; may be yellow, pleochroic with $\varepsilon > \omega$.

Optical properties. Uniaxial −ve or +ve.
In the gehlenite–åkermanite series, $\omega = 1.632–1.669$, $\varepsilon = 1.640–1.658$. $\delta = 0–0.011$, gehlenite being −ve, åkermanite +ve. The change from +ve to −ve takes place at $Ge_{48}Ak_{52}$ which mineral is isotropic. Na

affects these properties so that the RI may be as low as $\varepsilon = 1.616$; Fe increases the RI which may be as high as $\omega = 1.7$.
Interference colors may be anomalous blue.
Straight extinction and either (−ve) slow along, or (+ve) fast along the cleavage.

Occurrence. Rather rare, but may be abundant in some undersaturated lavas. Also found in thermally metamorphosed and metasomatized carbonate rocks.

Distinguishing features. Tabular crystals, uniaxial character, moderate relief, anomalous interference colors, and occurrence are distinctive. May be confused with vesuvianite or zoisite, but both these minerals have a higher RI, and zoisite is biaxial. The peg structure of inclusions perpendicular to (001) is very characteristic.

B. Cyclosilicates

In the cyclosilicates, the $[SiO_4]$ tetrahedral groups are linked together to form rings. A hexagonal ring $[Si_6O_{18}]$ is the dominant structural feature of the minerals beryl, cordierite, and tourmaline; axinite contains a hexagonal ring joined with four other tetrahedra to form unique $[B_2Si_8O_{30}]$ groups; eudialyte (eucolite) has both trigonal and nonagonal rings.

No. 26. BERYL (−ve) $Be_3Al_2[Si_6O_{18}]$

The precious variety *emerald* contains Cr.
Hexagonal. Imperfect {0001} cleavage. Usually six-sided prismatic euhedral crystals.

Color in thin section. Colorless.

Color in detrital grains. Usually blue-green, pleochroic with $\omega > \varepsilon$.

Optical properties. Uniaxial −ve.

$$\omega = 1.568\text{–}1.608,\ \varepsilon = 1.564\text{–}1.600.$$

$$\delta = 0.004\text{–}0.009.$$

Prismatic crystals have straight extinction and are length fast.

Occurrence. Principally in granite pegmatites, but also in metasomatized schists and carbonate rocks.

Distinguishing features. Low to moderate relief, uniaxial −ve character, and low δ distinctive. May be confused with quartz or apatite, but quartz has a lower RI and is +ve, and apatite has a higher RI.

No. 27. CORDIERITE (−ve and +ve) $(Mg,Fe)_2Al_3[Si_5AlO_{18}]$

See Figure 9.18.
Orthorhombic. Imperfect {010} cleavage. Anhedral or euhedral short prismatic pseudohexagonal crystals made up of sector twins (Figure 9.18). Twinning common; either sector twins or multiple twinning (often intersecting sets) with composition planes approximately parallel to {110} or {130}. The sector twins may radiate from a central point at intervals of 30°, 60° or 120° (Figures 9.18 and 9.19), and have been interpreted by Zeck (1972) to be the result of transformation from a high-temperature hexagonal state.

Color in thin section. Usually colorless; may have yellow pleochroic haloes around inclusions of zircon and apatite.

Color in detrital grains. Weakly pleochroic with X = colorless or pale yellow, Y and Z = violet or blue.

Optical properties. Biaxial −ve or +ve. $2V_x = 42°–104°$. $r < v$ (weak).

$$\alpha = 1.522–1.560,\ \beta = 1.524–1.574,\ \gamma = 1.527–1.578.$$

$$\delta = 0.005–0.018.\ X = c,\ Y = a,\ Z = b.$$

RI shows a general increase with Fe content; Selkregg and Bloss (1980) demonstrate a simple relationship of RI to the ratio Fe + Mn/Fe + Mn + Mg after heating cordierites for 6 hr at 800°C.

Orientation diagrams. *(001) section* (Figure 9.19): bisectrix figure; δ' (very low) = 0.003; may show sector twins or intersecting sets of lamellar twins; orientations shown in Figure 9.19a,b are for twins with composition planes ca. {110}—interchange Y and Z positions for ca. {130} composition planes.
Prismatic sections may display simple or multiple twin lamellae parallel to the c-axis.

Occurrence. Most commonly found in thermally metamorphosed pelitic sediments, often as ovoid poikiloblastic grains. Also in pelitic rocks regionally metamorphosed under high geothermal gradients, and in metasomatized(?) gneisses with anthophyllite. Present in some igneous rocks, especially contaminated gabbros, and in some granite pegmatites. Alters to pinite, a mixture of sericite and chlorite, or a yellow-brown amorphous (isotropic) substance.

Distinguishing features. The sector twinning, yellow pleochroic haloes, and pinite alteration are very distinctive, but not always present. In their absence, cordierite may be confused with quartz and plagioclase feldspar. Quartz is uniaxial, but the distinction from plagioclase may be very difficult, especially if the multiple lamellar twinning is present; however, the twins in cordierite frequently occur in patches and wedge-

Figure 9.18 Sector twinning in cordierite from a dacite (crossed-polarized light). From Zeck (1972), and reprinted by permission from *Nature* (*Phys. Sci.*), vol. 238, no. 81, pp. 47–48; copyright © 1972 Macmillan Journals Limited.

out. Lack of cleavage suggests cordierite, but it may be necessary to observe the pleochroism of cordierite in thick sections to distinguish the two. A staining method to distinguish cordierite from quartz and plagioclase has been described by Gregnanin and Viterbo (1965); the thin section is etched with HF vapor and stained with a solution of HCl

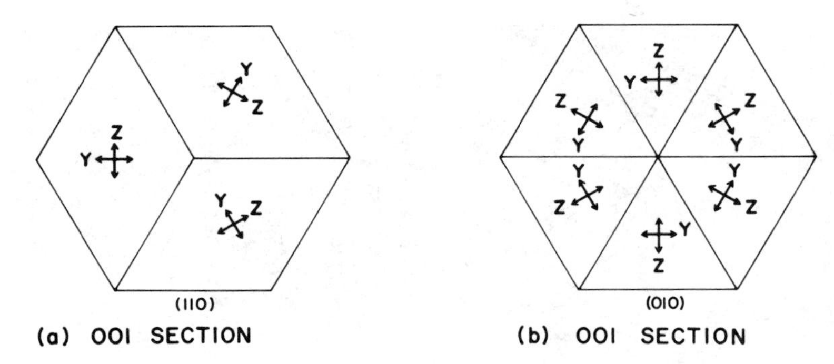

Figure 9.19 Orientation diagrams for cordierite with sector twinning; ca. {110} composition planes.

and potassium ferricyanide which turns cordierite azure but leaves quartz and plagioclase colorless.

No. 28. TOURMALINE (−ve)

$Na(Mg,Fe,Mn,Li,Al)_3Al_6[Si_6O_{18}](BO_3)_3(OH,F)_4$

Common Fe-tourmaline is known as *schorl,* Mg-tourmaline as *dravite,* and alkali-tourmaline as *elbaite*.
Trigonal. No good cleavage. Usually elongate prismatic crystals with a hexagonal or triangular (convexly curved sides) cross section.

Color in thin section. Variable, and strongly pleochroic with $\omega > \varepsilon$; schorl is black, brown, green, or blue; dravite is brown, yellow or colorless; elbaite is colorless or pink. Often color-zoned. The maximum absorption perpendicular to the prism length is very distinctive.

Color in detrital grains. Variable as above, but deeper colors.

Optical properties. Uniaxial −ve.

$$\omega = 1.635\text{–}1.675;\ \varepsilon = 1.610\text{–}1.650.$$

$$\delta = 0.017\text{–}0.035.$$

RI and δ increase with Fe content. An unusual Fe^{3+}-rich tourmaline with $\omega = 1.735$ and $\varepsilon = 1.655$, and an authigenic tourmaline with $\omega = 1.633$ and $\varepsilon = 1.621$ have been reported.
Prismatic sections have straight extinction, and are length fast.

Occurrence. Schorl is an accessory in many granites, pegmatites, schists, and gneisses, and is an important constituent of some metasomatic rocks. It is also a common detrital mineral. Dravite is found in metamorphosed and metasomatized carbonate rocks. Elbaite is found in granite-pegmatites in association with other Li minerals such as lepidolite.

Distinguishing features. *Schorl*—the moderate relief and δ, strong color and pleochroism, lack of cleavage, and crystal shape are very distinctive. It may be confused with (001) sections of biotite or (100) sections of hornblende. Both these minerals differ from tourmaline in other orientations in which they display good cleavages, and a maximum absorption parallel to their length; hornblende is biaxial. *Dravite*—may be confused with chondrodite with which it may be associated. Dravite is distinguished by its uniaxial character. *Elbaite*—the moderate δ distinguishes elbaite from other colorless uniaxial minerals of similar relief.

No. 29. AXINITE (−ve, rarely +ve)

$(Mn,Fe^{2+},Mg)_2(Ca,Mn)_4(Al,Fe^{3+})_4[B_2Si_8O_{30}](OH)_2$

Ferroaxinite has Ca > 1.5, Fe > Mn, *manganaxinite* Ca > 1.5, Mn > Fe, and *tinzenite* Ca < 1.5, Mn > Fe (Lumpkin and Ribbe, 1979). Triclinic, $\alpha = 92°$, $\beta = 98°$, $\gamma = 77°$. Distinct {100} and a number of other very poor cleavages. Usually euhedral wedge-shaped crystals with {010} and {011} well developed.

Color in thin section. Colorless, or yellow and violet pleochroic with $Y > X > Z$.

Optical properties. Biaxial −ve, rarely +ve. $2V_x = 63°–109°$. $r < v$ (distinct).

$$\alpha = 1.656–1.694,\ \beta = 1.660–1.701,\ \gamma = 1.668–1.705.$$

$$\delta = 0.007–0.014.$$

X is approximately normal to $(\bar{1}11)$ and Z parallel to $[0\bar{1}1]$.
Inclined extinction to crystal outlines and cleavages in most sections.
$2V_x$ increases with Mg, and RI decreases with Ca and Mg.

Occurrence. In thermally metamorphosed and metasomatized carbonate rocks, and in metasomatized igneous rocks. Also in regionally metamorphosed rocks, especially in vein assemblages.

Distinguishing features. The crystal shape, high relief, low δ, large −ve $2V$, and inclined extinction are distinctive.

No. 30. EUDIALYTE (+ve), EUCOLITE (−ve)

$Na_{12}(Ca,Re)_6(Fe,Mn,Mg)_3Zr_3[Si_3O_9]_2[Si_9O_{24}(OH)_3]_2$

Trigonal. Distinct {0001}, and imperfect $\{10\bar{1}0\}$ and $\{11\bar{2}0\}$ cleavages. Crystals vary from tabular to strongly prismatic. Concentric or hourglass zoning common.

Color in thin section. Colorless, pink or pale-yellow; may be pleochroic with $\varepsilon > \omega$.

Optical properties. Uniaxial +ve (eudialyte) or −ve (eucolite).

$$\omega = 1.567\text{–}1.660;\ \varepsilon = 1.572\text{–}1.655.$$

$$\delta = 0.000\text{–}0.008.$$

Zoned crystals may change from +ve to −ve with an intermediate isotropic zone. Generally −ve varieties have the higher RI. May be anomalously biaxial. The sign of elongation varies not only according to optic sign but also according to whether the crystals are tabular or prismatic.

Occurrence. In nepheline syenites, alkaline volcanics, and associated rocks, usually as a late, possibly pneumatolytic mineral.

Distinguishing features. Uniaxial character, color, zoned birefringence, and occurrence are distinctive.

C. Inosilicates

In the inosilicates, the $[SiO_4]$ tetrahedral groups are linked together to form chains. Minerals characterized by a single-chain structure $[SiO_3]$ and described here are: the pyroxenes; wollastonite; pectolite; rhodonite; aenigmatite; sapphirine. A double-chain structure, $[Si_4O_{11}]$, characterizes the amphiboles.

The Pyroxenes

Crystallography

The pyroxenes may belong to either the monoclinic or the orthorhombic systems. In the monoclinic pyroxenes, the crystallographic angle β varies between 105° and 110°. Most pyroxenes form rather stumpy prismatic crystals, though occasionally, as in the case of aegirine, the crystals are more elongate. Euhedral pyroxene crystals are characterized by an eight-sided cross section, and all pyroxenes have two cleavages parallel to the c-axis that are almost at right angles to each other (Figure 9.20).

The three following characteristic orientations may be observed in thin sections (Figure 9.21): sections at right angles to the c-axis which display two cleavages (and an eight-sided cross section in euhedral grains); prismatic sections displaying a single cleavage; prismatic sections parallel to (100) or (010), to which both cleavages are at too acute an angle to be always visible, especially in slightly thick sections.

Terminology

On the basis of the crystal system to which they belong, the pyroxenes are subdivided into clinopyroxenes (monoclinic) and orthopyroxenes.

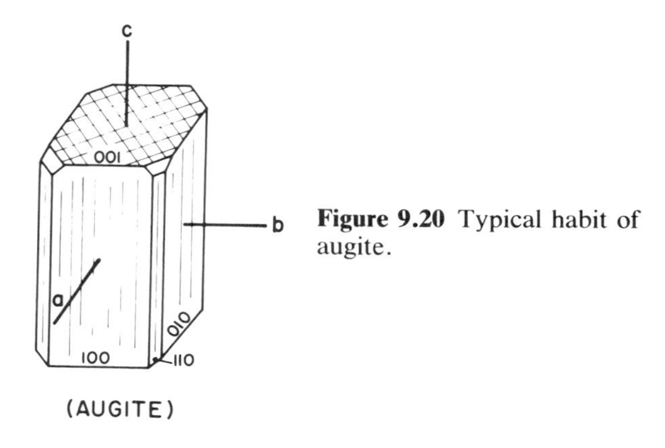

Figure 9.20 Typical habit of augite.

Clinopyroxenes. The most important clinopyroxenes can be considered as members of the system $CaMgSi_2O_6$–$CaFeSi_2O_6$–$MgSiO_3$–$FeSiO_3$ (Figure 9.22). Mg may be completely replaced by Fe^{2+} so that there is a continuous sequence of pyroxenes possible from $CaMgSi_2O_6$ to $CaFeSi_2O_6$, and from $MgSiO_3$ to $FeSiO_3$. There is more limited solid solution possible between the Ca-rich and Ca-poor pyroxenes, especially between the Mg-rich members. Subcalcic augite often forms under disequilibrium

Figure 9.21 Sections of augite: center, eight-sided section with two cleavages; top right and bottom right, sections with traces of cleavage in one direction; bottom left, section with no trace of cleavage. Basalt, Pahau River, New Zealand (view measures 3.4 × 2.1 mm). Plane-polarized light.

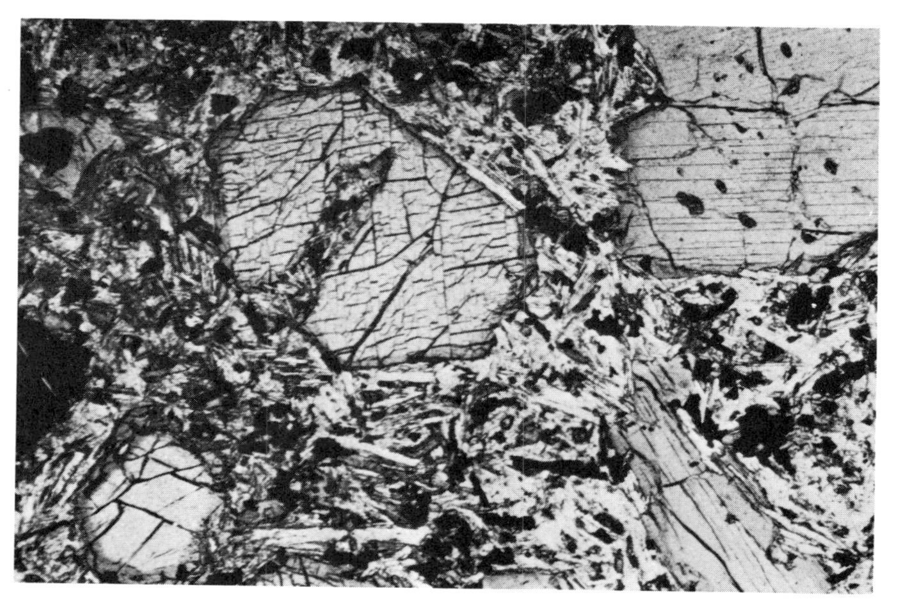

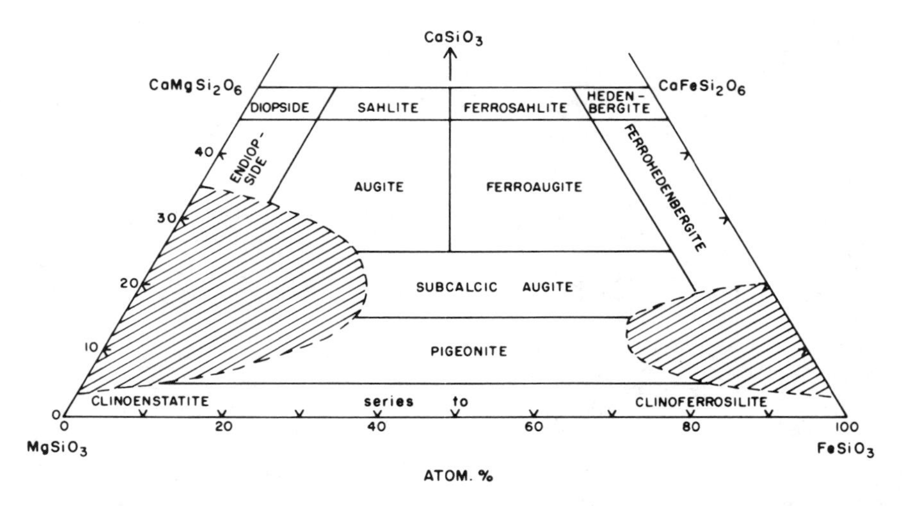

Figure 9.22 Terminology of clinopyroxenes in the system $CaMgSi_2O_6$–$CaFeSi_2O_6$–$MgSiO_3$–$FeSiO_3$. The shaded areas represent areas of immiscibility and lacking naturally occurring examples. Based on Poldervaart and Hess (1951) and Deer et al. (1963).

conditions of rapid crystallization in volcanic rocks, commonly being found there as groundmass or quench material (Ewart, 1976). Pyroxenes belonging to the series clinoenstatite–clinoferrosilite are not commonly recognized, and they are not described in this book. Recent work summarized in Deer et al. (1978) has shown that they may be produced by rapid chilling or by shear deformation of orthopyroxene; they may be the stable low-temperature forms although there is usually insufficient energy difference to drive their formation from orthopyroxene. Pigeonites are the common Ca-poor clinopyroxenes of intermediate Mg–Fe composition. They are metastably preserved in many volcanic rocks, but may invert on slow cooling to orthopyroxenes of similar composition exsolving excess Ca as augite in the process.

The crystallization of the Mg-rich pyroxenes in this system is represented diagrammatically in Figure 9.23. Solid solution is incomplete between $CaMgSi_2O_6$ and $MgSiO_3$, even at high temperatures, so that a melt of intermediate composition will crystallize both a Ca-rich and a Ca-poor pyroxene whose compositions are represented by the solvus curve.

The changes in composition of the two coexisting pyroxenes that crystallize under the influence of iron enrichment during crystal fractionation in the well-known Skaergaard intrusion are illustrated in Figure 9.24. The essential features are: initial crystallization of a diopsidic augite and Mg-rich orthopyroxene; at a composition of about $En_{70}Fs_{30}$, pigeonite crystallizes (with augite) in place of the orthopyroxene; with further fractionation, both the augite and pigeonite become more Fe-rich until eventually

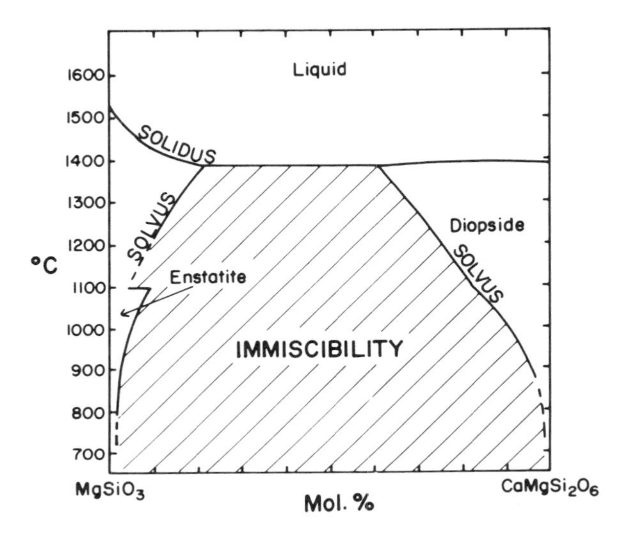

Figure 9.23 Solidus and solvus curves in the system $CaMgSi_2O_6$–$MgSiO_3$. Based on Boyd and Schairer (1964).

pigeonite ceases to crystallize. Very similar compositional trends are known from a large number of rock suites similarly affected by iron enrichment. In some volcanics, a different trend from diopside towards subcalcic augite and pigeonite has been observed (Kuno, 1955), this being regarded as typical of rapid cooling from high temperatures. Nakamura

Figure 9.24 Crystallization trends and coexisting pyroxene pairs (tied circles) in the Skaergaard intrusion. Based on Brown and Vincent (1963).

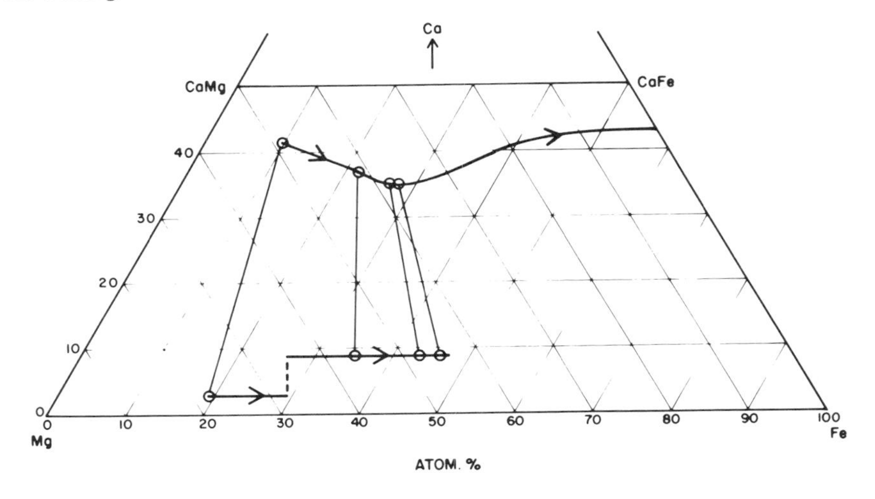

and Kushiro (1970) have described rocks in which the three pyroxenes pigeonite, orthopyroxene, and augite all coexist.

Other less common clinopyroxenes that cannot be represented by the chemical substitutions shown in Figure 9.22 include omphacite, fassaite, aegirine, jadeite, and spodumene. Omphacite resembles diopside, but is distinguished by substitution of Na for some of the Ca, and Al for some of the Mg. Fassaite also resembles diopside, but there is a significant replacement of both Mg and Si by Al. Aegirine has the formula $NaFe^{3+}[Si_2O_6]$ and aegirine-augite has a composition intermediate between this and members of the diopside–hedenbergite series: crystallization trends in this system are discussed by Deer et al. (1978). Jadeite has the formula $NaAl[Si_2O_6]$, and spodumene is a Li-pyroxene with the composition $LiAl[Si_2O_6]$.

Orthopyroxenes. The series of pyroxenes from clinoenstatite to clinoferrosilite represented in Figure 9.22 occur only rarely, and their place is taken in nature by a series of orthopyroxenes from enstatite to orthoferrosilite. Pigeonites commonly invert to orthopyroxenes on cooling. The nomenclature of the orthopyroxenes is given in Figure 9.26.

Exsolution phenomena

The two coexisting pyroxenes that initially crystallize in basic igneous rocks (Figure 9.24) may be metastably preserved on rapid cooling. However, the composition of each pyroxene will change during slow cooling in a way prescribed by the solvus curve of Figure 9.23. A Ca-poor pyroxene will become even more Ca-poor, and in so doing will exsolve a Ca-rich pyroxene as exsolution lamellae (see also Figure 9.54 and the accompanying text for details of a similar process in the feldspars). The Ca-rich pyroxene that originally crystallizes will exsolve a Ca-poor pyroxene. Poldervaart and Hess (1951) believed that a clinopyroxene exsolved from a clinopyroxene (e.g., pigeonite from augite and vice versa) would form exsolution lamellae parallel to (001); a clinopyroxene exsolved from orthopyroxene (or vice versa) would form exsolution lamellae parallel to (100). Although the latter situation is still valid, Robinson et al. (1971) and Jaffe et al. (1975) have shown that clinopyroxene/clinopyroxene composition plane boundaries can have a variety of orientations close to either (001) or (100). They show that exsolution lamellae have orientations representing the best dimensional fit between the crystal lattices of pigeonite and augite, and depending on the particular relative *a* and *c* crystal axes dimensions the lamellae may lie within either the obtuse or acute angle β up to about 20° from either (001) or (100). The entire range of possible orientations is shown in Figure 9.25. Note that the *b* and *c* crystal axes of all coexisting lamellae are parallel, and for clinopyroxene/clinopyroxene

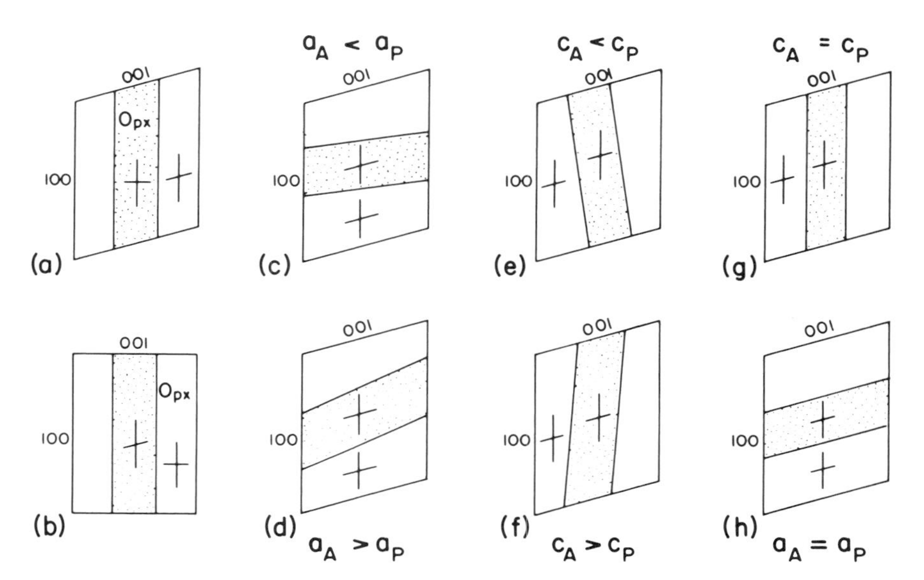

Figure 9.25 Possible orientations for exsolution lamellae in the pyroxenes (lamellae are represented diagrammatically by the dotted areas): **(a)** and **(b)** orthopyroxene/clinopyroxene pairs; **(c)–(h)** clinopyroxene/clinopyroxene pairs. The orientations of *a*- and *c*-axes are shown for host and lamellae by the crossed lines, and the relative values of *a* and *c* cell dimensions for augite (*A*) and pigeonite (*P*) are shown. Based on the work of Robinson et al. (1971) and Jaffe et al. (1975).

pairs the *a* axes are nearly parallel too (the angle β differs by only 3° or so), even though the lamellae are not necessarily on rational planes. The lamellae are often clearly visible because optical directions, of course, differ from lamella to lamella. Studies of exsolution lamellae are potentially fruitful areas for elucidating geological history. For example, pigeonite may initially exsolve augite, but if it inverts to orthopyroxene, further exsolution will be strictly on (100); similarly, augite may initially exsolve pigeonite but on further cooling exsolve orthopyroxene; and as Robinson et al. (1971) point out, changes in lattice parameters that occur during changes in pressure and temperature affect the angular relationships of lamellae and may thus record the pressure–temperature history of the pyroxene.

No. 31. ORTHOPYROXENE (+ve and −ve) $(Mg,Fe)[SiO_3]$

See Plates 3 and 4.

The terminology of the orthopyroxenes is given in Figure 9.26. Orthorhombic. Distinct {210} cleavages at 88° to each other. Parting and exsolution lamellae commonly developed parallel to {100} (Figure 9.25).

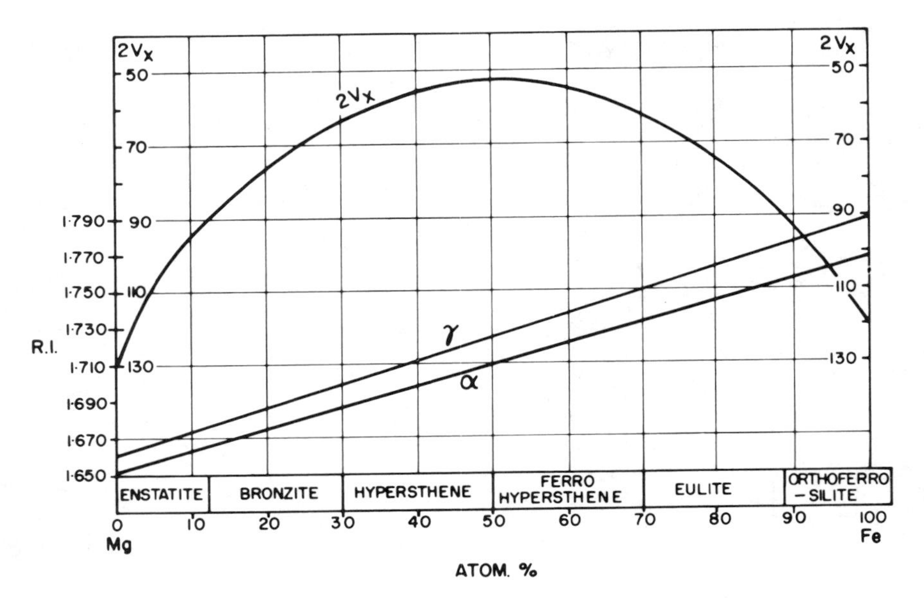

Figure 9.26 Terminology of the orthopyroxenes, and variation of RI and $2V$ with composition. Based on Leake (1968).

Often euhedral stubby prismatic crystals with eight-sided crosssections; also anhedral.

Color in thin section. Colorless, or pale-pink and green pleochroic with X = pink, Y = pale yellow, Z = green.

Color in detrital grains. As above but deeper colors.

Optical properties. Biaxial +ve or −ve. $2V_x = 45°–128°$. Dispersion about X is weak to distinct: $r < v$ from En_{100-85} and En_{50-01}, $r > v$ from En_{85-50}.

$$\alpha = 1.651–1.769,\ \beta = 1.653–1.771,\ \gamma = 1.658–1.788.$$

$$\delta = 0.007–0.019,\ X = b,\ Y = a,\ Z = c.$$

RI and δ increase and $2V_x$ varies with Fe content as shown in Figure 9.26. $2V$ determination with the U-stage (Chapter 6) combined with RI provides a good measure of composition.

Orientation diagrams. *(001) section* (Figure 9.27a): acute or obtuse bisectrix figure; δ' (very low or low) up to 0.010; symmetrical extinction to two distinct cleavages almost at right angles to each other; eight-sided sections typical; may display exsolution lamellae parallel to (100). *(100) section* (Figure 9.27b): flash figure; δ usually low but may be as

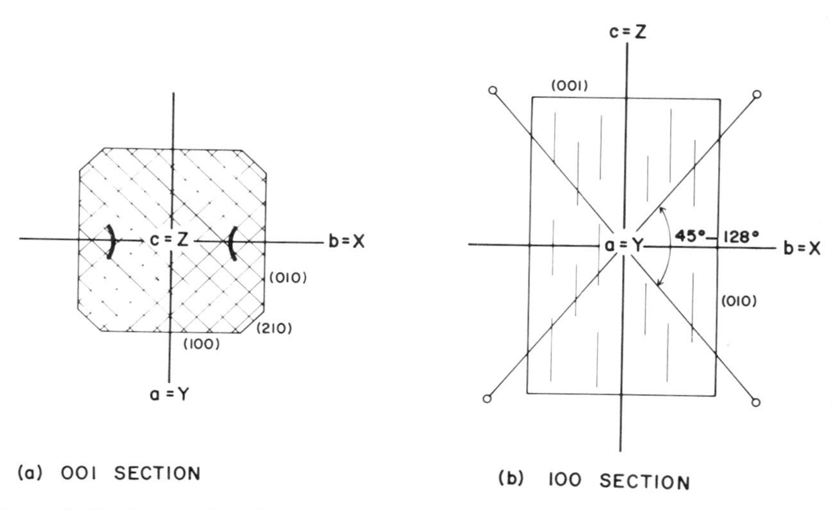

Figure 9.27 Orientation diagrams for orthopyroxene.

high as 0.019 in orthoferrosilite; straight extinction, but cleavage not always seen; may show marked pleochroism; if inverted from pigeonite may display exsolution lamellae on plane parallel to *b* but oblique to (001).

(010) sections are similar to (100) sections but provide bisectrix figures, and may display augite exsolution lamellae parallel to (100) or oblique to (100) and (001) if inverted from pigeonite.

Sections displaying a sharp single cleavage are oblique to the optical directions.

Optic-axis figures are obtained in sections oblique to the cleavages.

Occurrence. Mg-rich orthopyroxenes are characteristic of ultramafic rocks such as pyroxenites and harzburgite. Orthopyroxenes of varying Mg/Fe content are common in a wide range of basic igneous rocks, especially norite and andesite. Orthopyroxenes are also found in charnockites, granulites, and pelitic rocks thermally metamorphosed at high grades. Orthoferrosilite occurs in some metamorphosed Fe-rich sediments. The orthopyroxenes are commonly altered to serpentine, chlorite, or uralite (amphibole).

Distinguishing features. The two cleavages at right angles, straight extinction in prismatic sections, high relief, and low δ are characteristic. It may be difficult to distinguish from augite, especially when present in small quantities with augite. The best way to identify the orthopyroxene in these circumstances is to obtain numerous optic-axis figures. Augite always has a moderate +ve $2V$ whereas the orthopyroxene most commonly has a large $2V$ or is −ve. It may occasionally be confused with andalusite, especially in detrital grains, but andalusite is length fast, and euhedral crystals are different in shape.

No. 32A. DIOPSIDE–HEDENBERGITE (+ve) $Ca(Mg,Fe)[Si_2O_6]$

No. 32B. AUGITE (+ve) $(Ca,Mg,Fe,Ti,Al)_2[(Si,Al)_2O_6]$

No. 32C. SUBCALCIC AUGITE (+ve) $(Mg,Fe,Ca,Al)_2[(Si,Al)_2O_6]$

No. 32D. OMPHACITE (+ve) $(Ca,Na,Mg,Fe,Al)_2[Si_2O_6]$

No. 32E. FASSAITE (+ve) $Ca(Mg,Fe,Al)[(Si,Al)_2O_6]$

See Plates 1, 2, 13, and 14 and Figure 9.21.

The composition of these closely related minerals is discussed in the introductory section on the pyroxenes.

Monoclinic, β = 105°–106°. Distinct {110} cleavages at 87° to each other. Parting and exsolution lamellae of orthopyroxene commonly developed parallel to {100} (Figure 9.25). Often euhedral stubby prismatic crystals with eight-sided cross sections; also anhedral. Simple or lamellar twinning with {100} composition planes common. Oscillatory, hourglass, and sector zoning are often present, especially in Ti-rich titanaugite.

Color in thin section. Diopside, colorless; hedenbergite, pale green or green-brown; augite may be colorless or pale green, but a flesh color is particularly common; titanaugite has a characteristic purple-brown color; omphacite and fassaite are usually pale green; usually weak pleochroism.

Color in detrital grains. As above but deeper colors.

Optical properties. Biaxial +ve. $2V_z$ = 25°–83°. $r > v$ (weak to strong).

α = 1.659–1.743, β = 1.670–1.750, γ = 1.688–1.772.

δ = 0.018–0.033. $X \wedge a$ = 20°–34°, $Y = b$, $Z \wedge c$ = 35°–48° (higher values up to 70° have been reported for omphacite).

RI increases with Fe content. $2V$ increases with Ca and Na content. Figure 9.28 illustrates the variation of RI and $2V$ for the diopside–hedenbergite series, augite, and subcalcic augite (refer to Figure 9.22 for nomenclature). Fassaite has $2V_z$ = 51°–65°, and for omphacite, $2V_z$ = 58°–83°. The extinction angle $Z \wedge c$ increases with Fe content (Hess, 1949), but correlation of this angle with chemistry is not good. Fassaite has a stronger dispersion than most other pale green clinopyroxenes.

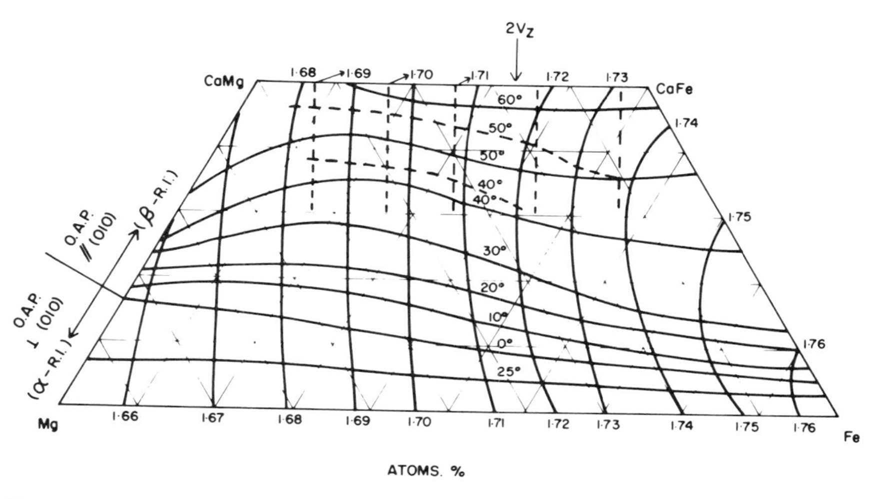

Figure 9.28 Variation of $2V$ and RI with composition in clinopyroxenes in the system $CaMgSi_2O_6$–$CaFeSi_2O_6$–$MgSiO_3$–$FeSiO_3$. Based on Hess (1949, Plate I), Muir (1951), and (dashed lines) Brown and Vincent (1963).

Orientation diagrams. *Section ⊥ c* (Figure 9.29a): close to an optic-axis figure; δ' low or very low; extinction symmetrical to two good cleavages almost at 90° to each other, but extinction may be indistinct due to proximity of an optic axis; eight-sided section characteristic; may display twinning and exsolution lamellae of orthopyroxene ∥ (100).
(010) section (Figure 9.29b): flash figure; δ (moderate) up to 0.033; inclined extinction; the trace of the prismatic cleavage is not always visible; twinning and exsolution lamellae of orthopyroxene ∥ (100), and lamellae of pigeonite close to either (100) or (001) may be visible.
(100) section (Figure 9.29c): close to an optic-axis figure; δ' low or very low; straight extinction; trace of cleavage is not always visible; exsolution lamellae of pigeonite close to (001) may be visible.
Cleavage fragments (and some sections showing a good single cleavage) lie parallel to (110). The extinction angle $Z' \wedge c$ is smaller by 10° or so than the angle $Z \wedge c$ in (010) sections.

Occurrence. Diopside–hedenbergite occurs in metamorphic rocks, diopside being characteristic of thermally metamorphosed and metasomatized carbonate rocks. Hedenbergite is formed in metamorphosed Fe-rich sediments. Members of the diopside–hedenbergite series also occur in basic igneous rocks, hedenbergite in syenites and, more rarely, granites. Augite is the typical pyroxene of many igneous rocks, and is an essential mineral of basalts and gabbros. Subcalcic augite is found in the groundmass of basalts and andesites and sometimes as the outer

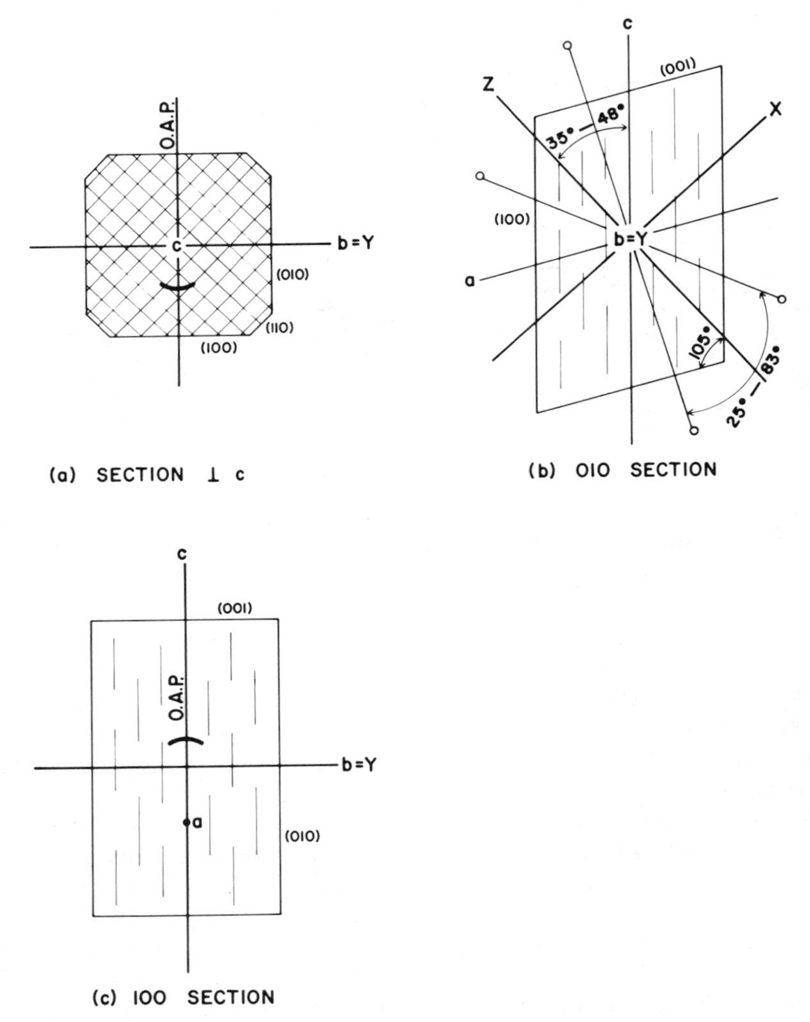

Figure 9.29 Orientation diagrams for the clinopyroxenes 32A–E.

zone of zoned augite crystals (Kuno, 1955). Omphacite is the typical pyroxene of eclogites and associated rocks. Fassaite is found in metamorphosed limestones in association with spinel. The common alteration product of diopside and common augite is uralite (tremolite, actinolite, or hornblende).

Distinguishing features. The clinopyroxenes described above are distinguished by their two cleavages at 87°, their moderate δ, the moderate extinction angle in (010) sections, and their +ve character. They are not easily confused with any other mineral. However, it is easy to overlook small quantities of orthopyroxene (usually −ve or a large $2V$) associated with augite. Individual species of clinopyroxene are less easy to distinguish without accurate $2V$, RI measurements and chemical analysis. Omphacite and fassaite are characterized by a higher $2V$, occurrence, and a pale-green color. Diopside is colorless, whereas augite is commonly flesh colored. Titaniferous augite is purple-brown in color. Subcalcic augite has a small but larger $2V$ than pigeonite (No. 33). The related pyroxenes jadeite (No. 35) and spodumene (No. 36) have distinctive occurrences, and aegirine (No. 34) is green pleochroic, and fast along.

No. 33. PIGEONITE (+ve) $(Mg,Fe,Ca)(Mg,Fe)[Si_2O_6]$

See Figure 9.22 for the compositional boundaries of pigeonite. Monoclinic, $\beta = 108°$. Distinct {110} cleavages at 87° to each other. Usually as very small crystals in the groundmass of volcanic rocks. Twinning with {100} composition plane.

Color in thin section. Colorless or pale pink or green; pleochroism weak or absent.

Optical properties. Biaxial +ve. $2V_z = 0°–30°$. $r <$ or $> v$ (weak to distinct)

$\alpha = 1.682–1.722$, $\beta = 1.684–1.722$, $\gamma = 1.705–1.751$.

$\delta = 0.023–0.029$. $X = b$ or $X \wedge a = 19°–26°$, $Y \wedge a = 19°–26°$ or $Y = b$, $Z \wedge c = 37°–44°$.

RI increases with Fe content. For the variation of RI and $2V$ with composition see Figure 9.28.

Orientation diagrams. Similar to augite (Figure 9.29), but OAP is most commonly ⊥ (010), and because of the very small $2V$ about Z, optic-axis figures are not obtained on sections ⊥ c or // (100).

Occurrence. Most common in rapidly chilled lavas, usually in the groundmass of andesites and dacites. Less commonly as phenocrysts in volcanic and hypabyssal rocks. Also found as exsolution lamellae in some augite. Pigeonite that has inverted to orthopyroxene may be detected by the presence of exsolution lamellae of augite on or close to the relict (001) plane of pigeonite.

Distinguishing features. Very similar to the other clinopyroxenes (No. 32), but distinguished from them by its very small $2V$.

No. 34A. AEGIRINE (ACMITE) (−ve) $NaFe^{3+}[Si_2O_6]$

No. 34B. AEGIRINE-AUGITE (+ve and −ve) $(Na,Ca)(Fe^{3+},Fe^{2+},Mg,Al)[Si_2O_6]$

Monoclinic, β = ca. 107°. Distinct {110} cleavages at 87° to each other. Aegirine is usually euhedral or subhedral with a very elongate or acicular prismatic habit. Aegirine-augite forms less elongate crystals. Twinning with {100} composition planes. Zoned crystals are common.

Color in thin section. Pale to dark yellowish-green; pleochroism weak to strong with $X > Y > Z$; acmite is distinguished from aegirine on the basis of color, and is brown or brownish-green in thin section.

Optical properties. Biaxial −ve or +ve. $2V_x$ = 60°–70° (aegirine), $2V_x$ = 70°–110° (aegirine-augite). $r > v$ (distinct to strong).

$$\alpha = 1.700\text{–}1.776,\ \beta = 1.710\text{–}1.820,\ \gamma = 1.730\text{–}1.836.$$

$$\delta = 0.030\text{–}0.060.\ X \wedge c = 0°\text{–}20°,\ Y = b,\ Z \wedge a = 7°\text{–}37°.$$

The variation in optical properties with composition is shown in Figure 9.30.

Figure 9.30 Variation of $2V$, $X \wedge c$, and RI with composition in aegirine-augite and aegirine. Based on Deer et al. (1966).

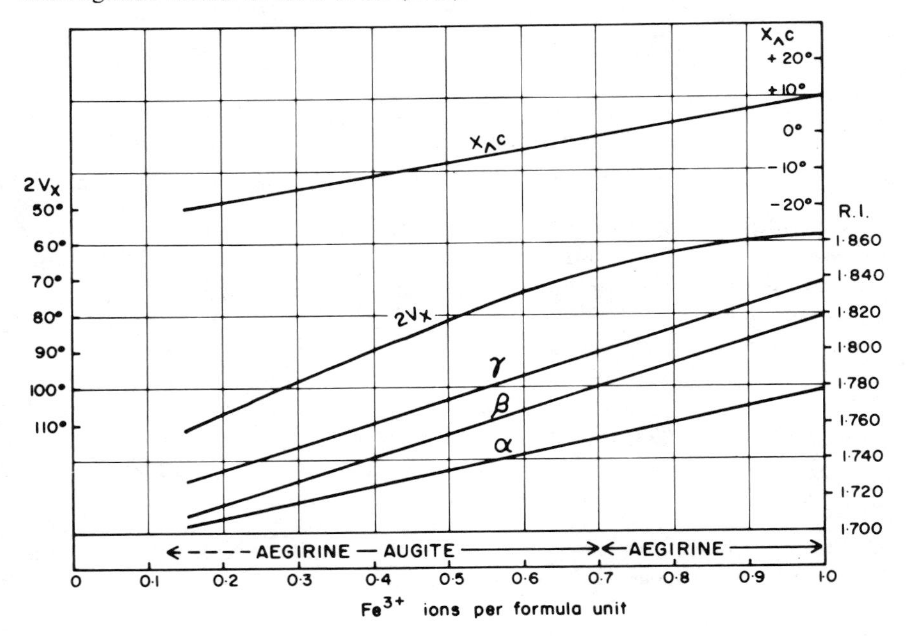

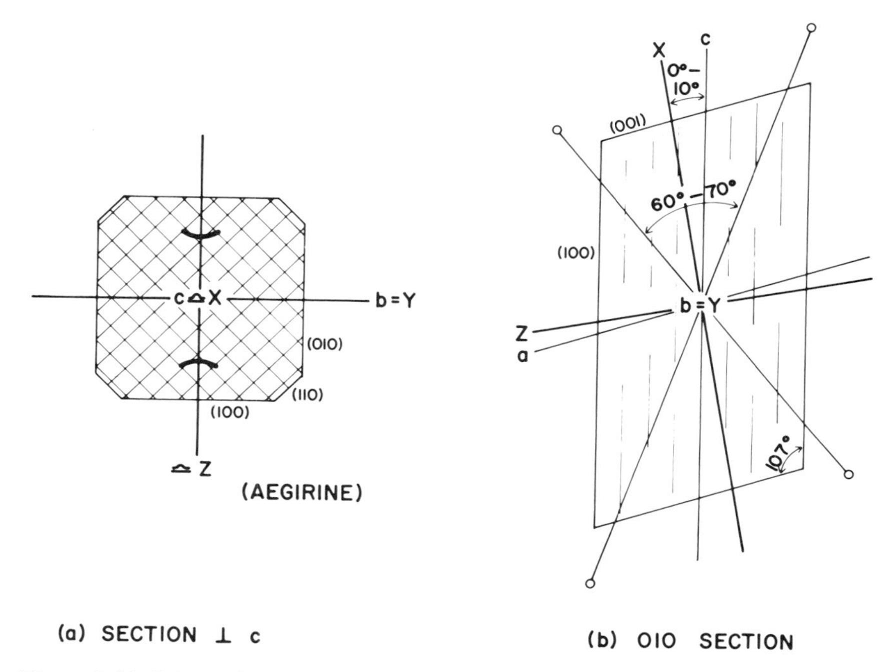

Figure 9.31 Orientation diagrams for aegirine.

Orientation diagrams. *Section* ⊥ *c* (Figure 9.31a): acute bisectrix figure for aegirine, obtuse bisectrix figure for some aegirine-augite; δ' (moderate) up to 0.020; symmetrical extinction to two distinct cleavages almost at right angles to each other; eight-sided sections typical.
(010) section (Figure 9.31b): flash figure; δ (high) = 0.030–0.060, highest for aegirine; small extinction angle $X \wedge c$ and fast along; cleavage oblique to section and may not be visible; may show marked pleochroism.
(100) section: similar to the (010) section, but will provide an obtuse or acute bisectrix figure; straight extinction and fast along.

Occurrence. The typical pyroxene of alkaline igneous rocks including syenites, nepheline-syenites, trachytes, and phonolites. Occasionally occurs in metamorphosed Na-rich rocks (e.g., spilite) in association with glaucophane or riebeckite. Aegirine has also been noted as an authigenic mineral.

Distinguishing features. The pyroxene habit and cleavages are distinctive but may be difficult to observe in strongly prismatic crystals. Distinguished from other pyroxenes by its color, prismatic habit, and fast-along character. The fast-along character distinguishes aegirine and aegirine-augite from similarly colored amphiboles.

No. 35. JADEITE (+ve) $NaAl[Si_2O_6]$

See Plates 23 and 24.
Monoclinic, β = 107°. Distinct {110} cleavages at 87° to each other. Usually anhedral and granular or fibrous. Twinning with {100} composition planes.

Color in thin section. Colorless.

Optical properties. Biaxial +ve. $2V_z$ = 67°–90°. $r < v$ (distinct).

α = 1.640–1.681, β = 1.645–1.684, γ = 1.652–1.692.

δ = 0.008–0.015. $X \wedge a$ = 13°–38°, $Y = b$, $Z \wedge c$ = 30°–55°.

Orientation diagrams. Similar to Figure 9.29, but a lower δ; euhedral crystals not common.

Occurrence. Restricted to metamorphic rocks formed under unusually high pressures. Common associates are albite, lawsonite, and glaucophane. Precious jade is composed of jadeite, and forms in association with the metamorphism of ultramafic rocks.

Distinguishing features. The typical pyroxene cleavages and the restricted occurrence are distinctive. It may be confused with omphacite in eclogites but differs from omphacite (and other clinopyroxenes) in its lower δ. It may be confused with some amphiboles, but differs in its larger extinction angle, its pyroxene cleavages, and the complete lack of color.

No. 36. SPODUMENE (+ve) $LiAl[Si_2O_6]$

Monoclinic, β = 110°. Distinct {110} cleavages at 87° to each other. Often euhedral as very large tabular crystals with {100} prominent and elongate parallel to *c*. Parting parallel to {100}. Twins with {100} composition plane common.

Color in thin section. Colorless.

Optical properties. Biaxial +ve. $2V_z$ = 55°–70°. $r < v$ (weak).

α = 1.648–1.663, β = 1.655–1.669, γ = 1.662–1.679.

δ = 0.014–0.027. $X \wedge a$ = 2°–7°, $Y = b$, $Z \wedge c$ = 22°–27°.

Orientation diagrams. Similar to Figure 9.29, but extinction angle $Z \wedge c$ is smaller.

Occurrence. Characteristic of Li-rich granite-pegmatites in which the spodumene crystals may be very large. Alters to micas and albite or clay minerals.

Distinguishing features. The pyroxene habit and cleavages, and occurrence are distinctive. Differs from other clinopyroxenes in its smaller extinction angle $Z \wedge c$.

No. 37. WOLLASTONITE (−ve) $Ca[SiO_3]$

Parawollastonite is a polymorph of wollastonite.
Triclinic, α = ca. 90°, β = 95°, γ = 103° (parawollastonite is monoclinic with β = 95°). Perfect {100} and distinct {001} and {$\bar{1}$02} cleavages. Crystals are often euhedral to subhedral as tablets parallel to {100} and/or elongated parallel to *b*. Multiple twinning with {100} composition planes and [010] twin axis common.

Color in thin section. Colorless.

Optical properties. Biaxial −ve. $2V_x$ = 38°–60°. $r > v$ (weak to distinct).

α = 1.616–1.640, β = 1.627–1.650, γ = 1.631–1.653.

δ = 0.013–0.015. $X \wedge c$ = ca. 30°–44°, $Y \wedge b$ = 3°–5° ($Y = b$ for parawollastonite), $Z \wedge a$ = ca. 35°–45°.

Orientation diagrams. *Sections parallel to b and the length of crystals* (Figure 9.32a): may be bisectrix, optic-axis, or off-centered figures, but all indicate the OAP is across the good cleavage traces and length of crystals; δ' (low or very low) up to 0.012; straight or inclined extinction up to 5° (always straight for parawollastonite) and may be either length fast or slow.
(010) section (Figure 9.32b): flash figure; δ (low) up to 0.015; three cleavages may be visible; inclined extinction to all cleavages; this section may not be easily observed in fibrous wollastonite.

Figure 9.32 Orientation diagrams for wollastonite.

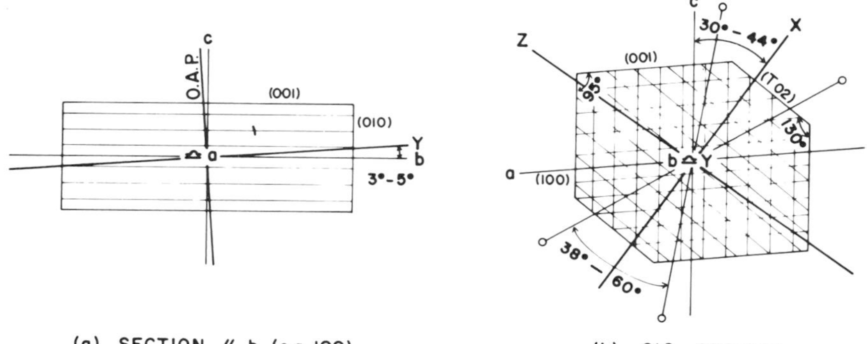

Occurrence. Wollastonite (and much less commonly parawollastonite) occurs in thermally metamorphosed and metasomatized carbonate rocks. It also occurs more rarely in alkaline undersaturated igneous rocks such as ijolites and phonolites.

Distinguishing features. The elongate crystals, the OAP across the cleavages, the −ve character and occurrence are distinctive. May be confused with tremolite, pectolite, and diopside, all of which commonly occur with wollastonite. However, the low δ and the position of the OAP distinguishes wollastonite from all three. In addition, pectolite and diopside are +ve, and tremolite has a larger $2V$. Sillimanite differs in being consistently length slow with straight extinction.

No. 38. PECTOLITE (+ve) $Ca_2NaH[SiO_3]_3$

Triclinic, $\alpha = 90°$, $\beta = 95°$, $\gamma = 102°$. Perfect {100} and {001} cleavages. Crystals elongate or fibrous parallel to b.

Color in thin section. Colorless.

Optical properties. Biaxial +ve. $2V_z = 50°–63°$. $r > v$ (weak).

$\alpha = 1.595–1.610$, $\beta = 1.605–1.615$, $\gamma = 1.632–1.645$.

$\delta = 0.030–0.038$. $X \wedge c = 5°–11°$, $Y \wedge a = 10°–16°$, Z is close to b.

Orientation diagrams. *Sections parallel to b and length of crystals* (Figure 9.33a): flash or obtuse bisectrix figure; δ' (moderate to high) up to 0.038; almost straight extinction; trace of cleavage parallel to b often visible.
(010) section (Figure 9.33b): acute bisectrix figure; δ' (very low) up to 0.005; inclined extinction to two cleavages at 95° to each other.

Figure 9.33 Orientation diagrams for pectolite.

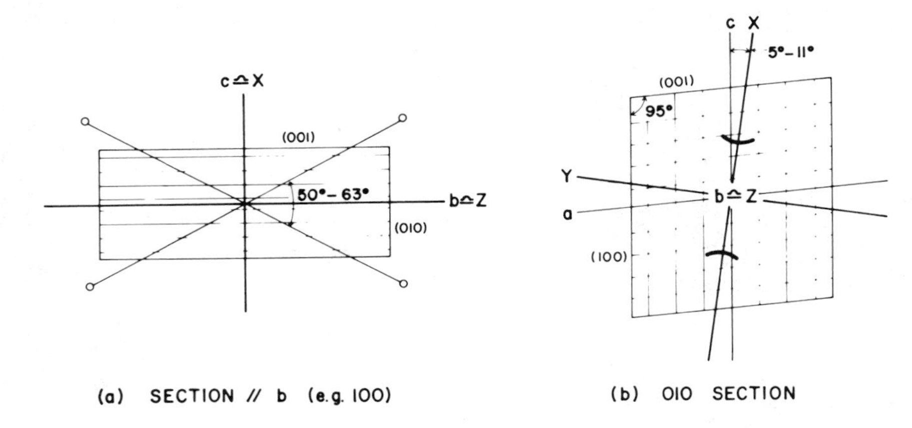

Occurrence. Most commonly found in hydrothermal veins and in cavities in basic igneous rocks. Also found in thermally metamorphosed carbonate rocks, and more rarely in undersaturated alkaline igneous rocks such as nepheline-syenite.

Distinguishing features. The elongate crystals, length-slow character, high δ and occurrence are characteristic. May be associated with prehnite and wollastonite, both of which superficially resemble pectolite. However, prehnite is length fast, and wollastonite is sometimes length fast, is −ve, and has a lower δ.

No. 39. RHODONITE (+ve) (Mn,Fe,Ca)[SiO_3]

Triclinic, $\alpha = 86°$, $\beta = 93°$, $\gamma = 111°$. Perfect {100} and {001} and distinct {010} cleavages. Euhedral to anhedral crystals, commonly tabular parallel to {010}. Some workers choose a different unit cell and index crystal planes differently.

Color in thin section. Colorless to pale pink; pleochroism weak.

Optical properties. Biaxial +ve. $2V_z = 61°–87°$. $r < v$ (weak).

$$\alpha = 1.711–1.738,\ \beta = 1.716–1.741,\ \gamma = 1.724–1.751.$$

$$\delta = 0.011–0.017.\ X \wedge c = \text{ca. } 45°,\ Z \wedge b = \text{ca. } 20°.$$

Orientation diagrams. *(010) section* (Figure 9.34a): since $b \wedge Z = 20°$, sections displaying two sharp cleavages at 93° provide off-centered

Figure 9.34 Orientation diagrams for rhodonite.

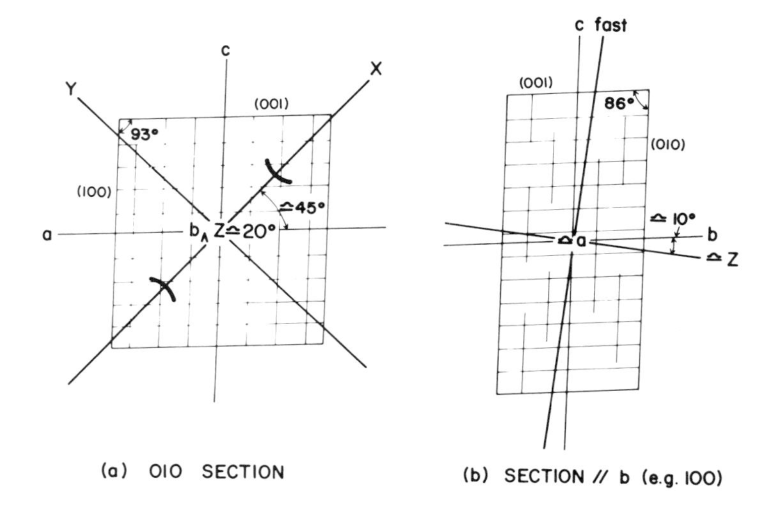

acute bisectrix figures; δ' (very low) up to 0.005; almost symmetrical extinction.

Sections parallel to b (Figure 9.34b): figures vary from flash to off-centered obtuse bisectrix figures; δ' (low) up to 0.017; may display perfect {001} or {100} cleavages and less distinct cross-cleavage {010}; generally inclined extinction, and the slow direction is across the length of tabular crystals.

Occurrence. Found in ore deposits of manganese (often associated with metasomatic activity). Alters to pyrolusite (black MnO_2) or rhodochrosite.

Distinguishing features. The association with Mn-ore deposits, the pale-pink color and triclinic nature are characteristic. It may be confused with clinopyroxene, but its lower δ is distinctive. The RI and δ are similar to those of kyanite, but the cleavages of kyanite are much more distinct.

No. 40. AENIGMATITE (+ve) $Na_2Fe_5^{2+}TiO_2[Si_6O_{18}]$

Triclinic, $\alpha = 105°$, $\beta = 97°$, $\gamma = 125°$. Distinct {010} and {001} cleavages. Euhedral crystals prismatic parallel to c. Simple and multiple twinning on $(0\bar{1}1)$ common.

Color in thin section. Red or brown, often almost opaque; strongly pleochroic with $Z > Y > X$.

Optical properties. Biaxial +ve. $2V_z = 27°–55°$. $r < v$ (strong).

$$\alpha = 1.790–1.81,\ \beta = 1.805–1.826,\ \gamma = 1.87–1.9.$$

$$\delta = 0.07–0.08.\ Z \wedge c = \text{ca. } 45°.\ \text{OAP is close to } (010).$$

Occurrence. In alkaline plutonic and volcanic rocks, usually as a groundmass phase in the latter.

Distinguishing features. Its dark, almost opaque color, prismatic habit, very strong pleochroism and δ, and occurrence are distinctive. May be confused with other opaque minerals and some dark amphiboles; the latter have a lower RI.

No. 41. SAPPHIRINE (−ve or +ve) $(Mg,Fe,Al)_8O_2[(Al,Si)_6O_{18}]$

Monoclinic, $\beta = 110°$ or 125°. Imperfect {010} cleavage. Anhedral, or tablets parallel to (010). Lamellar twins with {010} composition planes.

Color in thin section. Weakly pleochroic from colorless, yellow or pink to blue or blue-green, usually with $Z > Y > X$.

Optical properties. Biaxial −ve (more rarely +ve). $2V_x = 40°–114°$. $r < v$ (strong).

$\alpha = 1.701–1.731$, $\beta = 1.703–1.741$, $\gamma = 1.705–1.745$.

$\delta = 0.004–0.015$. $X \wedge a = 42°–45°$, $Y = b$, $Z \wedge c = 6°–9°$.

RI increases and $2V_x$ decreases with Fe content.

Occurrence. Usually found in high-grade regionally and thermally metamorphosed Al-rich rocks, and often associated with spinel, corundum, and cordierite.

Distinguishing features. High relief, low δ, blue color and lack of cleavage are distinctive. May be confused with corundum, blue-amphibole, kyanite, tourmaline, or dumortierite, but corundum and tourmaline are both uniaxial, blue-amphibole and kyanite have good cleavages, and tourmaline and dumortierite are more strongly pleochroic.

The Amphiboles

The amphiboles are a group of minerals characterized by the linkage of the $[(Si,Al)O_4]$ tetrahedra to form double chains of general composition $[(Si,Al)_4O_{11}]_n$. This structure is reflected in two cleavages parallel to the *c*-axis at an angle of approximately 54° to each other. The structure allows very extensive ionic replacement, and the resulting variation in chemistry is so great that no completely satisfactory classification has been devised. The International Mineralogical Association recently published its proposals for amphibole classification and nomenclature (Leake, 1978). The details of this are so complex that they cannot be dealt with completely in this book. The approved nomenclature is based solely on crystal chemistry, and a major problem is that physical properties (such as optics) are insufficient to characterize amphiboles precisely.

The recommendations involve a subdivision into four main amphibole groups. All are monoclinic except for some orthorhombic members of group 1. A summary of the proposals is:

GROUP 1A: Orthorhombic iron–magnesium–(manganese) amphiboles, including:

No. 42A. ANTHOPHYLLITE $(Mg,Fe^{2+})_7[Si_8O_{22}](OH,F)_2$

No. 42B. GEDRITE $(Mg,Fe^{2+})_5Al_2[Si_6Al_2O_{22}](OH,F)_2$

GROUP 1B: Monoclinic iron–magnesium–(manganese) amphiboles, including:

No. 43A. CUMMINGTONITE $(Mg,Fe^{2+})_7[Si_8O_{22}](OH)_2$

No. 43B. GRUNERITE $Fe^{2+}_7[Si_8O_{22}](OH)_2$

Group 2: Calcic amphiboles, including:

No. 44A. TREMOLITE $Ca_2Mg_5[Si_8O_{22}](OH,F)_2$

No. 44B. FERRO-ACTINOLITE $Ca_2Fe^{2+}_5[Si_8O_{22}](OH,F)_2$

(Amphiboles intermediate between tremolite and ferro-actinolite are called *actinolite*.)

No. 44C. HORNBLENDE

composed of varying proportions of:

Tremolite–Ferro-actinolite (as above)
Tschermakite $Ca_2(Mg,Fe^{2+})_3(Al,Fe^{3+})_2[Si_6Al_2O_{22}](OH,F)_2$
Pargasite $NaCa_2(Mg,Fe^{2+})_4Al[Si_6Al_2O_{22}](OH,F)_2$
Hastingsite $NaCa_2(Fe^{2+},Mg)_4Fe^{3+}[Si_6Al_2O_{22}](OH,F)_2$
Kaersutite $NaCa_2(Mg,Fe^{2+})_4Ti[Si_6Al_2O_{22}](O,OH)_2$
Edenite $NaCa_2(Mg,Fe^{2+})_5[Si_7AlO_{22}](OH,F)_2$

No. 44D. OXYHORNBLENDE

Calcic amphiboles with $(OH + F + Cl) < 1.00$.

GROUP 3: Sodic-calcic amphiboles, composed of varying proportions of:

No. 45. KATOPHORITE $NaCaNa(Mg,Fe^{2+})_4(Al,Fe^{3+})[Si_7AlO_{22}](OH)_2$

as well as:

Richterite $NaCaNa(Mg,Fe^{2+})_5[Si_8O_{22}](OH)_2$
Winchite $CaNa(Mg,Fe^{2+})_4(Al,Fe^{3+})[Si_8O_{22}](OH)_2$
Barroisite $CaNa(Mg,Fe^{2+})_3(Al,Fe^{3+})_2[Si_7AlO_{22}](OH)_2$

GROUP 4: Alkali amphiboles, including:

No. 46A. GLAUCOPHANE $Na_2(Mg,Fe^{2+})_3Al_2[Si_8O_{22}](OH)_2$

No. 46B. RIEBECKITE $Na_2(Fe^{2+},Mg)_3Fe_2^{3+}[Si_8O_{22}](OH)_2$

(Amphiboles intermediate between glaucophane and riebeckite are called *crossite*.)

No. 46C. ECKERMANNITE $Na_3(Mg,Fe^{2+})_4Al[Si_8O_{22}](OH,F)_2$

No. 46D. ARFVEDSONITE $Na_3(Fe^{2+},Mg)_4Fe^{3+}[Si_8O_{22}](OH,F)_2$

The international commission (Leake, 1978) recommends that end-member names should not be used if identification is based solely on optics. Hence *tremolitic amphibole* should be used rather than *tremolite*. *Hornblende* can be used as a general term for unspecified members of the calcic amphibole group.

As a family, the amphiboles are distinguished from other minerals by their two cleavages at 54°, their moderate to high RI, moderate to high birefringence, and their biaxial character. Many amphiboles are strongly colored and pleochroic. Identification of individual species is less easy and may not be feasible with optics alone because there are considerable overlaps in their properties. In order to compare them more easily and avoid unnecessary repetition, the details are compiled below under two main headings—crystallography and optical properties. Typical occurrences, habits, and distinguishing features are then itemized separately for each species to assist identification.

Crystallography

All the amphiboles are monoclinic except anthopyllite and gedrite which are orthorhombic. For the monoclinic members, the crystallographic angle β varies within the small range 103°–109°.

Crystal Habit

Crystal habit varies from stubby prismatic (not common) to elongate prismatic (very common) to fibrous or asbestiform (common). Amphiboles that may be fibrous or asbestiform include anthophyllite, gedrite, and members of the cummingtonite–grunerite, tremolite–ferroactinolite, and glaucophane–riebeckite series. *Crocidolite, montasite,* and *amosite* are familiar names for asbestiform riebeckite, grunerite, and anthophyllite, but the international commission recommends they not be used.

Euhedral prismatic crystals are generally four- or six-sided in cross section with the {110} form dominating, with or without {010} (Figure 9.35).

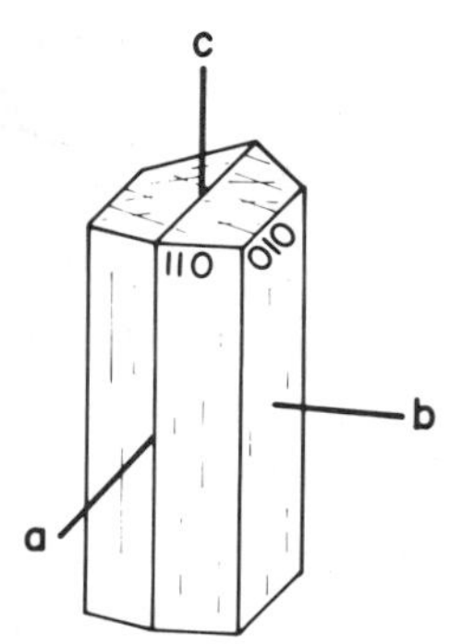

Figure 9.35 Typical habit of a prismatic amphibole crystal.

Cleavage

All amphiboles have perfect {110} or {210} (for orthorhombic amphibole) cleavages. The angle between the two cleavages is always in the range 54°–56°, and *b* bisects the acute angle. Therefore, three principal types of section are encountered in thin section (Figures 9.36, 9.39, and 9.40):

Figure 9.36 Sections of hornblende: bottom right, section with two cleavages; center left, section with no cleavage visible; top left and right, sections with traces of cleavage in one direction. Rotoroa Igneous Complex, New Zealand (view measures 0.7 × 0.45 mm). Plane-polarized light.

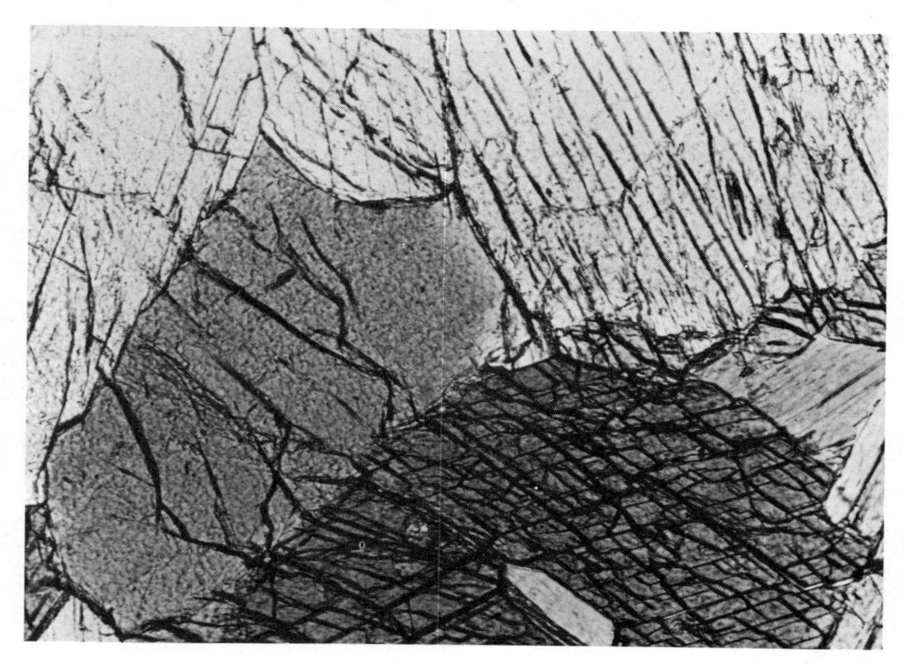

sections approximately at right angles to the c-axis that display two cleavages; sections parallel to c and close to (010) or (110) that display the traces of either or both cleavages in one direction parallel to the c-axis; sections parallel to (100) that display *no* cleavage in that both cleavages are at too acute an angle to the section to be visible.

Twinning

Simple or lamellar twinning with {100} composition planes is common in all the monoclinic amphiboles, but is not possible in orthorhombic anthophyllite and gedrite.

Optical Properties

There is as yet no certain way of correlating optics with amphibole chemistry except in a rather general way. The nature of the problem becomes clear when one realizes that Leake (1978) lists altogether 51 amphibole end members. Transitions between these create an immense variation leading to an almost impossible task as far as optical specification is concerned. Hornblende itself is a complex of 16 end members. The sodic-calcic amphiboles are particularly ill defined, and only the relatively well-known katophorite is described here. Nevertheless, various general relationships such as an increase of RI with Fe exist, and color is particularly useful in limiting one's options. Below I give an outline of the broad ranges of various optical properties.

Refractive Index and Birefringence

Figure 9.37 gives the commonly reported values for various amphiboles. Refractive indices for each series increase with iron content.

2V and the Position of the OAP (Figure 9.38)

Most amphiboles are −ve, but Fe-rich anthophyllite, most gedrite, cummingtonite, and some Mg-rich hornblendes are +ve. A simple correlation with chemistry is not, however, possible. $2V$ is difficult to determine in some of the alkali amphiboles because of their high dispersion and intense color. In orthorhombic amphiboles the OAP is parallel to (010) and $Z = c$ (Figure 9.39). A variety of optical orientations are possible for the monoclinic amphiboles, and these have been categorized as G (after glaucophane), C (after crossite), O (after osannite, a name for riebeckite recommended by Leake, 1978, for extinction), and R (after riebeckite). These are discussed by Borg (1967) who points out that the optical orientations must not be confused with the minerals after which they are

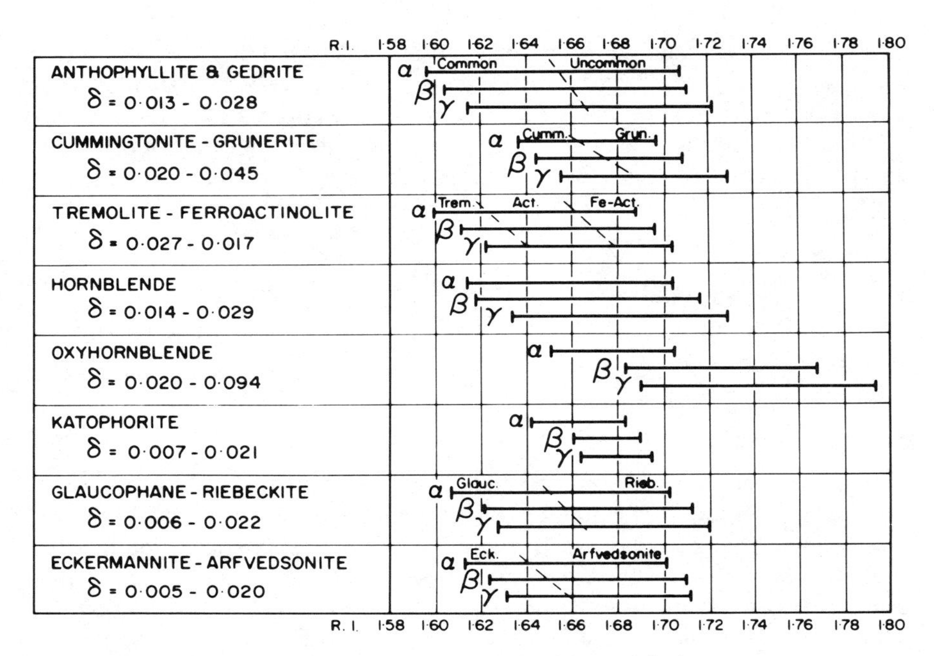

Figure 9.37 Ranges of RI and birefringence in the amphiboles.

named. For example, riebeckites have been described not only with an *R* orientation, but with *O* and *C* orientations too. The most common orientation is *G* in which the OAP is (010) and *Z* makes a small angle with *c*. If the OAP is (010) but *X* makes the smaller angle with *c* the orientation is *R*. The *C* and *O* orientations have the OAP perpendicular to (010); $Z = b$ and *Y* makes a small angle with *c* in the *C* orientation whereas *X* does in the *O* orientation. These four orientations are simplified here into two groups: (1) the OAP is (010) and $Y = b$ (Figure 9.40); (2) the OAP is perpendicular to (010) and $Z = b$ (Figure 9.41).

Orientation Diagrams and Extinction Angles

Note: cleavage fragments of any amphibole will lie parallel to the *c*-axis and one of the cleavages, and will always provide off-centered interference figures.

(1) *Orthorhombic amphiboles* (Figure 9.39). Basal sections (Figure 9.39a) have symmetrical extinction. All prismatic sections (Figure 9.39b,c) have straight extinction with *Z* parallel to *c*. Maximum interference colors are shown by (010) sections which display the traces of the cleavages parallel to the *c*-axis (Figure 9.39c). (100) and (001) sections provide bisectrix interference figures.

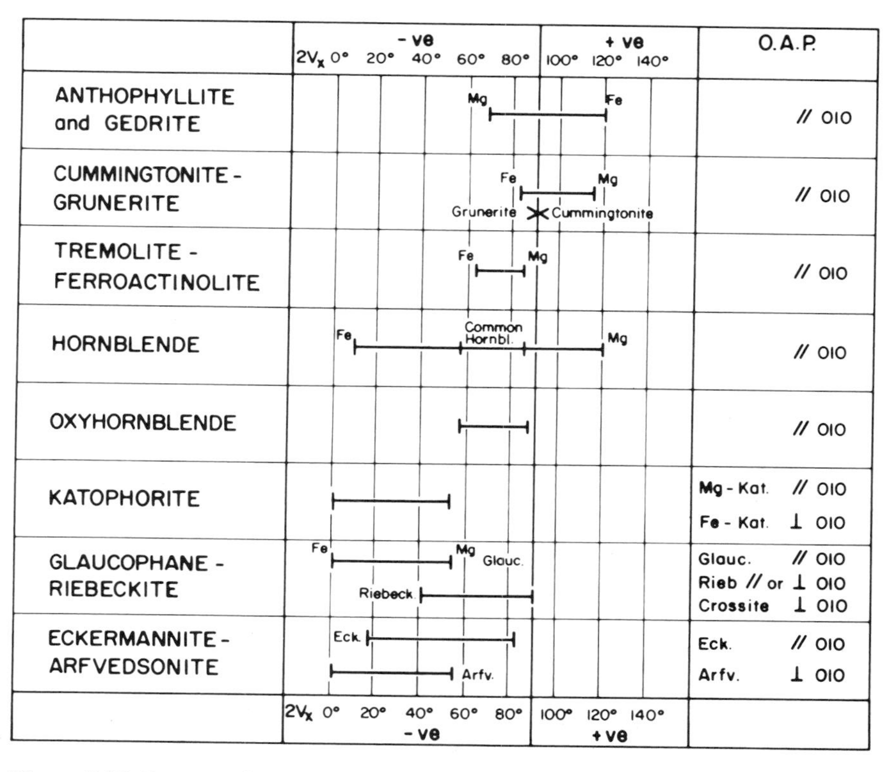

Figure 9.38 Ranges of $2V$ and orientation of the OAP in the amphiboles.

(2) *Monoclinic amphiboles with OAP // (010)* (Figure 9.40). Sections to which c is perpendicular have symmetrical extinction (Figure 9.40a). Sections approximately parallel to (100) which display no cleavage (Figure 9.40b) often provide good optic-axis interference figures, and display very little pleochroism. In (010) sections (Figure 9.40c), it is possible to measure the extinction angle $Z \wedge c$ (Table 9.1) It is important to choose the (010) section carefully; being the XZ plane, this section will display the highest interference color for the particular amphibole.

(3) *Monoclinic amphiboles with OAP ⊥ (010)* (Figure 9.41). It is customary to measure the extinction angle $X \wedge c$ (Table 9.1), although in practice this may be difficult because of the strong dispersion which results in poorly defined extinction in many amphiboles with this orientation.

Color and Pleochroism

The colors and pleochroic schemes for the amphiboles in thin section are summarized in Figure 9.42. The pleochroic scheme is usually $Z > Y > X$,

Table 9.1 Extinction Angles in the Monoclinic Amphiboles

Mineral	$Z \wedge c$ (Figure 9.40c), OAP // (010)		$X \wedge c$ (Figure 9.41) OAP ⊥ (010)
Cummingtonite	15°(Fe)–21°(Mg)		
Grunerite	10°(Fe)–15°(Mg)		
Tremolite	15°(Fe)–21°(Mg)		
Actinolite	10°(Fe)–15°(Mg)		
Hornblende	12°–34°		
Oxyhornblende	4°–14°		
Mg-katophorite	8°–16°		
Fe-katophorite			36°–70°
Glaucophane	4°–14°		
Riebeckite	ca. 98°	or	3°–21°
Crossite			ca. 80°
Eckermannite	30°–50°		
Arfvedsonite			0°–30°

or $Z = Y > X$, or $Z > Y = X$, but is $X > Y > Z$ for eckermannite–arfvedsonite, and some riebeckite, and $Z < Y > X$ for Mg-katophorite. The colors are darker in detrital grains.

Dispersion

Optic-axis dispersion is weak with $r<$ or $>v$ in the amphiboles of group 1. The more common calcic amphiboles likewise have variable but generally weak dispersion, but the more darkly colored varieties such as hastingsites and oxyhornblendes may have strong dispersion, again $r<$ or $>v$. The more strongly colored sodic-calcic and alkali amphiboles often have very strong dispersion. In katophorite, $r > v$ (OAP ⊥ 010) or $r < v$ (OAP = 010), and in riebeckite, eckermannite, arfvedsonite, and crossite, $r > v$ (about X), usually strong or very strong. Glaucophane has a weaker dispersion $r < v$.

Occurrence, Common Habits, and Distinguishing Features of Particular Amphiboles

All amphiboles may occur as detrital grains in sedimentary rocks. Listed below are their occurrences in igneous and metamorphic rocks. Minerals likely to be confused with amphiboles are discussed separately in the next section.

Anthophyllite and gedrite occur in a wide range of metamorphic rocks—often with cordierite in gneisses, and ascribed by some to metasomatism; they form with talc during the regional metamorphism of ultrabasic rocks. Sometimes grow as rims around pyroxene during retrograde

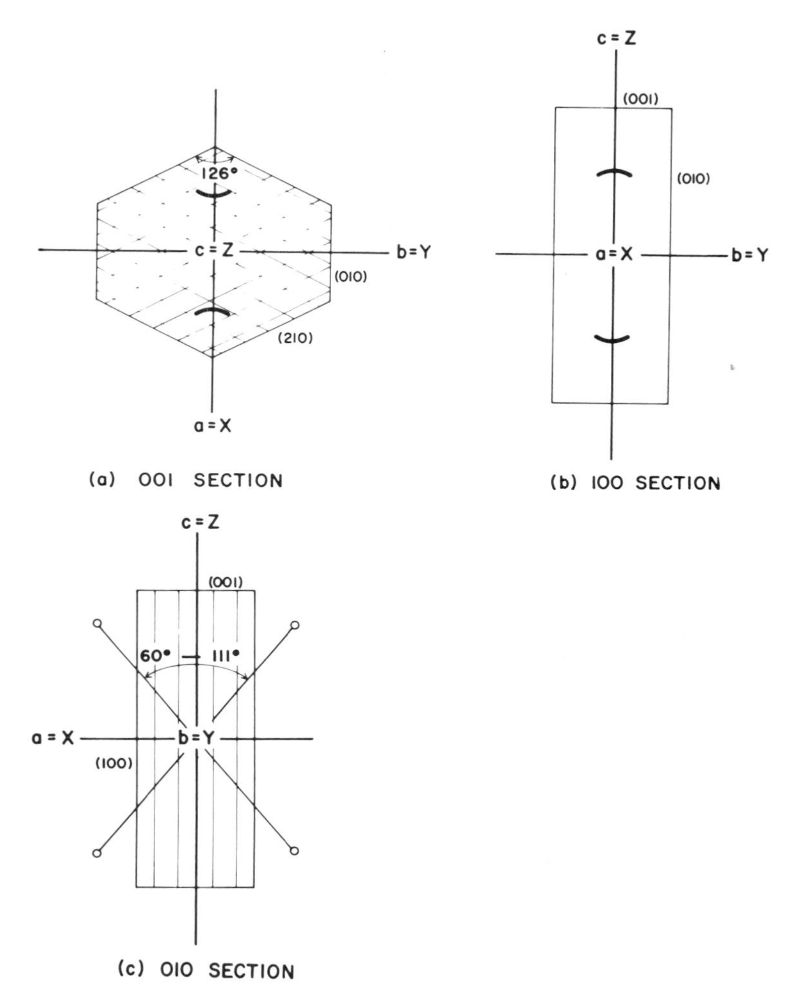

Figure 9.39 Orientation diagrams for the orthorhombic amphiboles.

metamorphism. Usually strongly prismatic or asbestiform. May alter to talc, or serpentine. Characteristics are: pale colored or colorless, usually elongate or fibrous crystals, and restricted to metamorphic rocks (and as detrital grains). May be confused with cummingtonite–grunerite or tremolite–ferro-actinolite. Distinguished as follows: straight extinction in (010) sections; generally lower δ than cummingtonite–grunerite; lack of twinning.

Cummingtonite–grunerite occurs almost exclusively in metamorphic rocks. Cummingtonite may be present together with anthophyllite or hornblende in amphibolites; it rarely occurs in some volcanic rocks (da-

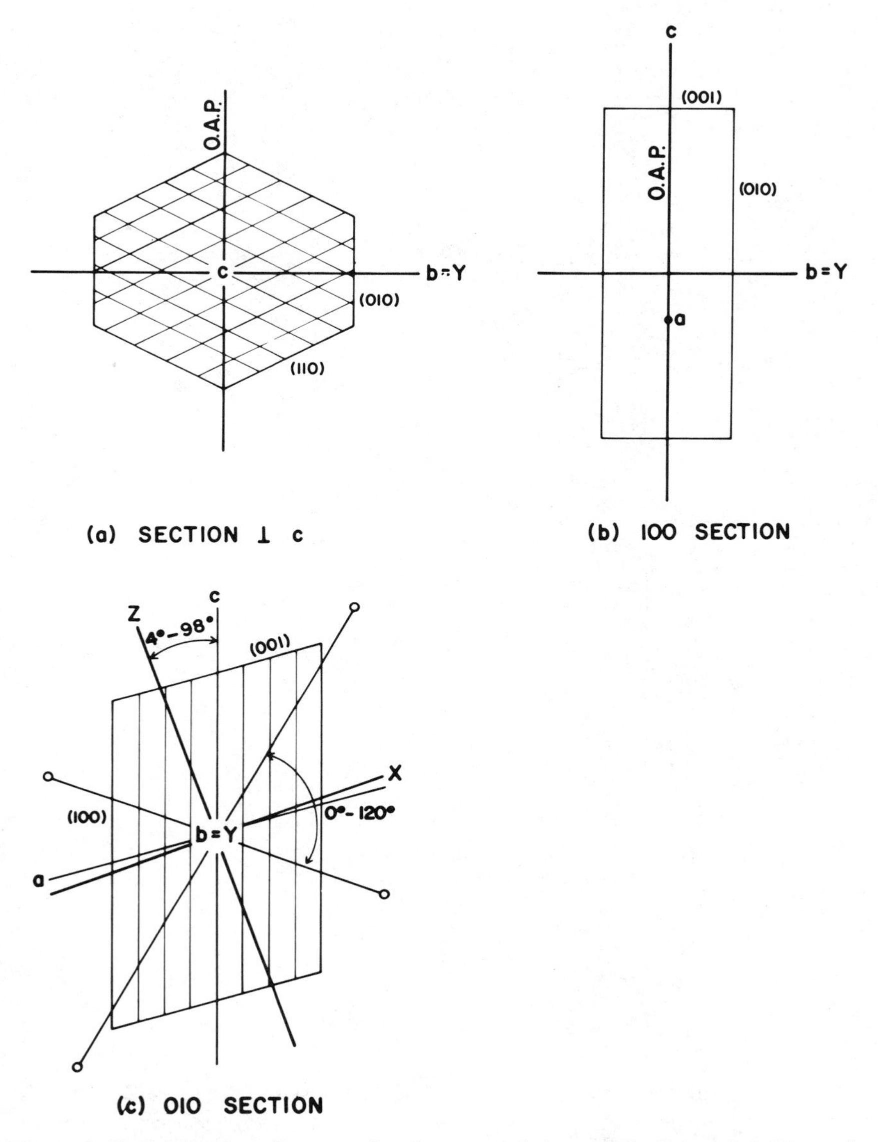

Figure 9.40 Orientation diagrams for the monoclinic amphiboles with OAP parallel to (010).

cites). Grunerite is found in metamorphosed Fe-rich sediments. Usually strongly prismatic, fibrous or asbestiform. May alter to talc or serpentine. Characteristics are: pale colored or colorless, usually elongate or fibrous crystals, and usually in metamorphic rocks (or as detrital grains). May be confused with anthophyllite, gedrite, and tremolite–ferro-actinolite. Distinguished as follows: cummingtonite is +ve whereas tremolite is −ve;

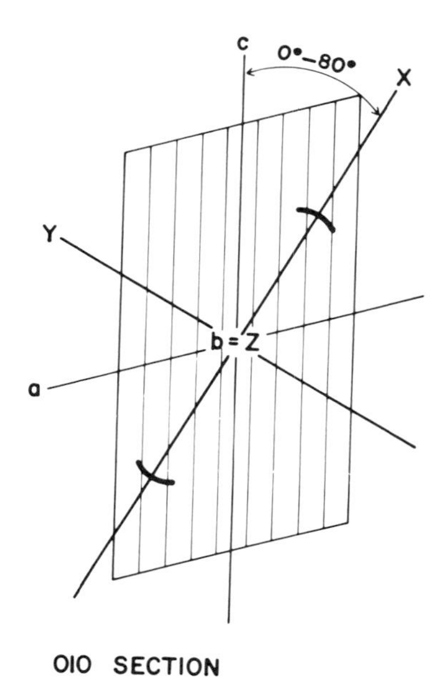

Figure 9.41 Orientation diagram for monoclinic amphiboles with OAP perpendicular to (010).

Figure 9.42 Pleochroic schemes and possible colors in the amphiboles.

	COLORLESS	YELLOW-BROWN	RED-BROWN	GREEN	BLUE-GREEN	BLUE
		Pale —— Dark	Pale —— Dark	Pale —— Dark	Pale —— Dark	Pale —— Dark
ANTHOPHYLLITE and GEDRITE	X Y Z	X Y— Z—		X Y— Z—	Y Z—	
CUMMINGTONITE — GRUNERITE	X Y Z (Mg-rich)	X Y— Z—		Z		
TREMOLITE — FE-ACTINOLITE	X Y Z (Mg-rich)	X Y—		X Y— Z——		
HORNBLENDE	X	X Y—— Z——	Y— Z——	X Y—— Z——	X Y—— Z——	
OXYHORNBLENDE		X — Y—— Z	Y—— Z			
KATOPHORITE		—X—	—Y— —Z—	—Z—	—Y—	
GLAUCOPHANE	X	X		X	X	Y (violet) Z —
RIEBECKITE		Z		Z	Z	X — Y —— Z ——
ECKERMANNITE — ARFVEDSONITE		——Y ——Z			X ——Y ——Z	

grunerite has a higher RI and δ than actinolite; inclined extinction, generally higher δ, and twinning distinguish the series from anthophyllite and gedrite.

Tremolite–ferro-actinolite (see Plates 33 and 34) is the typical amphibole formed during greenschist facies or low-grade thermal metamorphism of basaltic or impure carbonate rocks. *Nephrite* (also known as *jade* or *greenstone*) is ornamental material composed essentially of very fine-grained tremolite-actinolite (Leake, 1978, recommends these terms be abandoned). Some *uralite* is tremolite–actinolite formed by the replacement of pyroxene in basic and intermediate igneous rocks. Usually strongly prismatic, fibrous or asbestiform. May alter to chlorite. Characteristics are: colorless (tremolite) to moderate green (ferro-actinolite), usually elongate or fibrous crystals, and restricted occurrences in metamorphic rocks, as an alteration product of pyroxene, or as detrital grains. May be confused with anthophyllite, gedrite, and cummingtonite–grunerite. Distinguished as follows: tremolite is −ve whereas cummingtonite is +ve; actinolite has a lower RI and δ than grunerite; inclined extinction and twinning distinguish the series from anthophyllite and gedrite; green actinolitic amphiboles cannot be distinguished optically from some other common hornblendes.

Hornblende (see Figure 9.36 and Plates 3–6) has a very widespread occurrence. It is the typical amphibole formed during amphibolite facies or high-grade thermal metamorphism of basaltic or impure carbonate rocks. Abundant in many plutonic and volcanic igneous rocks, especially diorites and granites. Hornblendes are usually Mg rich in basic rocks, and Fe rich in acid rocks. Some *uralite* is hornblende formed by the replacement of pyroxene in basic igneous rocks. Hornblende may form intergrowths with spinel in corona structures around olivine and pyroxene in basic igneous rocks, or intergrowths with plagioclase around garnet in eclogites. Usually stubby to long prismatic crystals. May alter to biotite and chlorite. Some Fe-rich varieties may be replaced by iron oxides and siderite. Characteristics are its moderately strong green, blue-green, or brown color, and lack of a strongly elongate or prismatic habit. Hornblendes may be difficult to distinguish from strongly colored grunerite, though grunerite has a higher birefringence. Particular types of hornblende are not generally easy to distinguish, but Mg-rich varieties may be +ve and tend to be pale-colored whereas Fe-rich varieties are usually darker and have a smaller $2V_x$.

Oxyhornblende occurs as phenocrysts in a wide variety of volcanic rocks, usually as short to moderately long prismatic crystals. It is characterized by an intense brown to red-brown color. It has a higher RI and larger $2V$ than katophorite.

Glaucophane–riebeckite. Glaucophane (see Plates 19–22) is characteristic of the greenschist facies and the higher-pressure lawsonite–glaucophane facies of regional metamorphism. Riebeckite occurs in alkaline and acid plutonic and volcanic rocks, and also in metamorphosed Fe-rich rocks where it may be fibrous or asbestiform. It is also reported as a diagenic mineral. The pale-blue/violet color of most glaucophane is diagnostic. Riebeckite is characterized by deep blue colors and its length-fast nature. Riebeckite may be difficult to distinguish from arfvedsonite and eckermannite although it has a more widespread distribution than those two minerals. *Crossites* (intermediate forms between glaucophane and riebeckite) are supposedly characterized by the C (crossite) optical orientation with Y close to c. However, not all crossites have this orientation, and some riebeckites also have the C orientation.

Eckermannite–arfvedsonite is almost restricted to alkaline plutonic igneous rocks, usually as moderately elongate prismatic crystals. It is distinguished from most other amphiboles by its intense blue-green color, pleochroic scheme, and optical orientation, though this may be difficult to determine because of high dispersion. It may be difficult to distinguish from riebeckite. Eckermannite has $Y = b$, whereas arfvedsonite has $Z = b$.

Katophorite is rare, and restricted to basic alkaline igneous rocks. Mg-rich katophorite is distinguished from other red-brown amphiboles by its smaller $2V$, and Fe-rich katophorite has its OAP perpendicular to (010).

Minerals Likely to Be Confused with Amphiboles

The general properties of the colored amphiboles are quite distinctive and confusion with other minerals is not likely. Red-brown biotite is sometimes confused with red-brown amphibole (they often occur together), but the single cleavage, mottled extinction, and pseudo-uniaxial interference figure of the biotite enables the distinction to be made. (100) sections of hornblende are sometimes mistaken for uniaxial tourmaline, but these sections usually provide good biaxial interference figures. Aenigmatite is distinguished by its higher RI.

The colorless amphiboles may be confused with zoisite, sillimanite, or wollastonite. However, all three minerals lack the characteristic amphibole cleavages, zoisite often has anomalous interference colors, and wollastonite and some zoisite have their OAP across the cleavage or length of the crystals. Fibrous anthophyllite (or gedrite) and sillimanite may be very difficult to distinguish, although their occurrences are usually different. Sillimanite has a slightly higher RI than the common range of anthophyllite and gedrite.

Fibrous amphiboles have a higher RI than fibrous serpentine.

D. Phyllosilicates

In the phyllosilicates, each $[SiO_4]$ tetrahedral group shares three of its oxygen atoms with three other tetrahedra to form a continuous sheet of general formula $[Si_2O_5]$. Phyllosilicates described here are: the micas;

the chlorites and serpentines; pyrophyllite; talc; stilpnomelane; the brittle micas margarite and clintonite; prehnite; apophyllite; and the clay minerals.

The Micas

The micas are characterized by a composite structure of two [Si_2O_5] sheets between which are sandwiched Fe, Mg, Al, or Li cations. K or Na atoms lie between, and (OH) ions within, the composite sheets. The sheets may be stacked upon one another in a number of ways resulting in a variety of polymorphs, but a study of these is beyond the scope of this book. The "brittle micas" (margarite and clintonite) have a similar structure with Ca atoms lying between the composite sheets.

The micas described below are muscovite (including sericite, phengite, and fuchsite), paragonite, the Li-mica lepidolite, biotite–phlogopite, and glauconite. Characteristics common to them all include: a single perfect cleavage parallel to the sheet structure; *X* approximately perpendicular to (001)—hence all micas have a slow direction parallel to the cleavage traces in thin section; biaxial −ve, and sometimes pseudo-uniaxial—hence interference colors are always lower in basal sections than side sections; mottled extinction (Figure 9.43) caused by the microscopic

Figure 9.43 Mottled extinction in biotite, Constant Gneiss, New Zealand (view measures 3.3 × 2.1 mm). Crossed-polarized light.

buckling of cleavage planes during thin-section making. However, these characteristics, even in combination, *do not* distinguish micas from all other minerals.

No. 47. MUSCOVITE (−ve) $K_2Al_4[Si_6Al_2O_{20}](OH,F)_4$

See Plates 7–10, 19–22, 25, 26, 29, and 30.

Varieties include *phengite* (Si replaces Al in the sheet structure and Mg or Fe replaces some of the Al sandwiched between the composite sheets), *fuchsite* (Cr-rich mica), and *sericite* (fine-grained muscovite). Monoclinic, β usually = 95°. Perfect {001} cleavage. Often subhedral tabular crystals with {001} well developed; sometimes euhedral tablets with {001}, {110}, and {010} forms.

Color in thin section. Usually colorless, but fuchsite may be pale to dark emerald green with $X < Y \leqslant Z$.

Color in detrital grains. Usually colorless.

Optical properties. Biaxial −ve. $2V_x = 30°–47°$ (but as low as 0° in some phengite). $r > v$ (distinct).

$$\alpha = 1.552\text{–}1.578,\ \beta = 1.582\text{–}1.615,\ \gamma = 1.587\text{–}1.617.$$

$$\delta = 0.036\text{–}0.049.\ X \wedge c = 2°\text{–}4°,\ Y \wedge a = 1°\text{–}3°,\ Z = b.$$

RI and δ increase with Fe and Cr, and $2V$ decreases with Mg and Fe. Hence phengite and fuchsite are high RI muscovites.

Orientation diagrams. *(001) section and cleavage fragments* (Figure 9.44a): acute bisectrix figure; δ' (low) up to 0.006; extinction positions often indistinct due to proximity of optic axes; extinction may be wavy due to bending of cleavages; cleavage *not* visible.

(010) section (Figure 9.44b): obtuse bisectrix (flash) figure; δ' (high) up to 0.037; good cleavage, slow along and almost straight extinction (mot-

Figure 9.44 Orientation diagrams for muscovite.

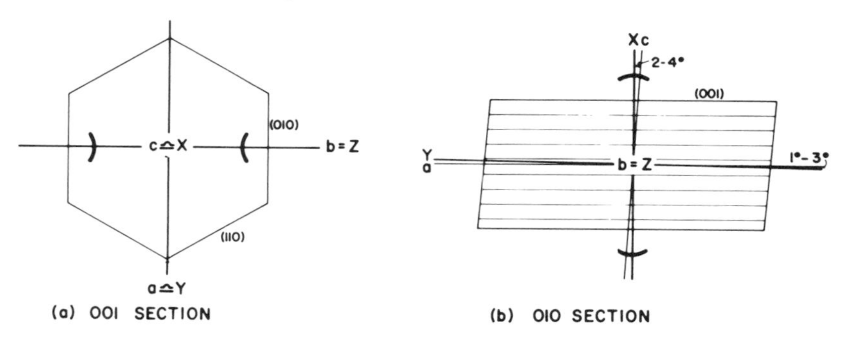

tled); if α has a low value, this section "twinkles" on rotation of stage.
(100) section: since Y ca. = Z, this section is almost identical to the (010) section; δ up to 0.049.
Optic-axis figures obtained in sections slightly oblique to (001), and in which cleavage is not visible.

Occurrence. Widespread in regionally metamorphosed rocks, especially those derived from pelitic sediments. At high grades, muscovite may break down to form sillimanite and K-feldspar; sillimanite frequently occurs as swarms of minute fibers embedded in muscovite. Phengite is common in high-pressure low-temperature metamorphic rocks. Muscovite is less common in thermally metamorphosed rocks. It is a common constituent of granites, pegmatites, and greisens, but is unstable in the volcanic environment of high temperatures and low pressures. It is resistant to weathering, and is therefore a common detrital mineral. It may also form during diagenesis. Many minerals (especially feldspar, nepheline, andalusite, kyanite, and cordierite) alter to fine-grained muscovite (sericite). Some sericite may, however, be paragonite or illite.

Notes. Sections of muscovite displaying cleavage and high interference colors are easily identified, but basal sections are often overlooked or misidentified. Loose grains of mucovite always lie on their cleavage, do not display high interference colors, and do not allow measurement of α. Rocks cut normal to a schistosity defined by muscovite will seldom display basal sections.

Distinguishing features (see Table 9.2). Perfect single cleavage, straight extinction (mottled), slow along, high δ, and lack of color are characteristic. Fuchsite is usually prominent as a green mineral in hand specimen. Muscovite is easily confused with phlogopite, talc, paragonite, the Li-micas, and pyrophyllite. Phlogopite and talc have a small $2V$, but may still be confused with some phengites. Talc and phlogopite have distinctive occurrences, and talc is soft and has a soapy feel in hand specimen. Pyrophyllite has a higher $2V$. Lepidolite may have a lower RI than muscovite, is often pink or purple in hand specimen, and occurs only in pegmatites and associated rocks. Paragonite is optically indistinguishable from muscovite. Chemical, X-ray, or percussion tests may be necessary to distinguish these minerals. There is a stain test to distinguish muscovite from paragonite (Laduron, 1971).

No. 48. PARAGONITE (−ve) $Na_2Al_4[Si_6Al_2O_{20}](OH)_4$

Monoclinic, $\beta = 95°$. Perfect {001} cleavage. Usually fine-grained aggregates of crystals with {001} dominating.

Color in thin section. Colorless.

Table 9.2 Summary of Distinctions Between Some of the Phyllosilicates

Mineral	Distinctions	
A. *With mottled extinction and high birefringence*		
Muscovite (47)	colorless; usually larger $2V$ than talc and phlogopite; smaller $2V$ than pyrophyllite; difficult to distinguish from paragonite and lepidolite; *var.* fuchsite is green in hand specimen	
Paragonite (48)	cannot be distinguished optically from muscovite	
Lepidolite (49)	similar to muscovite, but usually pink or violet in hand specimen, and restricted in occurrence	
Phlogopite and biotite (50)	brown or green pleochroic except for some colorless phlogopite which has a restricted occurrence; $2V$ very small; much higher δ than chlorite	
Talc (55)	colorless; usually smaller $2V$ than muscovite; distinctive occurrence and soapy feel in hand specimen	
Pyrophyllite (54)	colorless; distinguished from micas by its larger $2V$	
B. *With less pronounced mottled extinction and low birefringence*		
Chlorite (52)	green (or yellow) pleochroic; often anomalous interference colors	
Serpentine (53)	not so green or pleochroic as chlorite; always normal interference colors; some serpentine is fibrous	
C. *Without mottled extinction*		
Stilpnomelane (56)	similar to biotite	less perfect {001} cleavage than in the other phyllosilicates
Margarite (57) and clintonite (58)	δ moderate, between that of the normal micas and the chlorites	
Prehnite (59)	similar to muscovite, but fast along and +ve	

Optical properties. Biaxial −ve. $2V_x = 0°–46°$. $r > v$ (distinct).

$$\alpha = 1.564–1.580,\ \beta = 1.594–1.609,\ \gamma = 1.600–1.609.$$

$$\delta = 0.028–0.038.\ X \wedge c = \text{ca. } 5°,\ Y = \text{ca. } a,\ Z = b.$$

Orientation diagrams. Same as for muscovite (No. 47).

Occurrence. Occurs in schists and sediments. Some "sericite" may be paragonite. Because of the difficulty of distinguishing this mineral from muscovite, it may be more widespread than generally believed.

Distinguishing features (see Table 9.2). Same as for muscovite. Cannot be distinguished optically from muscovite, and chemical, X-ray, or staining methods (Laduron, 1971) must be used.

No. 49. LEPIDOLITE (−ve) $K_2(Li,Al)_{5-6}[Si_{6-7}Al_{2-1}O_{20}](OH,F)_4$

Fe may be present in small quantities.
Monoclinic, $\beta = 100°$. Perfect {001} cleavage. As large tabular crystals or fine-grained aggregates with {001} dominating.

Color in thin section. Colorless.

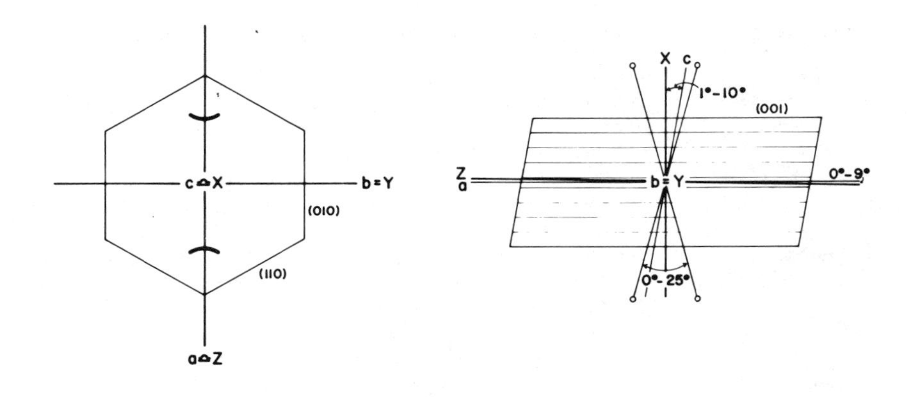

Figure 9.45 Orientation diagrams for biotite–phlogopite.

Color in detrital grains. Colorless; sometimes pink or lilac.

Optical properties. Biaxial −ve. $2V_x = 0°–58°$ (usually 30°–50°). $r > v$ (weak).

$$\alpha = 1.525–1.548,\ \beta = 1.548–1.585,\ \gamma = 1.551–1.587.$$

$$\delta = 0.018–0.039.\ X \text{ ca.} = c,\ Y = b,\ Z \text{ ca.} = a.$$

RI increases with Fe content.

Orientation diagrams. *(001) section and cleavage fragments:* similar to muscovite, but orientation of OAP is different (but same as for biotite—Figure 9.45a); orientation of OAP difficult to determine except in euhedral tablets or by percussion tests.
(010) and (100) sections: similar appearance to muscovite, but generally a slightly lower δ (up to 0.039).

Occurrence. Mainly in granite-pegmatites, aplites, and associated rocks.

Distinguishing features (see Table 9.2). Very similar to muscovite, and chemical or X-ray tests may be necessary to distinguish them. Lepidolite is, however, distinguished by its pink or purple color in hand specimen, and its restricted occurrence. It usually has a lower RI and δ than muscovite.

No. 50. BIOTITE (−ve) $K_2(Fe^{2+},Mg)_6(Fe^{3+},Al,Ti)_{0-2}[Si_{6-5}Al_{2-3}O_{20}]O_{0-2}(OH,F)_{4-2}$
and PHLOGOPITE (−ve) $K_2(Mg,Fe^{2+})[Si_6Al_2O_{20}](OH)_4$

See Figure 9.43 and Plates 7–12, 25–28, 31, 32, and 35–38.

Phlogopite forms a series with biotite. Phlogopite has Mg/Fe > 2/1.

Monoclinic, $\beta = 100°$. Perfect {001} cleavage. Often subhedral tabular with {001} dominating; euhedral crystals are tablets with {001}, {110}, and {010} forms. Twinning with {001} composition plane–twin plane is {110}.

Color in thin section. Biotite is strongly pleochroic, pale to dark brown or green with $X < Y \leqslant Z$. Phlogopite is colorless or pale brown or green. Reddish-brown varieties usually have a high Ti and Fe content, and green ones usually have a high Fe^{3+} content. Pleochroic haloes are common around inclusions of radioactive minerals such as zircon.

Color in detrital grains. As above but deeper colors.

Optical properties. Biaxial −ve. $2V_x = 0°–25°$ (usually <10°). $r < v$ (weak), rarely $r > v$.

$\alpha = 1.530–1.625$, $\beta = 1.557–1.696$, $\gamma = 1.558–1.696$.

$\delta = 0.028–0.078$. $X \wedge c = 1°–10°$, $Y = b$, $Z \wedge a = 0°–9°$. Occasionally $Z = b$, and Y ca. $= a$.

RI and δ increase generally with Ti and Fe, especially Fe^{3+}.

Orientation diagrams. *(001) section and cleavage fragments* (Figure 9.45a): acute bisectrix figure; δ' (very low) usually 0.001; since Y ca. $= Z$, pleochroism is weak or absent; cleavage not visible; extinction may be wavy due to bending of cleavages.
(010) section (Figure 9.45b): flash figure; δ (high or very high) up to 0.078; good cleavage, slow along, and almost straight extinction (mottled); strongly pleochroic with $Z > X$ in colored members; may display twin lamellae parallel to cleavage in members whose extinction angle is measurable.
(100) section: since Y ca. $= Z$, this section is identical to the (010) section, but always has straight extinction.
Optic-axis figures obtained on (001) sections in which cleavage is not visible.

Occurrence. *Biotite*—present in a very wide range of plutonic igneous rocks, but particularly common in granites. Also as phenocrysts (often showing signs of corrosion) in volcanic rocks. Fe content of biotite is generally higher in the more felsic igneous rocks. Very widespread in metamorphic rocks, especially metamorphosed pelites; forms at the onset of thermal metamorphism, but slightly later with the higher pressures of regional metamorphism. At high grades, it may break down to sillimanite, the sillimanite often being embedded in the biotite. It alters to chlorite, but nevertheless is common as a detrital mineral in sediments. *Phlogopite*—more restricted in occurrence. Most common in metamorphosed impure carbonate rocks, in ultrabasic rocks—especially kimberlite, and in some volcanics.

Notes. Sections of biotite and phlogopite displaying cleavage, pleochroism, and high interference colors, are easily identified, but basal sections are often overlooked or misidentified. Loose grains always lie on their cleavage, appear almost isotropic, and do not allow measurement of α. Rocks cut normal to a schistosity defined by biotite will seldom display basal sections.

Distinguishing features (see Table 9.2). Perfect single cleavage, almost straight extinction (mottled), length-slow nature, high δ, and pleochroism are characteristic. Biotite is not easily confused with other minerals, but basal sections may be confused with tourmaline (lacks cleavage) and hornblende (the confusing (100) sections of hornblende provide good figures indicating a moderate to large $2V$). Stilpnomelane is very similar to biotite, but has a less perfect {001} cleavage, does not display mottled extinction, and is "brittle" in hand specimen. Phlogopite may be confused with muscovite, talc, paragonite, pyrophyllite, and lepidolite. However, except for talc, phlogopite is distinguished by its usually smaller $2V$ and occurrence; talc is soft and soapy to the touch in hand specimen.

No. 51. GLAUCONITE (−ve)

$(K,Na,Ca)_{1.2-2.0}(Fe^{3+},Al,Fe^{2+},Mg)_4[Si_7AlO_{20}](OH)_4 \cdot n(H_2O)$

Monoclinic, $\beta = 100°$. Perfect {001} cleavage. As fine-grained aggregates in pellets.

Color in thin section. Green, pleochroic with $X < Y = Z$.

Optical properties. Biaxial −ve. $2V_x = 0°–20°$.

$\alpha = 1.585–1.616$, β ca. $= \gamma = 1.600–1.644$.

$\delta = 0.014–0.032$. $X \wedge c$ up to 10°, $Y = b$, Z ca. $= a$.

Orientation diagrams. Usually too fine grained to determine properties precisely.

Occurrence. Restricted to sediments. Forms as a diagenetic mineral under marine conditions, often as sand-sized pellets. A major constituent of greensands.

Distinguishing features. Its green color, fine-grained nature, and occurrence are distinctive. Distinguished from chlorite by its higher δ.

No. 52. CHLORITE (+ve or −ve) $(Mg,Fe,Al)_{12}[(Si,Al)_8O_{20}](OH)_{16}$

See Plates 11, 12, 25, 26, 29, and 30–32.

Mn or Cr may also be present.

Usually monoclinic with $\beta = 97°$. Perfect {001} cleavage. Euhedral–subhedral pseudohexagonal tabular crystals, scaly aggregates, or radiating aggregates, with {001} dominating. Twinning with {001} composition planes.

Color in thin section. Generally colorless, pale to dark green, pleochroic with $X = Y > Z$ (+ve varieties), or $Z = Y > X$ (−ve varieties); Mn-rich chlorites are usually orange, and Cr-rich chlorites, pink or violet in color.

Color in detrital grains. As above, but deeper colors in coarse material.

Optical properties. Biaxial +ve or −ve. $2V_z = 0°–70°$ or $2V_x = 0°–60°$. $r <$ or $> v$ (often strong).

$\alpha = 1.550–?$, $\beta = 1.550–1.70$, $\gamma = 1.550–?$.

$\delta = 0.00–0.020$. $X \wedge a$ (+ve) or $Z \wedge a$ (−ve) $< 3°$; $Y = b$, $Z \wedge \perp (001)$ (+ve) or $X \wedge \perp (001)$ (−ve) $< 3°$.

Terminology and correlation of chemistry with optics. The compositional and structural variation in the chlorites is wide ranging, and numerous varietal names have been adopted, as shown in Figure 9.46.

Figure 9.46 Terminology and variation of RI, birefringence, color, and optical sign with composition in the chlorites. Based on Saggerson and Turner (1982).

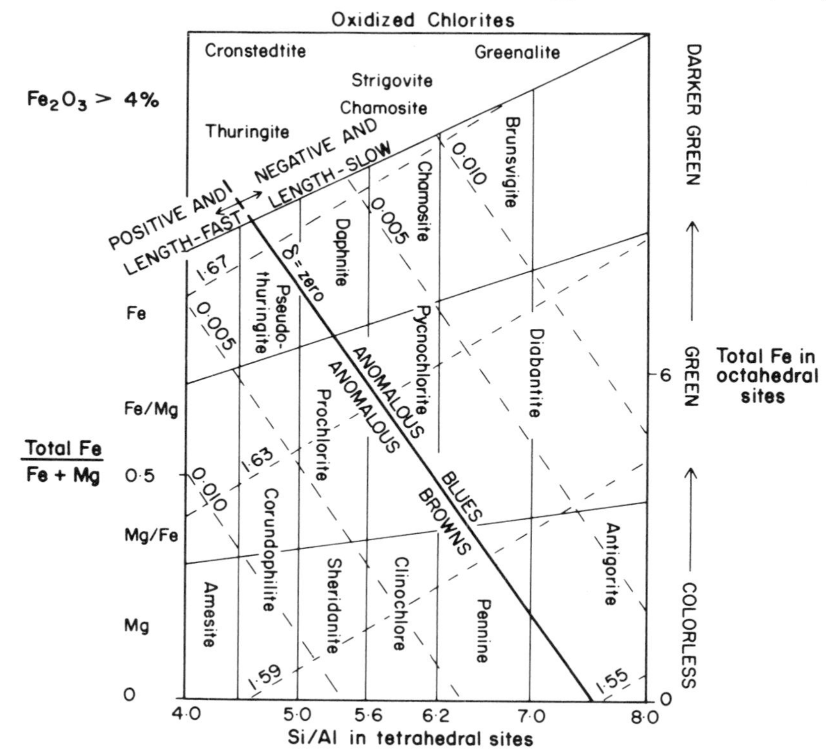

To classify the chlorites and use varietal name properly require both chemical and X-ray work (Phillips, 1964). However, Saggerson and Turner (1982), show how a combination of color and RI (both increase with Fe content), optic sign (+ve chlorites are Al–Mg rich, −ve chlorites Fe–Si rich), and birefringence (see Figure 9.46) enable one to specify approximately the type of chlorite. Note that +ve chlorites with a low birefringence usually display anomalous brown interference colors, whereas −ve low birefringence chlorites display anomalous blue colors. Note also that +ve chlorites are always length fast, −ve chlorites length slow.

Orientation diagrams. *(001) section and cleavage fragments* (Figure 9.47a): acute bisectrix figure; because of very low δ', sections may appear isotropic, and isogyres are difficult to observe; may display anomalous interference colors; pleochroism absent, and no cleavage visible.
(010) section (Figure 9.47b): flash figure; δ (very low to low) up to 0.020; often anomalous interference colors; good cleavage, fast along (+ve) or slow along (−ve); almost straight extinction (may be mottled); pleochroic with maximum absorption parallel to cleavage.
(100) section: almost identical to the (010) section.
Optic-axis figures obtained in basal or near-basal sections showing no cleavage or pleochroism—but isogyres usually difficult to observe.

Occurrence. Chlorite is widespread in metamorphic rocks of low grade, particularly chlorite-schists derived from pelites, and greenschists derived from basic igneous rocks. In both metamorphic and igneous rocks, chlorite is a common alteration product of pyroxene, amphi-

Figure 9.47 Orientation diagrams for chlorite.

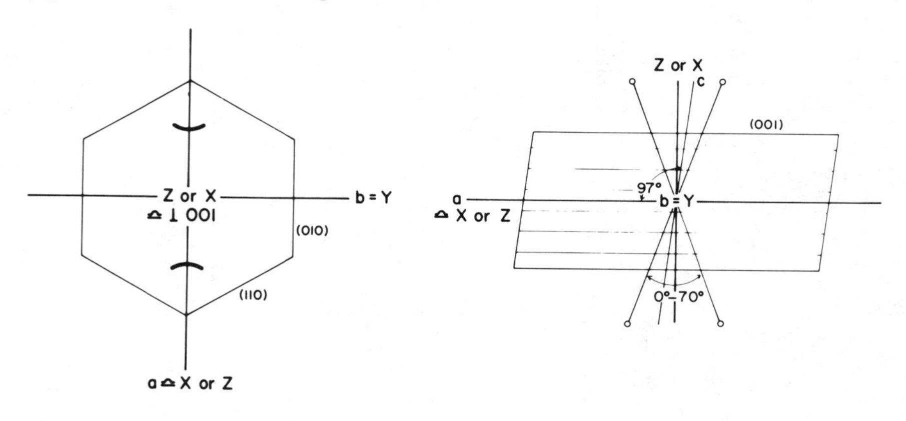

bole, and biotite, the composition of the chlorite being closely related to that of the original mineral. Chlorite is particularly characteristic of spilites and adinoles. Many amygdales in lavas are filled with chlorite, and chlorite is also a very common vein mineral, often associated with adularia and quartz. In sediments, chlorite is a very common detrital and authigenic mineral, often as mixed-layer structures, e.g., composite crystals made of chlorite and clay minerals. *Chamosite* is a name used for chlorite-forming ooliths in sedimentary iron formations; some chamosite is, however, now known to be a clay mineral. Mn-rich and Cr-rich chlorites are found in Mn- and Cr-rich rocks, respectively.

Distinguishing features (see Table 9.2). The good single cleavage, green pleochroism, low δ and anomalous interference colors are distinctive. The length-fast character of +ve chlorites distinguishes these varieties from all other sheet silicates except prehnite. Some pale-colored, low-RI chlorites with normal interference colors are essentially the same as serpentine (antigorite). Chloritoid may be mistaken for chlorite, but it has a higher RI, and a larger $2V$ and extinction angle. Glauconite has a much higher δ than chlorite.

No. 53. SERPENTINE (−ve) $Mg_3[Si_2O_5](OH)_4$

See Figure 9.48

Small amounts of Fe and Al are usually present.

As for the chlorites, an adequate characterization of the serpentine polymorphs is only possible using X-rays. Three varieties are normally distinguished: *chrysotile, lizardite,* and *antigorite*. Chrysotite is usually fibrous, lizardite platy, and antigorite fibrous or platy. Lizardite has a high Fe^{3+}/Fe^{2+} ratio. Antigorite is also a variety of chlorite.
Monoclinic, β ca. 90°–93°. Fine-grained aggregates; fibrous (asbestiform) usually parallel to a, or platy with a perfect {001} cleavage.

Color in thin section. Colorless to pale green.

Optical properties. Biaxial −ve. $2V_x$ ca. = 20°–60°. $r > v$ (weak) in antigorite.

$\alpha = 1.532–1.570$, $\gamma = 1.545–1.584$.

$\delta = 0.004–0.017$. X ca. $\perp$ (001) in antigorite and lizardite; Z is normally parallel to fiber length in chrysotile.

The lowest RI and highest δ are associated with chrysotile, the highest RI and lowest δ with antigorite and lizardite.

Orientation diagrams. *Platy serpentine:* very similar to −ve chlorite. Sections displaying cleavage have straight extinction and are slow along. (001) sections show no cleavage, and provide acute bisectrix

Figure 9.48 Serpentine (view measures 1.5 × 1.0 mm). Crossed-polarized light.

figures, but these are difficult to obtain due to the fine-grained nature of most serpentine. Interference colors are normal, and pleochroism weak or absent.
Fibrous serpentine: straight extinction, and usually length slow. Figures impossible to obtain because of fine-grained size of fibers.

Occurrence. Serpentine forms by the hydrothermal alteration of ferromagnesium minerals below 400–500° C, particularly olivine and pyroxene. Serpentine replaces olivine as an irregular anastomosing network (Figure 9.48), but more regular arrangements mimicking cleavage are formed from pyroxene and amphibole. Replacement is usually accompanied by the formation of small grains of opaque iron oxides. Ultramafic rocks, such as dunite, may be completely replaced, forming large rock masses of serpentinite. For a detailed discussion of serpentine textures see Wicks and Whittaker (1977).

Distinguishing features (see Table 9.2). The occurrence, platy or fibrous nature, low δ and RI are characteristic. The generally lower RI, weaker pleochroism and color, and lack of anomalous interference colors distinguish serpentine from most chlorite. Fibrous amphiboles have a higher RI.

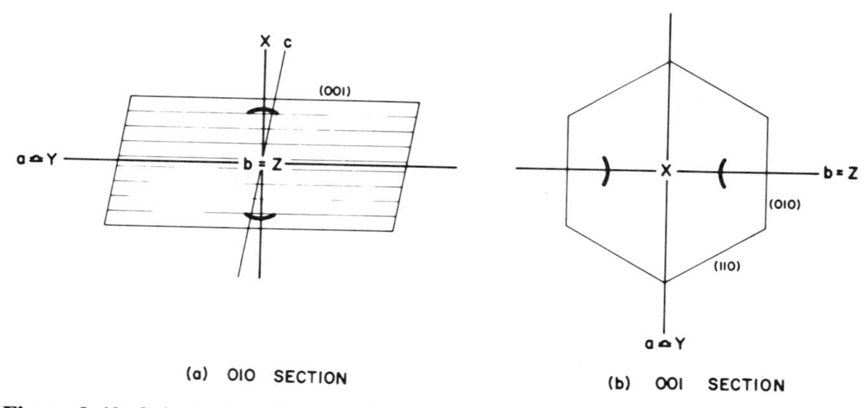

Figure 9.49 Orientation diagrams for pyrophyllite.

No. 54. PYROPHYLLITE (−ve) $Al_4[Si_8O_{20}](OH)_4$

Monoclinic (may also be triclinic), β = 100°. Perfect {001} cleavage. Subhedral tabular crystals parallel to {001}, or fine-grained aggregates, often as radiating groups.

Color in thin section. Colorless.

Optical properties. Biaxial −ve. $2V_x$ = 46°–62°. $r > v$ (weak).

α = 1.530–1.564, β = 1.586–1.592, γ = 1.596–1.601.

δ = 0.045–0.068. X ca. = ⊥ (001), Y ca. = a, Z = b.

Orientation diagrams. *(010) section* (Figure 9.49a): obtuse bisectrix figure; δ' (high) ca. 0.05; slow along good cleavage, and straight extinction.
(100) section: similar to (010) section, but flash figure.
(001) section (Figure 9.49b): acute bisectrix figure; δ' (low) ca. 0.01; no cleavage visible; extinction may be wavy due to bending of crystals.
Optic-axis figures are obtained in sections oblique to (001) that display no cleavage.

Occurrence. Not common. Found in low-grade schists, and as a sericitic hydrothermal alteration product of feldspars, kyanite, and andalusite.

Distinguishing features (see Table 9.2). Closely resembles muscovite and talc and other similar minerals. Muscovite and especially talc can be distinguished by their smaller $2V$, but with fine-grained material, chemical or X-ray tests may be necessary.

No. 55. TALC (−ve) $Mg_6[Si_8O_{20}](OH)_4$

Monoclinic, β = ca. 100°; more rarely triclinic. Perfect {001} cleavage. Usually in massive aggregates of flaky crystals with the {001} form dominating. Very soft, with a soapy feel in hand specimen.

Color in thin section. Colorless.

Optical properties. Biaxial −ve. $2V_x$ = 0°–30°. $r > v$ (weak).

$$\alpha = 1.539–1.550,\ \beta = 1.584–1.594,\ \gamma = 1.548–1.596.$$

$$\delta = 0.039–0.050.\ X \text{ ca.} = \perp (001),\ Y \text{ ca.} = a,\ Z = b.$$

Fe-rich varieties of talc are known with a higher RI (e.g., α = 1.580, γ = 1.615).

Orientation diagrams. As for muscovite (No. 47), but has a smaller $2V$.

Occurrence. As a hydrothermal replacement product of ultramafic rocks, especially serpentinite, in veins and as massive bodies (soapstone), often associated with magnesite, tremolite, anthophyllite, and chlorite. Also forms during the initial stages of the thermal metamorphism of impure dolomites.

Distinguishing features (see Table 9.2). The occurrence, perfect cleavage, straight extinction (mottled), length-slow nature, and high δ are characteristic, but talc is easily confused with muscovite, pyrophyllite, and other similar minerals. Talc has a smaller $2V$ than pyrophyllite and most muscovite, and a lower RI than most muscovite with a small $2V$ (phengite). The softness and soapy feel of talc in hand specimen are distinctive.

No. 56. STILPNOMELANE (−ve)

$(K,Na,Ca)_{0–1.4}(Fe,Mg,Al,Mn)_{5.9–8.2}[Si_8O_{20}](OH)_4(O,OH,H_2O)_{3.6–8.5}$

See Plates 31 and 32.
Triclinic, β = 96°. Perfect {001} and imperfect {010} cleavages. As thin platy crystals dominated by {001}, and with pseudohexagonal outlines.

Color in thin section. Pale to dark yellow, green or brown, pleochroic with $Z = Y > X$; Fe^{3+}-rich varieties are usually very dark brown, and Fe^{2+}-rich varieties, dark green in color.

Optical properties. Biaxial −ve. $2V_x$ = 0°–40° (usually ca. 0°).

$$\alpha = 1.543–1.634,\ \beta = \gamma = 1.576–1.745.$$

$$\delta = 0.033–0.111.\ X \text{ ca.} = \perp (001),\ Y \text{ ca.} = b,\ Z \text{ ca.} = a.$$

RI and δ increase with Fe^{3+} content.

Orientation diagrams. *(001) section* (Figure 9.50a): acute bisectrix figure, often pseudo-uniaxial; δ' (very low) ca. zero (appears isotropic); poor (010) cleavage parallel to OAP may be visible; maximum absorption in this section, no pleochroism.
(010) section (Figure 9.50b): flash figure; δ (high to very high) up to 0.111; good cleavage, straight extinction and length slow; pleochroism marked, with $Z > X$.
(100) section: similar to (010) section, but may show a poor cross-cleavage parallel to (010).

Occurrence. Relatively common as fine grains in low-grade schists and greenschists, especially those relatively rich in Fe and Mn. Also reported in xenoliths in igneous rocks.

Distinguishing features (see Table 9.2). The strong pleochroism, straight extinction, habit, high δ, and small $2V_x$ make stilpnomelane difficult to distinguish from green or brown biotite. However, the {001} cleavage of stilpnomelane is less perfect than that of biotite, and there may be, in addition, a very poor {010} cleavage. Mottled extinction is absent in stilpnomelane, and the flakes are not flexible as in biotite. Stilpnomelane may superficially resemble chlorite and chloritoid, but its δ is much higher.

No. 57. MARGARITE (−ve) $Ca_2Al_4[Si_4Al_4O_{20}](OH)_4$

A "brittle mica" with a muscovite-like structure, but with Ca instead of K.
Monoclinic, $\beta = 95°$. Perfect {001} cleavage. Subhedral platy crystals dominated by {001}. Twinning with {001} composition plane.

Figure 9.50 Orientation diagrams for stilpnomelane.

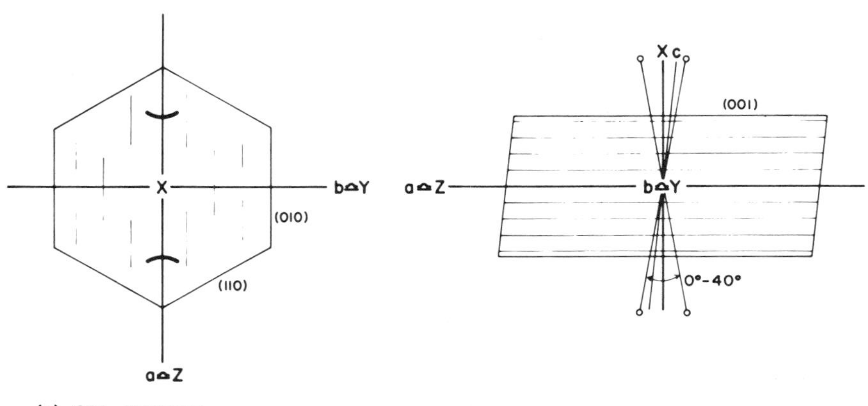

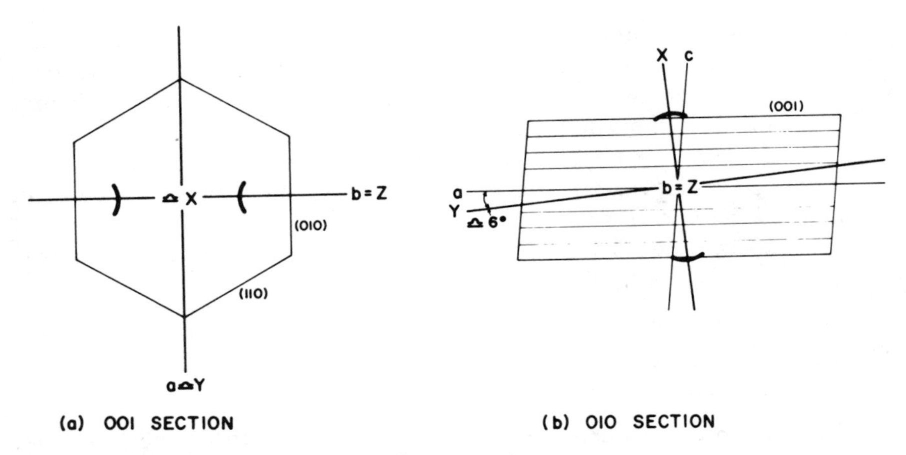

Figure 9.51 Orientation diagrams for margarite.

Color in thin section. Colorless.

Optical properties. Biaxial −ve. $2V_x = 40°–67°$. $r < v$ (distinct).

$\alpha = 1.630–1.638$, $\beta = 1.642–1.648$, $\gamma = 1.644–1.650$.

$\delta = 0.012–0.014$. $X \wedge \perp (001) =$ ca. 6°, $Y \wedge a =$ ca. 6°, $Z = b$.

Orientation diagrams. *(001) section* (Figure 9.51a): acute bisectrix figure; δ' (very low) = 0.002; indistinct extinction positions; no cleavage visible.
(010) section (Figure 9.51b): obtuse bisectrix figure; δ' (low) up to 0.012; good cleavage with slightly inclined extinction; slow along.
(100) section: similar to (010) section, but flash figure and straight extinction.
Optic-axis figures obtained in sections slightly oblique to (001).

Occurrence. Usually associated with corundum in emery deposits or veins, but also in some low-grade schists.

Distinguishing features (see Table 9.2). Margarite may be confused with muscovite, talc, and −ve chlorite. Muscovite and talc have a lower RI, a higher δ, and more flexible cleavages; −ve chlorite is distinguished by its green color, and usually by its lower δ.

No. 58. CLINTONITE (−ve) $Ca_2(Mg,Fe)_{4.6}Al_{1.4}[Si_{2.5}Al_{5.5}O_{20}](OH)_4$

A "brittle mica" with a biotite-like structure, but with Ca instead of K.
Monoclinic, $\beta = 100°$. Perfect {001} cleavage. Subhedral platy crystals with {001} dominating. Twinning with {001} composition plane.

Color in thin section. Colorless, pale green or brown, pleochroic with $Z = Y > X$.

Optical properties. Biaxial −ve. $2V_x = 2°–40°$. $r < v$ (weak).

$$\alpha = 1.643–1.649,\ \beta = 1.655–1.662,\ \gamma = 1.655–1.663.$$

$$\delta = 0.012–0.014.\ X \text{ ca.} = \perp (001),\ Y \text{ ca.} = a,\ Z = b.$$

The related mineral *xanthophyllite* has X ca. $= \perp$ (001), $Y = b$, Z ca. $= a$.

Orientation diagrams. Similar to those for margarite (No. 57).

Occurrence. Not common. Found in chlorite-schists with talc, and in metasomatically altered limestones.

Distinguishing features (see Table 9.2). Distinguished from the micas by its lower δ and less flexible cleavage. −ve chlorite usually has a lower δ, and chloritoid a higher RI.

No. 59. PREHNITE (+ve) $Ca_2Al[AlSi_3O_{10}](OH)_2$

See Plates 17 and 18.

Fe may substitute for Al.

Orthorhombic. Distinct {001} and poor {110} cleavages. Crystals are usually tabular parallel to (001) and are often arranged in radiating or sheaflike aggregates; also prismatic parallel to c. Very fine lamellar twinning on {100} may be present.

Color in thin section. Colorless.

Optical properties. Biaxial +ve. $2V_z = 65°–69°$. $r > v$ (weak); or $r < v$ (weak to strong).

$$\alpha = 1.611–1.632,\ \beta = 1.615–1.642,\ \gamma = 1.632–1.665.$$

$$\delta = 0.021–0.035.\ X = a,\ Y = b,\ Z = c.$$

Orientation diagrams. *(001) section* (Figure 9.52a): acute bisectrix figure; δ' (low) < 0.01; the poor {110} cleavages may be visible.
(010) section (Figure 9.52b): flash figure; δ (moderate to high) up to 0.035; distinct cleavage, straight extinction and fast along.
(100) section: straight extinction and fast along; similar to the (010) section, but a slightly lower δ', and provides an obtuse bisectrix figure.
Optic-axis figures are obtained on sections oblique to (001) showing no good cleavage.

Occurrence. Commonly found in veins and amygdales in basic volcanic rocks. Forms as a result of the hydrothermal alteration of Ca-bearing minerals such as plagioclase and amphibole. Also found in thermally

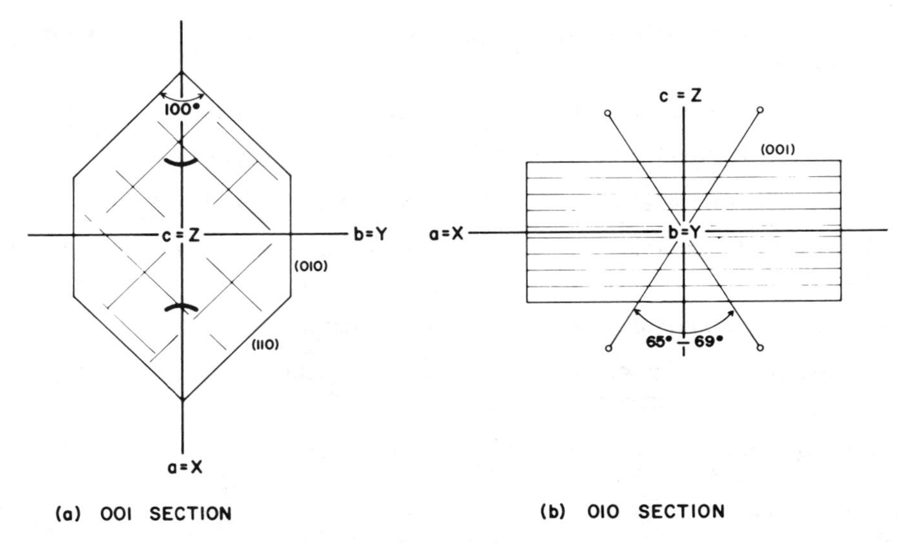

Figure 9.52 Orientation diagrams for prehnite.

metamorphosed impure carbonate rocks, and is a characteristic mineral in a wide variety of rocks affected by the low-grade prehnite–pumpellyite facies of regional metamorphism, and in veins cutting such rocks.

Distinguishing features (see Table 9.2). The moderate RI, high δ, good single cleavage, straight extinction, and +ve character are distinctive. It may be confused with muscovite, especially in metamorphic rocks, but is distinguished by its length-fast character, the lack of mottled extinction, and the less distinct cleavage.

No. 60. APOPHYLLITE (+ve or −ve) $KFCa_4[Si_8O_{20}]8H_2O$

Tetragonal. Perfect {001} and imperfect {110} cleavages. Often stubby euhedral prismatic crystals with {100} well developed; also anhedral.

Color in thin section. Colorless.

Optical properties. Uniaxial +ve, more rarely −ve.

$$\omega = 1.5335\text{–}1.5445,\ \varepsilon = 1.5352\text{–}1.5439.$$

$$\delta = 0.00\text{–}0.002.$$

Interference colors are often anomalous.
Occasionally biaxial +ve, with complex twinning.

Occurrence. Fairly common in cavities and amygdales in basic volcanic rocks, usually associated with zeolites.

Distinguishing features. The RI (approximately the same as that of Canada balsam), uniaxial character, anomalous interference colors, very low δ, and occurrence are characteristic. Associated zeolites have either a lower RI or a higher δ.

The Clay Minerals

Clay is used as a rock term, as a qualification of grain size, and also for a group of hydrous aluminium sheet-silicates characteristically found in clays, argillites, soils, and related rocks. Sheet silicates with a more widespread occurrence, such as chlorite and mica, may also be present in clay grain-size material.

Because of the very fine grain size of most clay minerals, they do not lend themselves to optical study. To study them properly requires the use of X-rays, differential thermal analysis, electron microscopy, and chemistry. The description of the clay minerals given here will therefore be brief. Further details can be found in books such as Grim (1968) and Brindley and Brown (1980).

No. 61. KAOLINITE GROUP

includes the following species:

No. 61A. Kaolinite $Al_2[Si_2O_5](OH)_4$

No. 61B. Dickite, a polymorph of kaolinite.

No. 61C. Nacrite, a polymorph of kaolinite.

No. 61D. Halloysite, a hydrated form.

No. 62. SMECTITE GROUP

includes the following species:

No. 62A. Montmorillonite $(\frac{1}{2}Ca,Na)_{0.66}(Al,Mg)_4[Si_8O_{20}](OH)_4 \cdot nH_2O$

No. 62B. Beidellite $(\frac{1}{2}Ca,Na)_{0.66}Al_4[(Si,Al)_8O_{20}](OH)_4 \cdot nH_2O$

No. 62C. Nontronite $(\frac{1}{2}Ca,Na)_{0.66}Fe^{3+}_4[(Si,Al)_8O_{20}](OH)_4 \cdot nH_2O$

No. 62D. Saponite $(\frac{1}{2}Ca,Na)_{0.66}Mg_6[(Si,Al)_8O_{20}](OH)_4 \cdot nH_2O$

No. 62E. Hectorite $(\frac{1}{2}Ca,Na)_{0.66}(Mg,Li)_6[Si_8O_{20}](OH)_4 \cdot nH_2O$

No. 63. ILLITE GROUP $K_{1-1.5}Al_4[Si_{7-6.5}Al_{1-1.5}O_{20}](OH)_4 \cdot (+H_2O?)$

The status of this group is uncertain. Illite is essentially a mica (also known as hydromuscovite). It differs from the micas in having less substitution of Al for Si and possibly being hydrated. Some "sericite" may be illite.

No. 64. VERMICULITE GROUP

$(Mg,Ca)_x(Mg,Fe)_6[(Si_{8-x},Al_x)O_{20}](OH)_4 \cdot 8H_2O$ where $x = 1\text{–}1.4$

No. 65. PALYGORSKITE-SEPIOLITE GROUP

includes the following species:

No. 65A. Palygorskite $(Mg,Al)_4[Si_8O_{20}](OH)_3(OH_2)_3 \cdot 4H_2O$

No. 65B. Sepiolite $(Mg,Al,Fe^{3+})_8[Si_{12}O_{30}](OH)_4(OH_2)_4 \cdot 8H_2O$

No. 66. ALLOPHANE

A term used for amorphous clay material.

Crystallography and optical properties. Most of the clay minerals are monoclinic, though kaolinite is triclinic and some palygorskite-sepiolite is orthorhombic. Table 9.3 summarizes the optical properties and crystal habits of the clay minerals—the data coming mainly from Grim (1968). It should be noted that properties such as RI may change during the drying of a sample, or as a result of absorption of immersion liquids during RI measurement. Identification is also hindered by the commonly mixed character of clay samples, and interlayering of different clay structures within individual crystals.

Occurrence. The factors governing the formation of particular clay-mineral species are exceedingly complex. There are, in general, three principal modes of occurrence: (1) as a result of hydrothermal alteration; (2) as a result of weathering, and in soils; (3) in sediments. Acid conditions appear to favor the formation of kaolinite, whereas alkaline conditions favor the smectites and illite. Illite is favored by high concentrations of K, and may be derived from muscovite. Smectites are a characteristic alteration product of basic volcanic rocks. Vermiculite is often coarser grained than the other clays, and forms pseudomorphs after biotite. Palygorskite and sepiolite are fibrous alteration products of Mg-rich minerals and rocks; they are often found in saline deposits and along cave walls. The clay minerals in a sediment reflect not only the parent soils and weathered rocks, but the conditions during and subsequent to deposition which may bring about diagenetic changes.

Distinguishing features. Apart from coarse-grained pseudomorphs of vermiculite after biotite, the clay minerals are characterized by their very fine grain size, and their occurrence in clays, soils, and weathered or altered rocks. Optical determination is not often possible. Members of the kaolinite group are distinguished from the other clay minerals by their low δ. Vermiculite and nontronite are colored and pleochroic. Vermiculite differs from biotite in its weaker pleochroism and smaller δ.

Table 9.3 Optical Properties and Crystal Habits of the Clay Minerals

Mineral	α	γ	δ	$2V$ and sign	Habit and other optics
Kaolinite (61A)	1.553–1.563	1.560–1.570	0.006–0.007	24°–50° (−ve)	{001} flakes; $X \wedge \perp (001) = 3°$, Z ca. $= b$
Dickite (61B)	1.560–1.562	1.566–1.571	0.006–0.009	52°–80° (+ve)	{001} flakes; $X \wedge c = 15°–20°$, $Z = b$
Nacrite (61C)	1.557–1.560	1.563–1.566	0.006	40° (−ve)	{001} flakes; $X \wedge \perp (001) = 10°–12°$, $Z = b$
Halloysite (61D)	mean value = 1.526–1.556		0.002–0.001	?	elongate tubes
Montmorillonite and Beidellite (62A,B)	1.48 –1.57	1.50 –1.60	0.020–0.030	0°–30° (−ve)	usually flake aggregates; X ca. $= \perp (001)$
Nontronite (62C)	1.565–1.60	1.60 –1.640	0.035–0.040	moderate −ve	flakes or rods; yellow, brown, green, pleochroic
Saponite (62D)	1.480–1.490	1.510–1.525	0.030–0.035	moderate −ve	flakes; X ca. $= \perp (001)$
Hectorite (62E)	1.485	1.516	0.031	small −ve	laths; X ca. $= \perp (001)$
Illite (63)	1.545–1.63	1.57 –1.67	0.022–0.055	small −ve	flakes; X ca. $= \perp (001)$
Vermiculite (64)	1.525–1.56	1.545–1.585	0.020–0.030	small −ve	flakes; X ca. $= \perp (001)$; green brown, pleochroic
Palygorskite (65A)	1.50 –1.52	1.54 –1.555	0.020–0.035	0°–60° (−ve)	fibers, length-slow; may be yellow, pleochroic
Sepiolite (65B)	1.49 –1.522	1.505–1.530	0.009–0.020	20°–70° (−ve)	fibers, length-slow; may be yellow, pleochroic
Allophane (66)	n = 1.468–1.512		isotropic		amorphous

Flakes of the clay minerals are always length slow. Palygorskite and sepiolite are characterized by their fibrous habit.

E. Tectosilicates

In the tectosilicates, all four oxygen atoms of each [SiO_4] tetrahedral group are shared with adjacent tetrahedra so that a framework of general formula [SiO_2] is formed. The silica minerals (including quartz) are composed solely of such frameworks, and have the formula SiO_2. In the other tectosilicates, some of the Si is replaced by Al, and other elements such as Na, K, and Ca balance the electrical charges in the formulae. Tectosilicates described here are: the silica minerals; the feldspars; the feldspathoids; the zeolites; and scapolite.

The Silica Minerals

SiO_2 occurs naturally as a number of polymorphs at which the most important and abundant is trigonal low quartz. At atmospheric pressure, low quartz rapidly transforms above 573° C to the closely similar hexagonal high quartz. High quartz always inverts to low quartz below 573° C. Above 870°, high quartz may transform to hexagonal tridymite, which in turn transforms above 1470° C to cubic cristobalite (Frondel, 1962). These transformations are sluggish and do not always take place, either with increasing or decreasing temperature. Hence, unstable low forms of tridymite and cristobalite may occur at low temperatures. The structures of tridymite and cristobalite are very open, and they readily accept impurities, and it is found that, in some cherts (Lancelot, 1973) and hydrothermal deposits (e.g., siliceous sinters), low forms of tridymite and cristobalite grow in the presence of impurities at low temperatures. At very high pressures (in excess of 20 kilobars) the polymorphs *coesite* and *stishovite* may form. These high-pressure polymorphs are found naturally in meteorite craters; they are not described here.

In addition to the above polymorphs, quartz may occur as the microcrystalline or fibrous form known as chalcedony; chalcedony differs from quartz in properties such as RI due to the admixture of varying quantities of water and other impurities. Opal is a form of hydrous silica which consists of cryptocrystalline low cristobalite or, in the precious variety, regularly packed spheres of amorphous silica.

No. 67. QUARTZ (+ve) [SiO_2]

See Figures 5.4 and 9.53 and Plates 7, 8, 16–18, 21–26, 29, and 30. Trigonal. No good cleavage. In volcanic rocks, quartz phenocrysts are euhedral bipyramidal hexagonal prisms (Figure 9.53a)—originally hexagonal high quartz; in quartz veins, crystals may be euhedral singly-

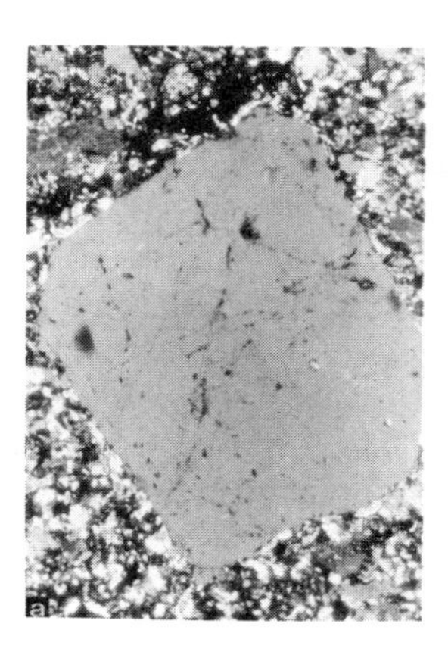
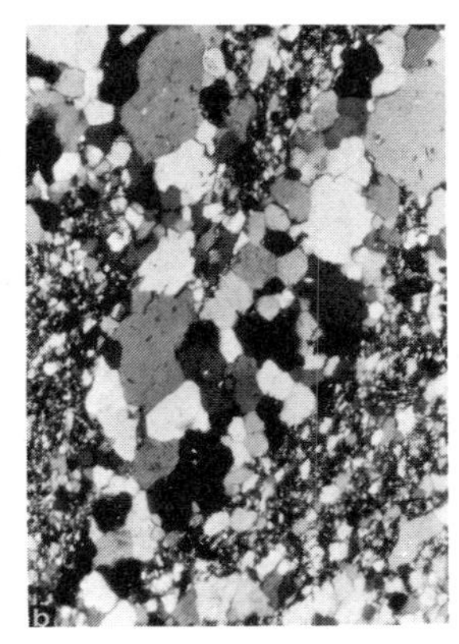

Figure 9.53 Quartz: **(a)** typical section of bipyramidal quartz in dacite (view measures 1.7 × 1.2 mm); **(b)** mosaic of recrystallized quartz grains in biotite schist, Taipo Valley, New Zealand (view measures 1.8 × 1.3 mm); **(c)** deformation bands and undulatory extinction, Greenland Group, New Zealand (view measures 1.8 × 1.3 mm). Crossed-polarized light.

terminated pseudohexagonal elongate prisms; quartz is commonly anhedral, filling the gaps between other crystals in igneous rocks or forming mosaics in metamorphic rocks (Figure 9.53b). Parallel twinning on the Dauphiné (twin axis = *c*-axis) and Brazil (twin plane = $\{11\bar{2}0\}$) laws very common, but cannot be observed in thin sections, since the *c*-axes of the twins are parallel (Frondel, 1962).

Color in thin section. Colorless.

Color in detrital grains. Usually colorless, but may be weakly colored in a wide variety of tints.

Optical properties. Uniaxial +ve (rarely anomalously biaxial).

$$\omega = 1.544, \varepsilon = 1.553.$$

$$\delta = 0.009.$$

Euhedral prismatic crystals have straight extinction, and are length slow. Bipyramidal crystals in volcanic rocks have symmetrical extinction.

Deformation and recrystallization. Quartz deforms easily by translation gliding and Dauphiné twinning (Tullis et al., 1973), especially in the presence of water which weakens the bonding in the SiO_2 structure. Translation gliding may take place on several systems, but most importantly parallel to the base {0001} and the prisms, and less importantly parallel to the rhombohedra. Translation allows bending, slippage, and rotation within the quartz grains. Evidence that translation has taken place may be provided by *deformation lamellae,* which are narrow, closely spaced, subplanar features with a slightly different RI from the host quartz. Frequently, however, such evidence is destroyed during subsequent recrystallization, which involves the nucleation of relatively strain-free grains and the growth of these at the expense of the deformed quartz; such recrystallization may result in the development of a mosaic characterized by strain-free grains with subplanar boundaries and triple-point junctions (Figure 9.53b). *Deformation bands* and *undulatory extinction* are very commonly observed in quartz (Figure 9.53c), and probably result from a process of polygonization, that is, the migration of dislocations (resulting from translation) so that they are concentrated in zones perpendicular to translation planes; in other words, they represent a partial recovery following deformation. Since basal slip is the most common translation system, deformation bands are most commonly parallel to the *c*-axis of quartz. Deformation and recrystallization often result in the development of a preferred orientation of the *c*-axes of quartz (Tullis et al., 1973; Lister et al., 1978; Shelley, 1980); a quantitative assessment of this is made using the U-stage or X-rays, but a qualitative assessment can be made in thin section by using the sensitive-tint plate, and observing the preferred orientation of the fast directions ($= \omega$, $\perp$ *c*-axis) in a quartz mosaic.

Occurrence. Quartz is highly resistant to weathering, and is an abundant detrital mineral being the prime constituent of many sandstones. In addition, the secondary diagenetic formation of quartz is common. Many cherts are made primarily of quartz. It is stable throughout the entire range of metamorphism, and is important in a wide variety of metamorphic rocks; it is the principal constituent of quartzite. It is found in a wide variety of igneous rocks, and is an essential component of some such as granite and rhyolite. Phenocrysts of quartz in volcanic rocks are usually bipyramidal, and originated as the high-temperature hexagonal form; such phenocrysts are commonly corroded. Quartz is often intergrown with feldspar in igneous rocks to form graphic and granophyric intergrowths and myrmekite (see under feldspars). It does not occur in igneous rocks containing feldspathoids. Quartz is also an important hydrothermal mineral occurring in veins.

Distinguishing features. Quartz is distinguished by its lack of color, cleavage, and visible twinning, by its low relief and δ, by its uniaxial +ve character, and by the lack of alteration. It is most commonly confused with the feldspars or cordierite. Both these minerals are commonly twinned or altered and in the case of feldspar, cleaved, but in the absence of these features, distinction can be difficult. The RI of cordierite and feldspar may be distinctive, but Ca-oligoclase, Na-andesine, and some cordierite have the same RI as quartz. The uniaxial +ve character of quartz is diagnostic, but if quartz and feldspar or cordierite occur together, it may be necessary to use U-stage (Chapter 6) or staining techniques (see under feldspars and cordierite) to quantitatively assess the *amount* of quartz present. Less commonly, beryl, scapolite, and nepheline may be confused with quartz, but all these minerals are uniaxial −ve, and beryl, nepheline, and most scapolite differ in their RI.

No. 68. CHALCEDONY (+ve) $[SiO_2]$ + <10% H_2O

Essentially microcrystalline quartz with loosely held water, but there may be some substitution of OH for O in the $[SiO_4]$ tetrahedra. Crystals are of quartz; microcrystalline and often fibrous, the fibers being elongate parallel to $[11\bar{2}0]$, or less commonly $[10\bar{1}0]$ or the *c*-axis [0001].

Color in thin section. Colorless.

Optical properties. Uniaxial +ve, but aggregates often give biaxial +ve interference figures.

$$\omega = 1.526\text{–ca. } 1.544,\ \varepsilon = 1.532\ (?)\text{–ca. } 1.553.$$

$$\delta = 0.005\text{–ca. } 0.009.$$

Optical properties may be difficult to observe due to fine grain size. Fibers are fast or, less commonly, slow along.

Occurrence. Chalcedony occurs filling vesicles and cavities in igneous rocks, in veins and sinter deposits, in cherts, and replacing a variety of rocks, particularly volcanics and those rich in carbonates. Well-known varieties of chalcedony include *agate* (banded), *jasper* (red), *flint* (massive, nodular) and *chert*. Chalcedony often grades into quartz within a specimen.

Distinguishing features. The occurrence, fibrous or microcrystalline nature, low relief and δ, and lack of alteration are distinctive. Microcrystalline chalcedony is very similar to the felsitic groundmass of acid volcanic rocks. A felsitic groundmass is composed, however, of alkali-feldspar and quartz, and the presence of these two minerals is easily detected by closing the substage diaphragm and observing Becke lines.

No. 69. TRIDYMITE (+ve) $[SiO_2]$ + Al, Na, etc. impurities?

High-temperature tridymite is hexagonal, low-temperature tridymite orthorhombic. No good cleavage. Crystals are usually very small hexagonal or pseudo-hexagonal {0001} plates. Wedge-shaped or sector twins common.

Color in thin section. Colorless.

Optical properties. Usually biaxial +ve. $2V_z = 30°–90°$.

$$\alpha = 1.468–1.479,\ \beta = 1.469–1.480,\ \gamma = 1.473–1.483.$$

$$\delta = 0.002–0.007 \text{ (usually } <0.005\text{)}.\ Z = c.$$

Platelike crystals are fast along.

Occurrence. In cavities, vesicles, the groundmass, or less commonly as phenocrysts in volcanic rocks, especially rhyolites, trachytes, and andesites. Also occurs in siliceous sinters and high-grade thermally metamorphosed sandstones. Some cherts contain tridymite (Oehler, 1973), and some workers (Lancelot, 1973) believe such deposits to form instead of quartz-cherts where impurities contributed by clays are readily available.

Distinguishing features. The low RI and δ, the small platelike crystals with twinning, and occurrence are distinctive. Tridymite has a slightly lower RI than cristobalite. Some of the zeolites (e.g., chabazite and heulandite) are similar, but usually have a higher RI and occur in association with basic igneous rocks.

No. 70. CRISTOBALITE (−ve) [SiO_2] + Al, Na, etc. impurities?

High-temperature cristobalite is cubic; low-temperature cristobalite is tetragonal. No cleavage. Euhedral crystals are usually octahedra; also spherulitic; in cherts, cristobalite may form spherules consisting of tridymite-like plates. Penetration twins may result during inversion from the high to low form.

Color in thin section. Colorless.

Optical properties. Uniaxial −ve or pseudo-isotropic.

$$\omega = 1.487,\ \varepsilon = 1.484.$$

$$\delta = \text{ca. } 0.003.$$

Occurrence. In cavities, vesicles, and the groundmass of volcanic rocks, especially rhyolites, trachytes, and andesites. Also found in siliceous sinters. In the presence of impurities contributed by clay minerals, cherts may develop cristobalite and tridymite (Lancelot, 1973, Oehler, 1973) instead of quartz at low temperatures.

Distinguishing features. The low RI, the very low δ, and occurrence are distinctive. The RI is slightly higher than that of tridymite. Some zeolites (e.g., chabazite) are similar, but they generally occur in basic volcanics.

No. 71. OPAL (isotropic) $[SiO_2] + <20\% H_2O$

Cryptocrystalline low cristobalite, or amorphous (precious varieties) (Jones et al., 1964).

Color in thin section. Usually colorless, but may be pale colored, especially brown.

Optical properties. Isotropic.
n = ca. 1.41–1.47, usually ca. 1.435–1.455,
RI increases with decrease in H_2O.

Occurrence. A secondary mineral in veins and cavities in igneous rocks, in sediments neighboring areas of volcanic activity, and in siliceous sinters. May be derived from organic remains, and the prime constituent of diatomite and some cherts. Also in other sediments, often as concretions or replacing fossils, and usually derived from siliceous organic remains, or as a result of weathering under arid conditions. Opal may grade into chalcedony and quartz.

Distinguishing features. The low RI (moderate relief), isotropic character, and lack of crystal shape and cleavage are distinctive.

The Feldspars

Feldspars are the most common rock-forming minerals in the earth's crust. They are extremely abundant in most igneous rocks, and the presence or absence of particular feldspar species is the basis of igneous rock classification. Feldspars are also important components of many metamorphic rocks and are common in some sediments, especially sandstones.

The feldspars are divided into two main solid solution series: (1) the alkali-feldspars, a series from $KAlSi_3O_8$ (Or) to $NaAlSi_3O_8$ (Ab); and (2) the plagioclase-feldspars, a series from $NaAlSi_3O_8$ (Ab) to $CaAl_2Si_2O_8$ (An). There is no solid-solution series between $KAlSi_3O_8$ and $CaAl_2Si_2O_8$. Rare Ba-feldspars (*hyalophane* and *celsian*) exist but will not be considered further.

The alkali- and plagioclase-feldspar series are subdivided further according to the particular proportions of the molecules Or, Ab, and An, and according to their structural state. The structural state is a reflection of thermal history, and allows feldspars to be subdivided into "high-" or "low-temperature" types. "High-temperature" feldspars are those which crystallized at high temperatures and cooled sufficiently rapidly (e.g., in some volcanic rocks) to preserve the structure and optics of high-temperature forms. "Low-temperature" feldspars either result from slow cooling of high-temperature forms (e.g., in some plutonic rocks) or crystallize initially at low temperatures (e.g., in some metamorphic rocks). Feldspars may also have "intermediate" structures and optics.

No. 72. ALKALI-FELDSPAR (−ve, also +ve) $(K,Na)[AlSi_3O_8]$

includes: No. 72A—microcline (−ve), triclinic; No. 72B—orthoclase (−ve), monoclinic; No. 72C—sanidine (−ve), monoclinic; No. 72D—anorthoclase (−ve), triclinic; No. 72E—adularia (−ve), monoclinic or triclinic; No. 72F—pericline (+ve), triclinic; and albite, normally discussed under plagioclase (No. 73A). See Plates 7, 8, 11, and 12 and Figures 9.60d and 9.72.

The terminology based on composition and structural state is explained in Figure 9.54. The boundaries between particular species are based partly on crystal symmetry, but are otherwise somewhat arbitrary. The solvus curve represents the limits of solid solution.

At low pressures and high temperatures (Figure 9.54a), there is complete solid solution between Or and Ab. The high-temperature forms may be metastably preserved by quick cooling. Slow cooling results in inversion to low-temperature forms, and may result in unmixing and the separation of a K-rich from a Na-rich feldspar if the solvus curve is intersected; this unmixing or exsolution is often represented by an intergrowth of the two feldspars known as "perthite" (Figure 9.72) "mesoperthite" or "antiperthite." At higher pressures (Figure 9.54b), less solid solution is possible, and melts of intermediate composition will crystallize two separate feldspars from the start; unmixing takes place in the same way as at low pressures if the solvus curve is intersected during slow cooling.

Figure 9.54 (a) Terminology of the alkali-feldspars and positions of solidus and solvus curves at low pressures (curves based on Tuttle and Bowen, 1958). **(b)** Solidus and solvus curves at P_{H_2O} = 5 kbars. Based on Morse (1970).

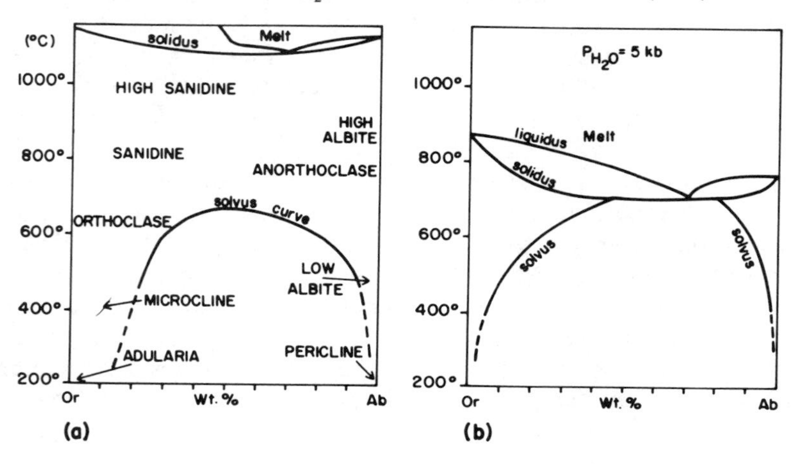

Adularia and pericline crystallize in veins from hydrothermal solutions, and usually have compositions closely approaching the two pure end members of the series.

No. 73. PLAGIOCLASE-FELDSPAR (+ve and −ve)

A series from $Na[AlSi_3O_8]$ to $Ca[Al_2Si_2O_8]$ subdivided according to the percentage proportions of the two molecules Ab and An (Table 9.4). See Plates 1–12, 27, 28, 35, and 36 and Figures 9.60, 9.69, and 9.73.

The plagioclase-feldspars are qualified as "high temperature," "intermediate," or "low temperature" (Figure 9.55). Complete solid solution at high temperatures breaks down at lower temperatures in a complicated and not fully understood way, but the most important feature is the peristerite solvus curve (Figure 9.55). For igneous rocks that cool slowly and contain oligoclase, this curve represents the unmixing of the oligoclase into a more sodic and a more calcic plagioclase which together form an intergrowth called peristerite; this intergrowth is too fine to be seen with the microscope, and in optical mineralogy the peristerite may be called low-temperature oligoclase. For metamorphic rocks in which feldspars first crystallize at low temperatures, the curve prescribes those compositions that are possible; hence during progressive metamorphism, early pure albite does not become progressively more calcic, but is succeeded abruptly by calcic oligoclase or andesine.

The feldspars are so important in petrology that it is necessary to specify the type as precisely as possible, and to do this a number of special methods have been devised. To employ these, the student must first understand the crystallography and twin laws of the feldspars.

Crystallography and Twin Laws

Feldspars belong to either the monoclinic (high sanidine, sanidine, orthoclase, and some adularia) or the triclinic crystal systems. There is little variation in the axial ratios and the crystallographic angles among the

Table 9.4 Subdivision of Plagioclase-Feldspar (mol.%)

	Ab (%)	An (%)
No. 73A. Albite	100–90	0– 10
No. 73B. Oligoclase	90–70	10– 30
No. 73C. Andesine	70–50	30– 50
No. 73D. Labradorite	50–30	50– 70
No. 73E. Bytownite	30–10	70– 90
No. 73F. Anorthite	10– 0	90–100

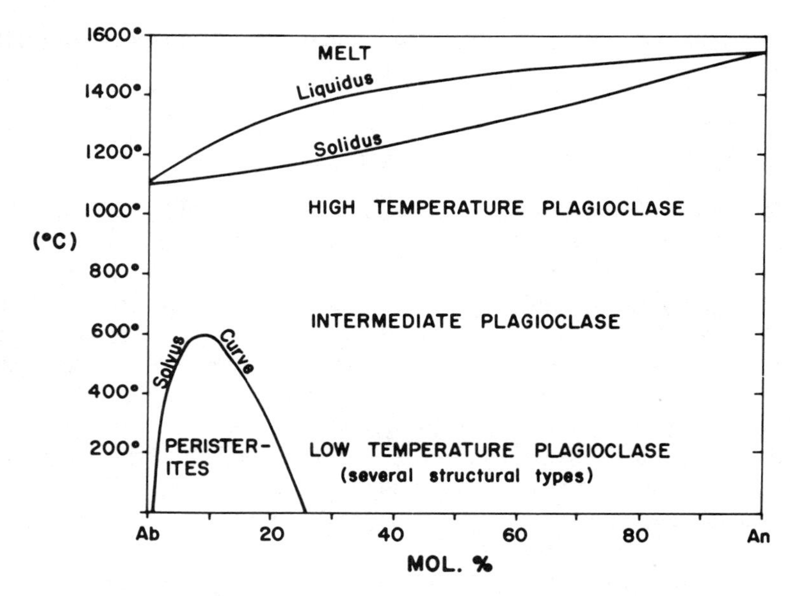

Figure 9.55 Terminology, liquidus, solidus, and peristerite solvus curves for the plagioclase-feldspars. Based on information from Bowen (1913) and Barth (1969).

feldspars. For example: monoclinic orthoclase has an axial ratio $a:b:c = 0.658:1:0.555$, and $\alpha = 90°$, $\beta = 116°00'$, $\gamma = 90°$; triclinic anorthite has an axial ratio $a:b:c = 0.635:1:0.557$, and $\alpha = 93°10'$, $\beta = 115°50'$, $\gamma = 91°15'$.

The crystal habits of all the feldspars are similar, varying from stubby prismatic to tabular (with (010) dominating) or lathlike (usually elongate parallel to *a*) (Figure 9.56). All feldspars have perfect {001} and distinct {010} cleavages.

Twinning is very common in all feldspars. The more important twin laws are summarized in Table 9.5. For details of the other less common

Figure 9.56 Common habits of feldspar.

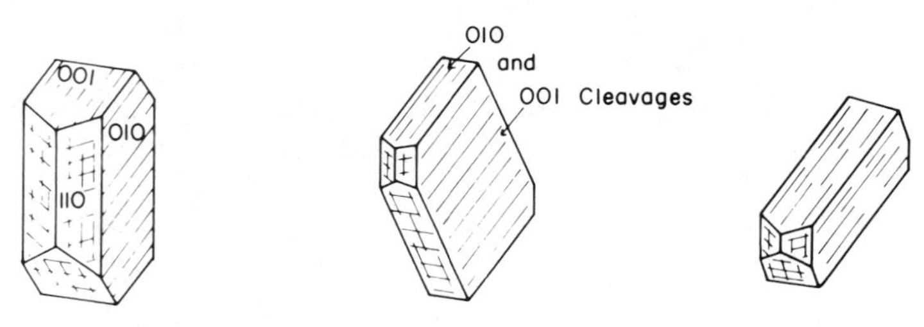

(a) STUBBY PRISM (b) 010 TABLET (c) LATH ELONGATE // a

Table 9.5 Common Twin Laws of the Feldspars

Name	Twin axis	Composition plane	Remarks
Normal twins			
Albite	⊥ (010)	(010)	often repeated[a]; very common; only in triclinic feldspar
Manebach	⊥ (001)	(001)	usually two individuals; not common
Baveno	⊥ (021)	(021)	usually two individuals; not common
Parallel twins			
Carlsbad	*c*	(010) or (100) or irregular	usually two individuals; very common
Pericline	*b*	rhombic section (see text)	often repeated[a]; very common; only in triclinic feldspar
Acline	*b*	(001)	often repeated[a]; not common; only in triclinic feldspar
Complex twins			
Albite-Carlsbad	⊥*c* in (010)	(010)	usually a small number of individuals; common; only in triclinic feldspar

[a] Repeated twinning is also called multiple or polysynthetic twinning. Feldspar crystals may display any combination of twin laws. Hence plagioclase may within one crystal display Carlsbad, albite, and pericline twinning. Twin laws are best determined using the universal-stage methods described by Slemmons (1962).

twin laws see Barth (1969) or Slemmons (1962). Figure 9.57 illustrates some of the more common twin relationships.

The rhombic section (the composition plane for pericline twinning) is defined as the section that contains b, and intersects (010) in a line which is perpendicular to b (Tunell, 1952). The position of the section is very sensitive to changes in the value of the crystallographic angle γ. For example, the rhombic section is close to $(10\bar{1})$ in microcline, but close to (001) for anorthoclase (Figure 9.58). In the plagioclase feldspars, the angle between a and the trace of the rhombic section on (010) varies both with composition and structural state (Figure 9.59).

Origin of Twinning in Feldspar

Feldspars may twin during growth (primary or growth twinning), during subsequent deformation (secondary deformation twinning), or during inversion from monoclinic to triclinic symmetry (secondary transformation twinning) (Vance, 1961, 1969). Carlsbad, Manebach, Baveno and albite-Carlsbad twins can only form during growth. Albite and pericline twins can be of growth, deformation, or transformation origin.

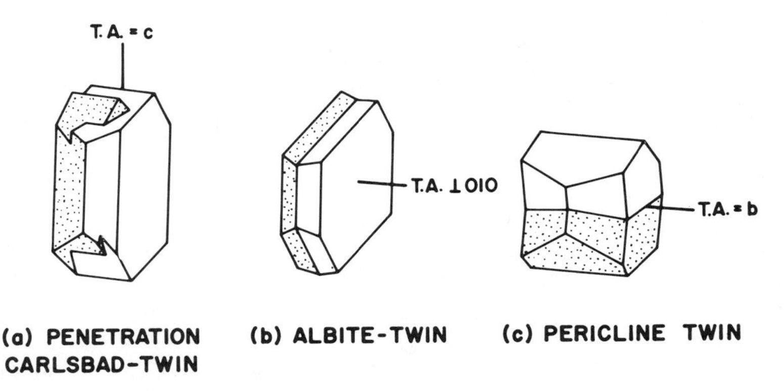

Figure 9.57 Common twin relationships in feldspar.

Growth twins are often made of only two individuals, but if the twins are repeated, the spacing is irregular (Figure 9.60a). Growth twins may be nucleated during growth, but may also result from synneusis or the colliding of crystals in a melt to join together in a twinned orientation (Figure 9.60b). Deformation twins are usually repeated and regularly spaced, and may be wedge shaped if the crystal is bent (Figure 9.60c). Transformation twins (on the albite and pericline twin laws) form in response to a change from monoclinic to triclinic symmetry in alkali-feldspar. The change commences at numerous sites in the crystal, and may be achieved by forming either of the albite or pericline twin orientations. The particular twin orientations first formed, spread and coalesce as the change of symmetry spreads through the crystal. The result is a characteristic pattern of interpenetrating albite and pericline twins called "cross-hatch" twinning (Figure 9.60d). Transformation twinning is characteristic of microcline and anorthoclase which have formed by inversion from orthoclase and high sanidine, respectively.

Figure 9.58 Orientation of albite and pericline twins in **(a)** microcline and **(b)** anorthoclase.

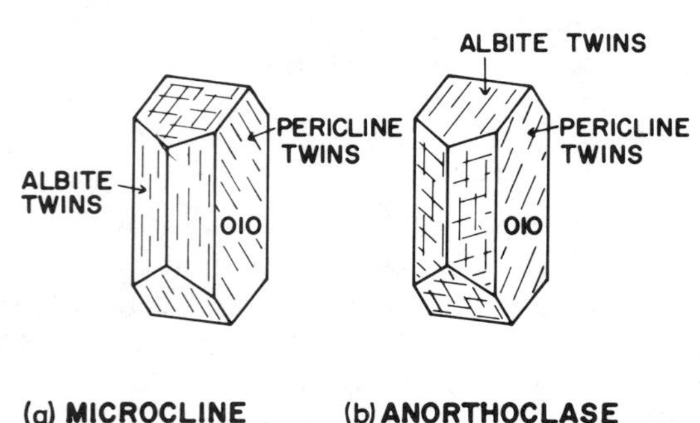

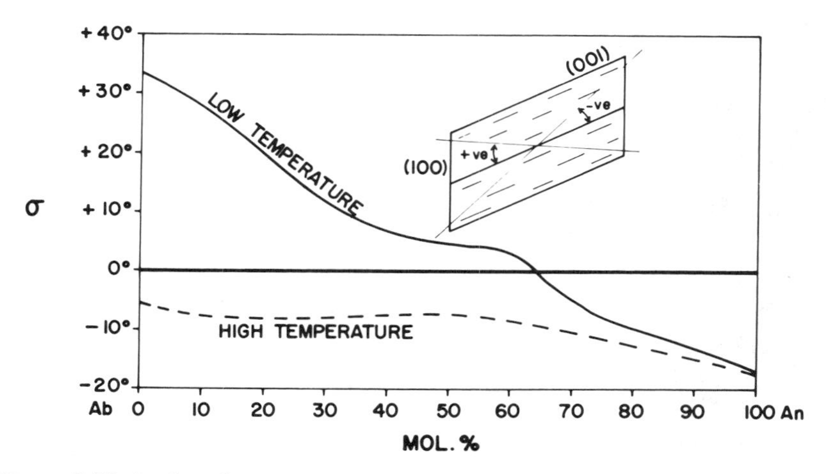

Figure 9.59 Angle σ between a and the trace of the pericline twin plane on (010) for high- and low-temperature plagioclase. Based on Starkey (1967).

Carlsbad twins are characteristic of magmatic rocks, and are very rare in metamorphic rocks. The reason for this is not clear since primary albite twins are common in metamorphic rocks. It may be that a high degree of supersaturation is necessary to nucleate Carlsbad twins, and that this is possible only during the initial growth stage in magmas; this may explain why only one central Carlsbad composition plane is formed in most crystals.

Identification of the Feldspars

The feldspars are distinguished from other minerals by the following general properties: RI in the range 1.518–1.590; δ in the range 0.005–0.013; $2V_x$ in the range 0°–105°; two cleavages {001} and {010}. Dispersion is generally weak with $r <$ or $> v$. Some of the zeolites, particularly scolecite and laumontite, have similar properties to the low RI feldspars, but have quite different crystal habits, occurrence, and orientation diagrams.

The procedures for identifying specific feldspars are described separately for thin sections and crushed-grain mounts. Staining methods are also described briefly.

Thin Sections

The procedure is as follows.

General approach

The first step is to estimate RI (Figure 9.61). This can be done *qualitatively* in thin section by comparison with the RI of Canada balsam and/or quartz (if present confirm by obtaining a uniaxial +ve interference figure), and by observing relief and Becke lines (Table 9.6).

Figure 9.60 **(a)** Growth twins in plagioclase, gabbro, Onawe, New Zealand (view measures 0.8 × 0.6 mm). **(b)** Twinning resulting from synneusis of two plagioclase crystals (note reentrant angles), Te Pua Andesite, New Zealand (view measures 1.7 × 1.3 mm). **(c)** Secondary deformation albite and pericline twinning in plagioclase (note bending and wedging out of twin lamellae), Constant Gneiss, New Zealand (view measures 1.8 × 1.3 mm). **(d)** "Cross-hatch" transformation twinning in microcline, granite, Bavaria (view measures 0.7 × 0.4 mm). Crossed-polarized light.

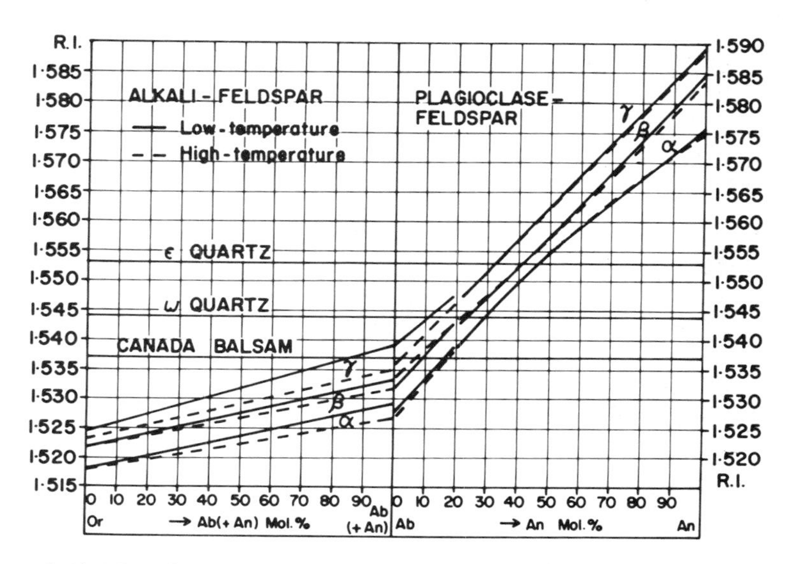

Figure 9.61 RI variation with feldspar composition. Based on Tröger (1971) and Smith (1958, Figure 3).

Having thus *estimated* the composition of the feldspar, more specific identifications are made as follows.

Alkali-feldspar

To determine the composition *accurately,* a precise RI measurement of crushed grains must be made; alternatively, chemical or X-ray methods can be used (Bowen and Tuttle, 1950; Fraser and Downie, 1964). How-

Table 9.6 Estimation of Feldspar Composition from Refractive Index

Feldspar	RI compared with CB and quartz	Becke line	Relief
Anorthite, bytownite and labradorite	distinctly above	strong	moderate
Andesine	just above quartz, distinctly above CB	strong	low
Calcic oligoclase	same as quartz, just above CB	weak	low
Sodic oligoclase	just below quartz, same as CB	very weak	low
Albite	distinctly below quartz, just below CB	weak	low
K-feldspar	distinctly below	strong	low

Note: The lower diaphragm of the microscope must be at least partly closed to observe the Becke lines.

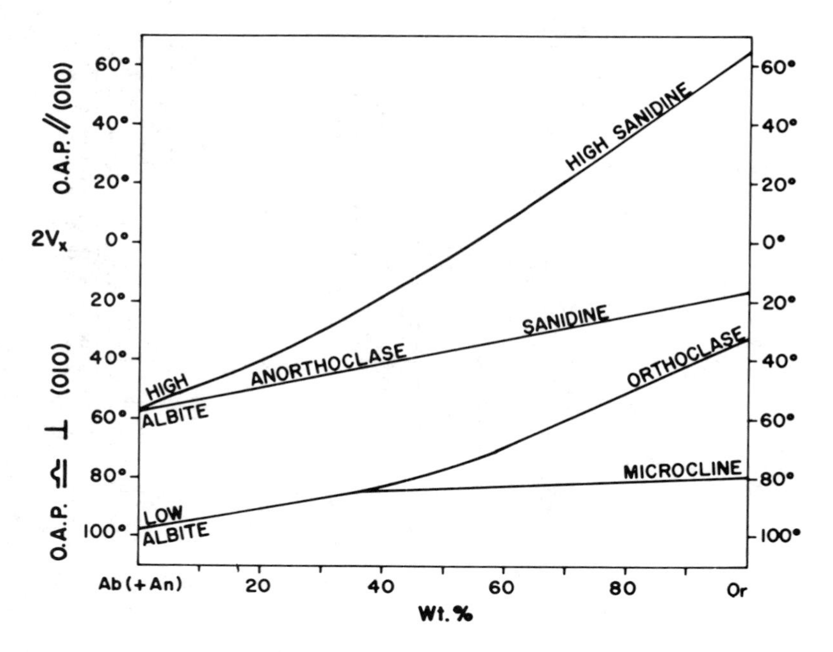

Figure 9.62 $2V$ variation with composition and structural state of alkali-feldspar. Based on Tuttle (1952) and MacKenzie and Smith (1956, Figure 1).

ever, the main types can be *roughly* determined using $2V$ and twinning characteristics. $2V$ varies with structural state and composition as shown in Figure 9.62.

For K-rich feldspars, a small $2V$ indicates sanidine, a moderate $2V$ orthoclase, and a high $2V$ microcline. K-rich high sanidine has the OAP parallel to (010), whereas the other alkali-feldspars have their OAP perpendicular to (010) (Figure 9.63). Albite and pericline twinning are impossible in orthoclase and sanidine, but microcline and anorthoclase are triclinic and normally display the transformation-type cross-hatched twinning. They are distinguished from each other by occurrence (anorthoclase in volcanics, microcline in granites and metamorphic rocks) and also by the orientation of the pericline twin plane (Figure 9.58); hence the cross-hatched effect is seen in the high-interference color sections (001) of microcline, but in the low-interference color sections (100) of anorthoclase.

Adularia is characterized by its occurrence in hydrothermal veins. It often has variable optics (from sanidine to microcline type) within a single crystal and may display cross-hatched twinning.

The sodic alkali-feldspars are best treated as plagioclase. Albite is distinguished from the other alkali-feldspars by its higher RI. Almost all albite crystallizes initially in the triclinic system, so that the albite and

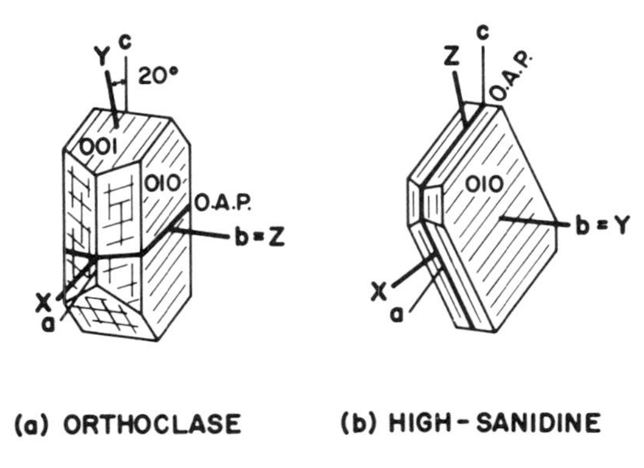

Figure 9.63 Optical orientations of the alkali-feldspars.

pericline twinning, if present, is *not* of the transformation type. Low albite is further characterized by being biaxial +ve. Pericline is found in hydrothermal veins as crystals elongate parallel to *b*.

Plagioclase-feldspar

Refractive index determination: for sodic plagioclase associated with quartz. Many rocks containing sodic plagioclase also contain quartz. Figure 9.61 shows that the RI of plagioclase from An_0 to An_{45} overlaps the RI of quartz or Canada balsam. By a careful selection of grains and examination of Becke line movements along quartz/plagioclase/Canada-balsam boundaries, an accurate determination of An% can be made. The method requires identification of plagioclase grains which contain X ($= \alpha$), Y ($= \beta$), or Z ($= \gamma$) so that those directions can be placed parallel to the polarizer vibration direction, and identification of the ω and ε directions in quartz. Details on how to do this are contained in Chapter 5; here I describe two simple examples to show how the method works. Find a low δ' grain of quartz (produces a uniaxial cross figure, and contains ω in all directions) which is adjacent to a high δ' grain of plagioclase (which contains α, the fast direction, and γ, the slow direction). Place the plagioclase in extinction so that α and γ are in turn parallel to the polarizer vibration direction, remove the analyzer, and examine Becke lines between the quartz and plagioclase. A direct comparison of α and γ with ω is made. By choosing adjacent grains of low δ' plagioclase (β in all directions) and high δ' quartz (ω the fast direction, ε the slow direction) β can be compared with ω and ε. To compare α and γ with ε normally requires two adjacent grains that extinguish at the same time. α, β, and γ should be compared with Canada balsam, too.

Extinction-angle methods. The plagioclase-feldspars are all triclinic, and the positions of X, Y, and Z vary considerably with composition

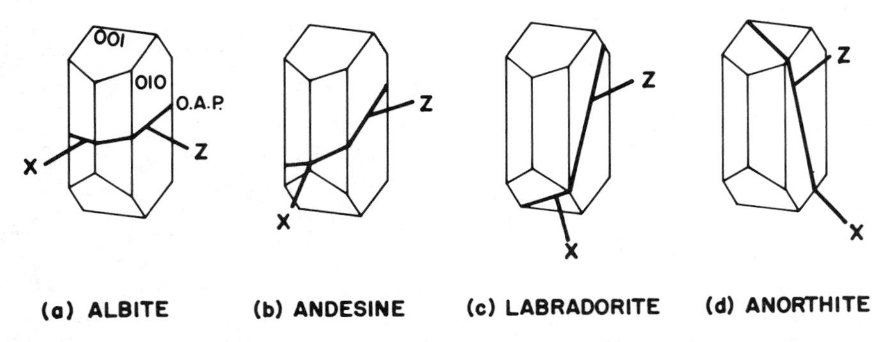

Figure 9.64 Variations in positions of *X, Z,* and the OAP in plagioclase.

(Figure 9.64) and structural state. $2V$ also varies with composition and structural state (Figure 9.65), and is not a useful property by itself. However, these two variable properties taken together can be used to determine both composition and structural state. Proceed as follows:

1. Estimate composition by measuring the extinction angles according to one of the three methods described below. These methods make use of the wide variation in the *X, Y,* and *Z* positions. Two curves are provided for each method; the high-temperature curve should be used for volcanic plagioclase, and the low-temperature curve for plutonic and metamorphic plagioclase. Detrital plagioclase may be of either type.
2. The choice of curve should then be checked by measuring $2V$ and referring to Figure 9.65, using the composition estimated by the

Figure 9.65 Variation of $2V$ with composition and structural state of the plagioclase-feldspars. Based on Smith (1958, Figure 2).

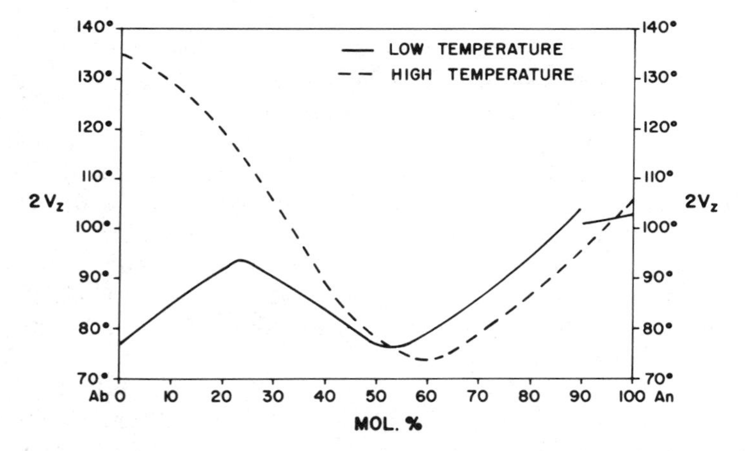

extinction-angle method. The estimation of composition should be judiciously modified to ensure that the same structural state is indicated by both $2V$ and extinction-angle values.

Cautionary notes. For strongly zoned plagioclase, it is generally necessary to use method 2, which enables the range of composition to be measured. Method 1 assumes that all plagioclase in one rock has the same composition; apart from the presence of zoning, this is normally the case for igneous and metamorphic rocks, but is an unjustified assumption for detrital plagioclase. Microlites in volcanics are usually elongate parallel to a, and for these, method 3 is the best.

Method 1. The Michel–Lévy method. The *maximum* extinction angle of the fast ray (α') onto (010) is measured in sections normal to (010). Sections normal to (010) are recognized by the presence of sharply bounded albite twins parallel to the (010) cleavage. Albite twins are recognized as the common twins that run along the length of euhedral/subhedral igneous plagioclase, and which are characterized by having even-illumination when the twin-planes are oriented E–W, N–S, and in the 45° position (Figure 9.66). Each twin orientation in turn is put into extinction, the fast direction identified, and the angle between the fast direction (α') and (010) measured. The two readings thus obtained are averaged. If the readings are not within 5° of each other they should be discarded. Readings must be made from at least seven suitably twinned crystals. The *maximum* reading obtained is selected, and the composition determined from Fig. 9.67. The ambiguity for readings less than 20° can be resolved by RI, since plagioclase more sodic than An_{20} has $\alpha <$ Canada balsam.

The method is not satisfactory for strongly zoned plagioclase, or detrital plagioclase, since the method assumes a restricted range of composition within the rock. There is a danger of confusing some pericline twins with albite twins in sodic plagioclase in metamorphic rocks where the grains lack euhedral outlines; in such cases it is best to abandon the Michel–Lévy technique and instead concentrate on RI determination.

Figure 9.66 Recognition of albite twinning by the even illumination in three positions.

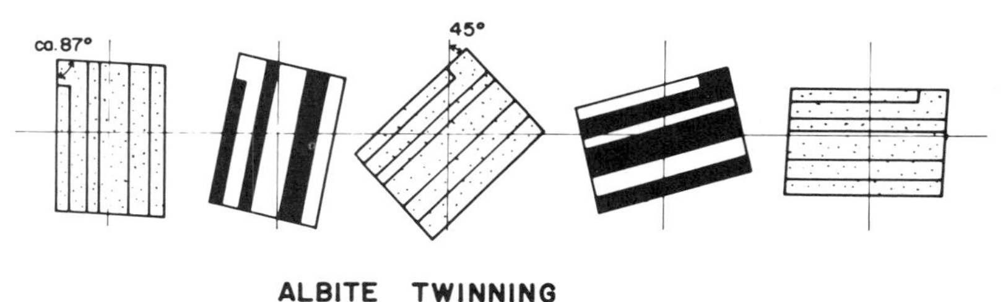

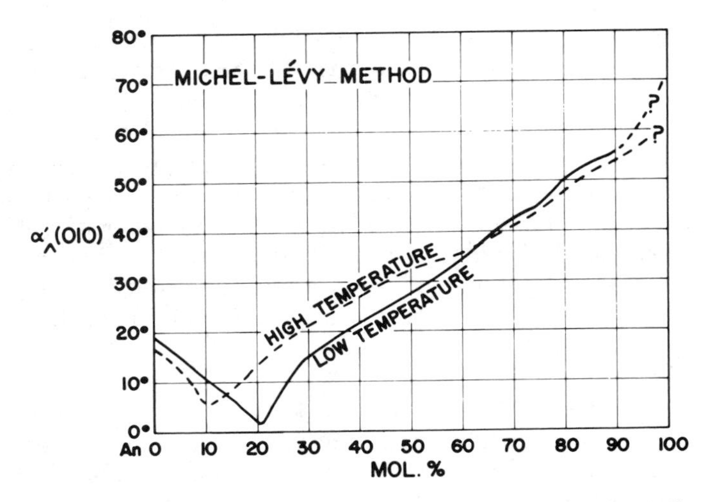

Figure 9.67 Michel–Lévy extinction-angle method for plagioclase. Curves are derived from Figure 9.70.

The Michel–Lévy method is easily adapted to the U-stage which enables one to measure the maximum extinction angle in the zone perpendicular to (010) quickly and accurately. Zoned crystals and microlites in volcanic rocks can be dealt with satisfactorily if the U-stage is used (see Chapter 6 for details).

Method 2. Extinction angle in section ⊥ a. The section ⊥ *a* is recognized by the presence of two cleavages almost at right angles to each other (Figure 9.68). Both cleavages should appear sharp (i.e., vertical) in

Figure 9.68 Extinction angles in plagioclase sections perpendicular to *a*. Curves are based on the data of Burri et al. (1967, p. 307).

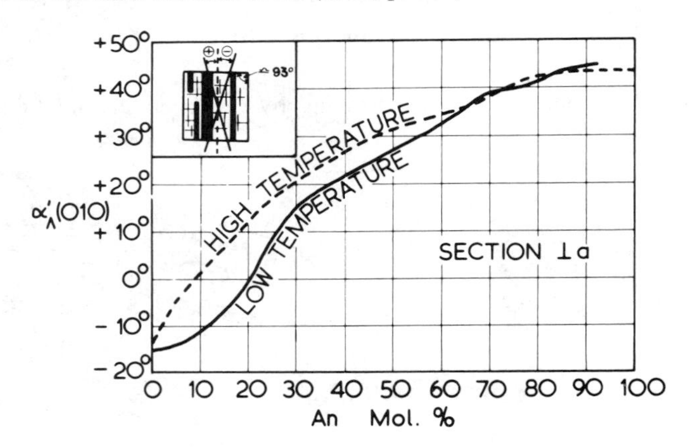

thin section. Albite twins will often be visible parallel to the (010) cleavage. The extinction angle between the fast ray (α') and (010) is measured (it will always be less than 45°), and the appropriate curve of Figure 9.68 referred to. The convention for +ve and −ve angles is shown in Figure 9.68, but the ambiguity for readings less than 20° can also be resolved by RI, since plagioclase more sodic than An_{20} has α < Canada balsam.

This method is particularly appropriate for strongly zoned plagioclase (range of composition equals the variation in extinction angle within one crystal). Unfortunately, it is not always possible to find a suitably oriented section, particularly for low-relief plagioclase whose cleavages tend to become invisible (due to low relief). It is not a very suitable method for discriminating amongst the highly calcic plagioclases inasmuch as the curve flattens out for these compositions.

Method 3. Using combined albite and Carlsbad twins. The extinction angle of (α') onto (010) is measured on each side of a Carlsbad twin. The presence of sharply bounded albite twins on each side of the Carlsbad twin enables the position of (010) to be accurately judged (Figure 9.69). The albite twins are recognized by their even illumination in the N–S, E–W, and 45° positions (Figure 9.66). The Carlsbad twin does not have even illumination in these positions, and remains visible, particularly in the 45° position.

Figure 9.69 Combined albite and Carlsbad twinning in several intergrown crystals of andesine (view measures 2.7 × 1.8 mm). Crossed-polarized light.

Measure and average the readings of the fast direction onto (010) from the albite twins (as in the Michel–Lévy method) for each side of the Carlsbad twin. Two (usually different) values are thus obtained, and the composition determined from Figure 9.70, using the solid curves for the larger angle and the dashed curves for the smaller angle. Positive and negative angles can only be determined in sections ⊥ *a* (Tobi and Kroll, 1975), but if such a section is found, the albite–Carlsbad method is superfluous and method 2 should be used. This means that readings greater than 20° are nearly always ambiguous, and for readings less than 20° there may be three possible solutions. Some ambiguities may be resolved by RI, but it is often necessary to make at least two determinations to resolve them.

Figure 9.70 Extinction-angle curves for combined albite and Carlsbad twinning in plagioclase. Based on Tobi and Kroll (1975). See text for explanation.

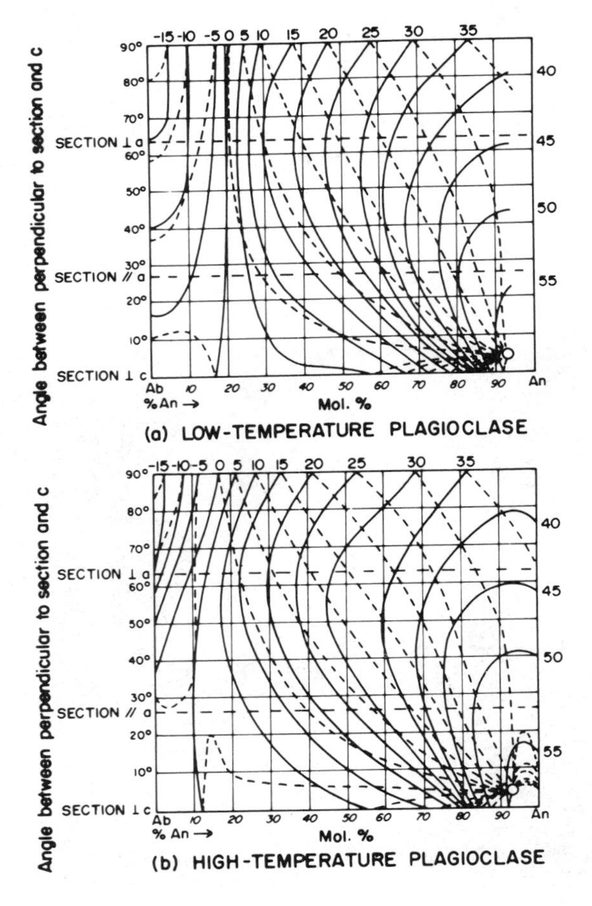

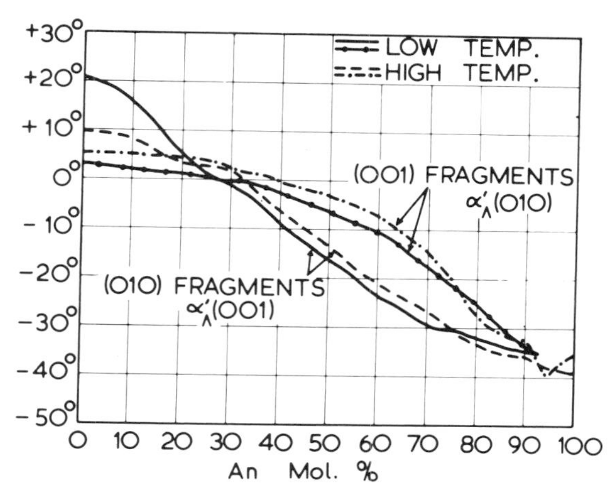

Figure 9.71 Extinction angles for (010) and (001) cleavage fragments. Curves are based on the data of Burri et al. (1967, p. 307).

Zoned crystals display a series of extinction angle pairs which correspond to values plotting along one of the horizontal lines of Figure 9.70. Microlites in volcanics are normally elongate parallel to *a*, and often Carlsbad twinned; this method is the best for determining their composition—use extinction values close to the horizontal line marked "section parallel to *a*" in Figure 9.70.

Crushed-Grain Mounts

An estimation of the composition of alkali-feldspar in crushed grains can be made from 2*V* and twinning characteristics, as described for thin sections. A more precise determination is possible by accurately measuring RI (Figure 9.61), using 2*V* to determine the structural state (Figure 9.62).

The composition of crushed grains of plagioclase can also be determined by accurately measuring RI (Figure 9.61), using 2*V* to specify the structural state (Figure 9.65). However, there may be difficulties in finding suitably orientated grains to measure α, β, or γ precisely, because the majority of crushed fragments will lie parallel to the {001} or less commonly, the {010} cleavages. An alternative to measuring RI precisely is to measure the extinction angles of α' onto the {010} cleavage trace in fragments lying on (001), or of α' onto the {001} cleavage trace in fragments lying on (010) (Figure 9.71). 2*V* must be measured to determine the structural state, and provides a check on the choice of curve in Figure 9.71, and hence allows an improvement in the accuracy of the determination. The ambiguous positive and negative angle values of Figure 9.71 are sorted most easily on the basis of RI.

Staining Methods

Many rocks contain two or three of the minerals quartz, alkali-feldspar, and plagioclase. For the purposes of modal analysis, it is necessary to identify every grain in a thin section (or rock slab) quickly, and this is not always easy with standard microscope procedures. To assist such analyses, various staining methods have been devised. The standard technique is to etch the specimen with HF, and stain K-feldspar yellow with sodium cobaltinitrite and plagioclase red by treating with $BaCl_2$ and K-rhodizonate or other dyes (Bailey and Stevens, 1960; Laniz et al., 1964). Modifications of this technique have been described recently by Wilson and Sedeora (1979) and Houghton (1980); Widmark (1979) describes a method for dealing with Ca-poor plagioclase. Lyons (1971) has described a more suitable procedure for rock-slab surfaces.

Occurrence and Alteration

The alkali-feldspars are essential constituents of many alkaline and acid igneous rocks, e.g., granites, syenites, nepheline-syenites, and their volcanic equivalents. In plutonic rocks, the alkali-feldspar is usually orthoclase or microcline, but in volcanics, the higher-temperature forms sanidine and anorthoclase are common. Orthoclase and microcline also occur in pegmatities, often in graphic intergrowth with quartz (see below). Orthoclase and microcline are formed during regional and thermal metamorphism, orthoclase being particularly characteristic of high-grade pelitic rocks. Plagioclase-feldspars are essential constituents of a wide range of igneous rocks. The more calcic plagioclases are typical of basic rocks, the more sodic of acid and alkaline types. Plagioclase is also common in metamorphic rocks. The lower grades of metamorphism are characterized by almost pure albite, and there is a sudden jump in composition (Figure 9.55 and accompanying text) to calcic oligoclase or andesine in the higher grades; this change coincides with the boundary between the greenschist and amphibolite facies of regional metamorphism. Calcic plagioclase is often abundant in thermally metamorphosed carbonate rocks. Plagioclase is unstable in the eclogite facies of metamorphism.

Adularia and pericline occur in hydrothermal veins, and both alkali- and plagioclase-feldspar (usually pure K and Na end members) may form during diagenesis.

Feldspars are prone to extensive weathering or alteration. The principal deuteric or hydrothermal alteration products are: sericite, composed of mica or illite—often fine-grained; saussurite (from plagioclase)—a mixture of zoisite, epidote, albite, chlorite, carbonates, and other minerals; clay minerals. Feldspars also weather readily to clay minerals, giving crystals a cloudy appearance. However, both alkali- and plagioclase-feldspars are abundant as detrital grains in some sediments, particularly sandstones.

Common Feldspar Textures and Intergrowths

Since any work with the feldspars must necessarily involve the very common textures and intergrowths, a brief description of these is given here. No attempt is made to provide an exhaustive account or to provide any more than brief suggestions as to their origin.

Perthite and Antiperthite

Perthite consists of discrete areas of sodic plagioclase within a host K-feldspar crystal. In antiperthite, the host is plagioclase. The term mesoperthite is used if the two feldspars are in more or less equal proportions. The enclosed feldspar may occur as lensoid areas variously described as rods, beads, or strings (Figure 9.72), or as irregular "flamelike" or blocky patches. Perthitic structures should not be confused with twinning from which they are easily distinguished by observing the contrast in RI using Becke lines. Rod, bead, string perthite, and antiperthite are normally considered to be the result of exsolution as prescribed by the solvus curve of Figure 9.54. Flame and patch perthite may be explained similarly, or may form by the replacement of one feldspar by another.

Figure 9.72 Perthite: sodic plagioclase (dark areas) in host K-rich feldspar; note cleavages run continuously from K-rich through Na-rich areas (view measures 3.3 × 2.1 mm). Crossed-polarized light.

Zoning

Zoning is present in some alkali-feldspars but is most common in plagioclase, to which the following descriptions apply.

Normal zoning is a progressive increase in Na content from the core to the rim of a plagioclase crystal. It is very common in igneous rocks, and is explained as a natural consequence of crystallization as defined by the solidus and liquidus curves of Figure 9.55 (see Turner and Verhoogen, 1960, or other petrology texts for a full explanation).

Reverse zoning is a progressive increase in Ca content from the core to the rim of a plagioclase, and is less common than normal zoning. In metamorphic rocks it may be explained by crystallization during a progressive increase in metamorphic grade.

Oscillatory zoning (Figure 9.73) consists of a sequence of normal zones which are separated by sharp reversals in composition. There is usually an overall normal trend in the zoning. Numerous explanations have ap-

Figure 9.73 Oscillatory zoning in broken plagioclase crystals, Te Pua Andesite, New Zealand (view measures 0.5 × 0.5 mm). Crossed-polarized light.

peared in the literature; for example, Bottinga et al. (1966) ascribe its development to a combination of crystal growth mechanisms and the relative rates of diffusion of Al and Si from the magma to the crystal faces whereas Loomis (1982) ascribes its development to periodic local convection. Well-developed oscillatory zoning is characteristic, if not diagnostic, of magmatic crystallization.

Myrmekite

This is a microscopic intergrowth of plagioclase and rods of quartz. The quartz rods are circular in cross section and are often arranged in a radiating and branching manner perpendicular to the outer boundary of the plagioclase (Figure 9.74a). The quartz in any one myrmekite has a single optical orientation. Myrmekite is almost ubiquitous in granites and granite-gneisses, and normally occurs along the grain boundaries of alkali-feldspar which it appears to replace. Interpretations of myrmekite include: replacement of K in alkali-feldspar by Na and Ca which causes a release of SiO_2 (Becke's hypothesis—see Ashworth, 1972); exsolution of Schwantke's molecule $Ca(AlSi_3O_8)_2$ and $NaAlSi_3O_8$ to the grain boundaries of the alkali-feldspar, the SiO_2 being released from Schwantke's molecule (Schwantke's hypothesis—see Ashworth, 1972); porphyroblastic growth of plagioclase including a recrystallizing groundmass of quartz (Shelley, 1973).

Granophyric Intergrowth

An intergrowth of quartz and K–Na-feldspar in which the quartz rods (which have a single optical orientation over large areas) are coarser than

Figure 9.74 **(a)** Myrmekite, Constant Gneiss, New Zealand (view measures 0.8 × 0.6 mm). **(b)** Granophyric intergrowth, Pepin Island, New Zealand (view measures 2.4 × 1.7 mm). Crossed-polarized light.

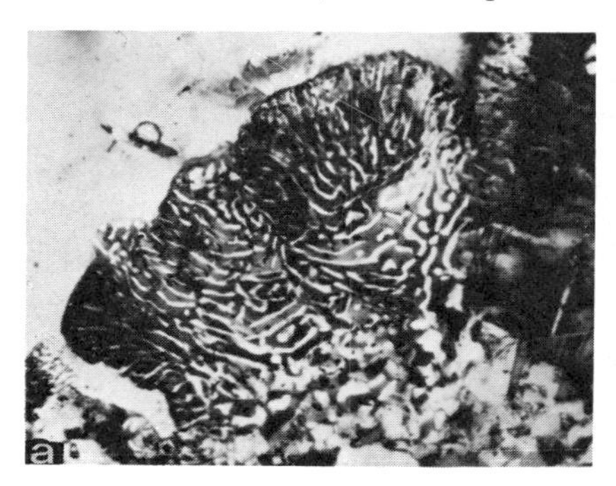

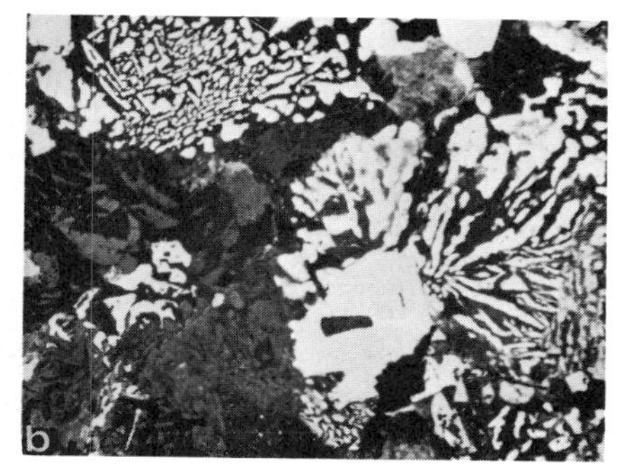

in myrmekite. The rods mostly have triangular cross sections and may radiate within the feldspar crystal (Figure 9.74b). It is generally interpreted (Barker, 1970) as resulting from the simultaneous crystallization of quartz and alkali-feldspar from magma at low pressures (hypersolvus conditions—see Figure 9.54a). Some superficial granite intrusions (granophyres) are composed mainly of this intergrowth.

Graphic Intergrowth

This is similar to granophyric intergrowth. However, the quartz rods are coarser, clearly visible in hand specimen, closely parallel to each other, and have cuneiform cross sections. The feldspar may be a K-rich alkali-feldspar or a sodic plagioclase. The intergrowth is generally interpreted (Barker, 1970) as the result of the simultaneous crystallization of quartz and feldspar at higher pressures (subsolvus conditions--see Figure 9.54b). It occurs in some granites, but is most common in pegmatites.

The Feldspathoids

As the name implies, the feldspathoids are similar in composition to the feldspars. They are related to them by the equation:

$$\text{Feldspathoid} + \text{silica} = \text{feldspar}$$

Feldspathoids and the silica minerals are antipathetic, although feldspathoids commonly coexist with the feldspars.

No. 74. LEUCITE (+ve, pseudocubic) $K[AlSi_2O_6]$

See Figure 9.75.
Tetragonal (pseudocubic), cubic above 625° C. Very poor {110} cleavages. Equant crystals; euhedral–subhedral trapezohedral crystals with eight-sided sections. Complex transformation-twin patterns on {110} very common (Figure 9.75).

Color in thin section. Colorless.

Optical properties. Uniaxial +ve or isotropic.

$$\omega = 1.508\text{–}1.511, \ \varepsilon = 1.509\text{–}1.511.$$

$$\delta = 0.00\text{–}0.001.$$

Small crystals may appear isotropic, but large crystals nearly always display a very weak δ and complex twin patterns.

Occurrence. As phenocrysts, or less commonly in the groundmass of silica-poor K-rich volcanic rocks, often with other feldspathoids such as

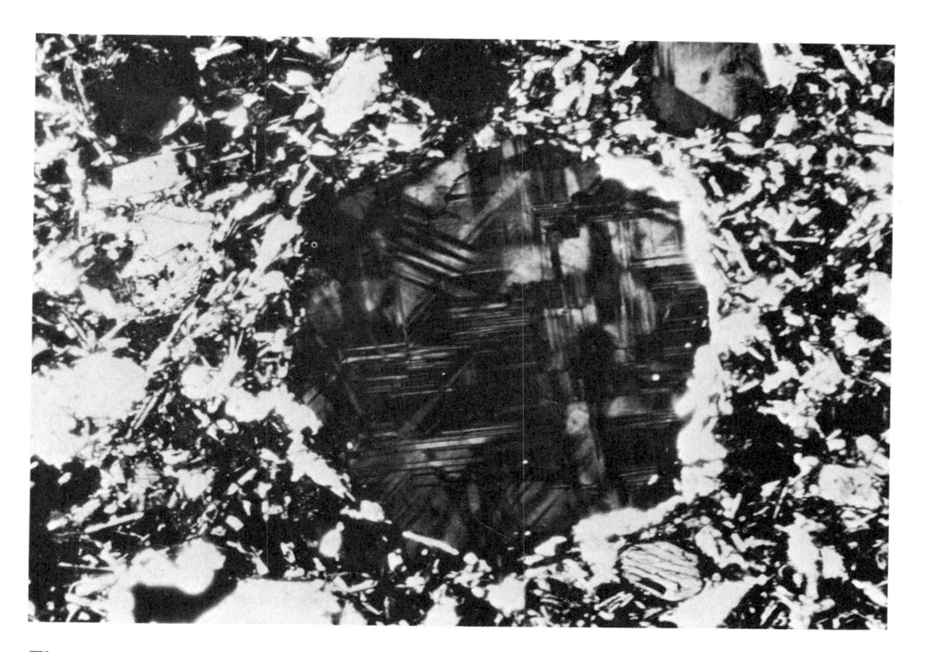

Figure 9.75 Leucite with weak birefringence and complex twin patterns, leucitophyre, Eifel, Germany (view measures 2.6 × 1.7 mm). Crossed-polarized light.

sodalite. May be replaced by a mixture of alkali-feldspar and nepheline (called pseudoleucite).

Distinguishing features. The low relief (RI < Canada balsam), very low δ, twin patterns, crystal shape (eight-sided sections), and occurrence are distinctive. It does not occur with quartz. Leucite may be confused with sodalite and analcime, but both these minerals lack the distinctive twinning; in addition, sodalite has a different crystal shape.

No. 75. NEPHELINE (−ve) $(K,Na)Na_3[Al_4Si_4O_{16}]$

Kaliophilite, the K-rich analogue of nepheline is rare. Somewhat less rare is the polymorph *kalsilite* ($KAlSiO_4$).
Hexagonal. Poor $\{10\bar{1}0\}$ cleavages. Euhedral–subhedral crystals, usually very stubby prisms with a hexagonal cross section and rectangular (almost square) side sections; also anhedral.

Color in thin section. Colorless.

Optical properties. Uniaxial −ve.

$$\omega = 1.529\text{–}1.547,\ \varepsilon = 1.526\text{–}1.542.$$

$$\delta = 0.003\text{–}0.007.$$

Square or rectangular sections of nepheline have straight extinction, and hexagonal sections provide uniaxial cross-interference figures; cleavage rarely seen.

Occurrence. In a wide range of silica-poor plutonic and volcanic igneous rocks, e.g., nepheline-syenite and phonolite. Also in nepheline-syenites and associated nepheine-bearing gneisses that are generally believed to have been produced by the metasomatic activity known as "fenitization." Nepheline is also formed in reaction zones between carbonates and basic igneous rocks. Kalsilite is present in some K-rich volcanic rocks. Nepheline alters easily to "sericite," zeolites, sodalite and cancrinite.

Distinguishing features. The RI close to Canada balsam, the low or very low δ, the uniaxial −ve character, lack of good cleavage, and alteration are distinctive. It does not occur with quartz. Often associated with alkali-feldspar with which it may be confused, but feldspar is biaxial and has good cleavages. Na-scapolite is similar, but has a higher δ and better cleavages. In fine-grained rocks, the presence of nepheline, or the *amount* of nepheline present, may be difficult to ascertain, and a stain test using phosphoric acid and methylene-blue dye may be used (Shand, 1939). Kalsilite and nepheline are best distinguished using X-rays.

No. 76. CANCRINITE (−ve)

$(Na,Ca,K)_{6-8}[Al_6Si_6O_{24}](CO_3,SO_4,Cl)_{1-2} \cdot 1\text{–}5H_2O$

Hexagonal. Perfect {1010} cleavages. Usually anhedral.

Color in thin section. Colorless.

Optical properties. Uniaxial −ve.

$$\omega = 1.490\text{–}1.528,\ \varepsilon = 1.488\text{–}1.503.$$

$$\delta = 0.002\text{–}0.025.$$

RI and δ decrease with increasing SO_4.
Crystals have straight extinction and are fast parallel to good cleavage traces.

Occurrence. Forms as a late-stage magmatic mineral in silica-poor plutonic igneous rocks such as nepheline-syenite. Also as a secondary mineral replacing nepheline and feldspar in such rocks.

Distinguishing features. The low RI, uniaxial −ve character, good prismatic cleavages, and occurrence are distinctive. Moderately birefringent cancrinite may superficially resemble muscovite, but cancrinite differs in its lower RI and length-fast nature. Scapolite has a higher RI and less distinct cleavage.

No. 77. SODALITE GROUP (isotropic)

Sodalite	$Na_8[Al_6Si_6O_{24}]Cl_2$
Nosean	$Na_8[Al_6Si_6O_{24}]SO_4$
Haüyne	$(Na,Ca)_{4-8}[Al_6Si_6O_{24}](SO_4,S)_{1-2}$

See Plates 13 and 14.
Cubic. Poor {110} cleavages. Euhedral–subhedral dodecahedral crystals with six-sided sections; also anhedral.

Color in thin section. Colorless, pale pink, blue or gray; sometimes color-zoned.

Optical properties. Isotropic.
n = 1.483–1.487 (sodalite), 1.461–1.495 (nosean), 1.493–1.509 (haüyne).
Crystals are often characterized by a zonal pattern of inclusions, some zones appearing dark and cloudy in thin section.

Occurrence. Members of the sodalite group are found in silica-poor igneous rocks such as nepheline-syenites and phonolites, often associated with nepheline and leucite. They also occur in metamorphosed and metasomatized carbonate rocks near igneous contacts.

Distinguishing features. The low RI, isotropic character, crystal shape (six-sided sections), and occurrence are distinctive. May be confused with analcime and leucite, but both these minerals, if euhedral, have eight-sided sections; in addition, leucite has a weak δ, and twinning. Anhedral analcime may be difficult to distinguish, although analcime has a different and better cleavage. Microchemical tests can be used to distinguish the members of the sodalite group, and to distinguish these from analcime (Deer et al., 1963, vol. 4, pp. 297 and 345).

The Zeolites

See Plate 15.

The zeolites are hydrated silicates of aluminium and the alkalies or alkaline-earths. When heated, they readily loose some or all of the water without destruction of the structural framework; this water can be readily reabsorbed. There are numerous zeolites, and only a few are described here.

No. 78. MORDENITE (+ve or −ve) $(Na_2,K_2,Ca)[Al_2Si_{10}O_{24}] \cdot 7H_2O$

No. 79. NATROLITE (+ve) $Na_2[Al_2Si_3O_{10}] \cdot 2H_2O$

No. 80. MESOLITE (+ve) $Na_2Ca_2[Al_2Si_3O_{10}]_3 \cdot 8H_2O$

No. 81. SCOLECITE (−ve) $Ca[Al_2Si_3O_{10}] \cdot 3H_2O$

No. 82. THOMSONITE (+ve) $NaCa_2[(Al,Si)_5O_{10}]_2 \cdot 6H_2O$

No. 83. HEULANDITE (+ve) $(Ca,Na_2)[Al_2Si_7O_{18}] \cdot 6H_2O$

(including *clinoptilolite*)

No. 84. STILBITE (−ve) $(Ca,Na_2,K_2)[Al_2Si_7O_{18}] \cdot 7H_2O$

No. 85. ANALCIME (isotropic) $Na[AlSi_2O_6] \cdot H_2O$

No. 86. WAIRAKITE (+ve or −ve) $Ca[Al_2Si_4O_{12}] \cdot 2H_2O$

No. 87. CHABAZITE GROUP (−ve or +ve) $Ca[Al_2Si_4O_{12}] \cdot 6H_2O$

(including *gmelinite* and *levyne*)

No. 88. LAUMONTITE (−ve) $Ca[Al_2Si_4O_{12}] \cdot 4H_2O$

No. 89. PHILLIPSITE (+ve) $(\frac{1}{2}Ca,K,Na)_3[Al_3Si_5O_{16}] \cdot 6H_2O$

Crystallography and optical properties. The principal properties of the zeolites are compiled in Table 9.7.

Orientation diagrams and habit. Mordenite, natrolite, mesolite, scolecite, and thomsonite are often fibrous. Mordenite (Figure 9.76a), natrolite (Figure 9.76b), mesolite (Figure 9.76c), and thomsonite (Figure 9.76e) have straight extinction; mordenite is always length fast, natrolite length slow, whereas mesolite and thomsonite have the OAP across the length, and may be length fast or slow. Scolecite (Figure 9.76d) has inclined extinction with $X \wedge c = 18°$.

Heulandite and stilbite are characterized by a single cleavage and platy habit with tabular or sheaflike crystals parallel to (010). Heulandite (Figure 9.76f) has the OAP across the cleavage whereas stilbite (Figure 9.76g) has the OAP parallel to (010).

Analcime is cubic, and euhedral crystals are eight-sided in thin section. It is often anhedral. Some analcime may display a weak birefringence. Wairakite, essentially a Ca-analcime, is pseudocubic, and usually has interpenetrating twin lamellae.

Members of the chabazite group are normally uniaxial −ve, but they may be +ve or anomalously biaxial. They crystallize most commonly with prominent $\{10\bar{1}1\}$ forms which resemble cubes. The members chabazite, gmelinite, and levyne are best sorted with X-rays.

Laumontite (Figure 9.76h) is characterized by three good cleavages seen together in (001) sections, and a small extinction angle $Z \wedge c$.

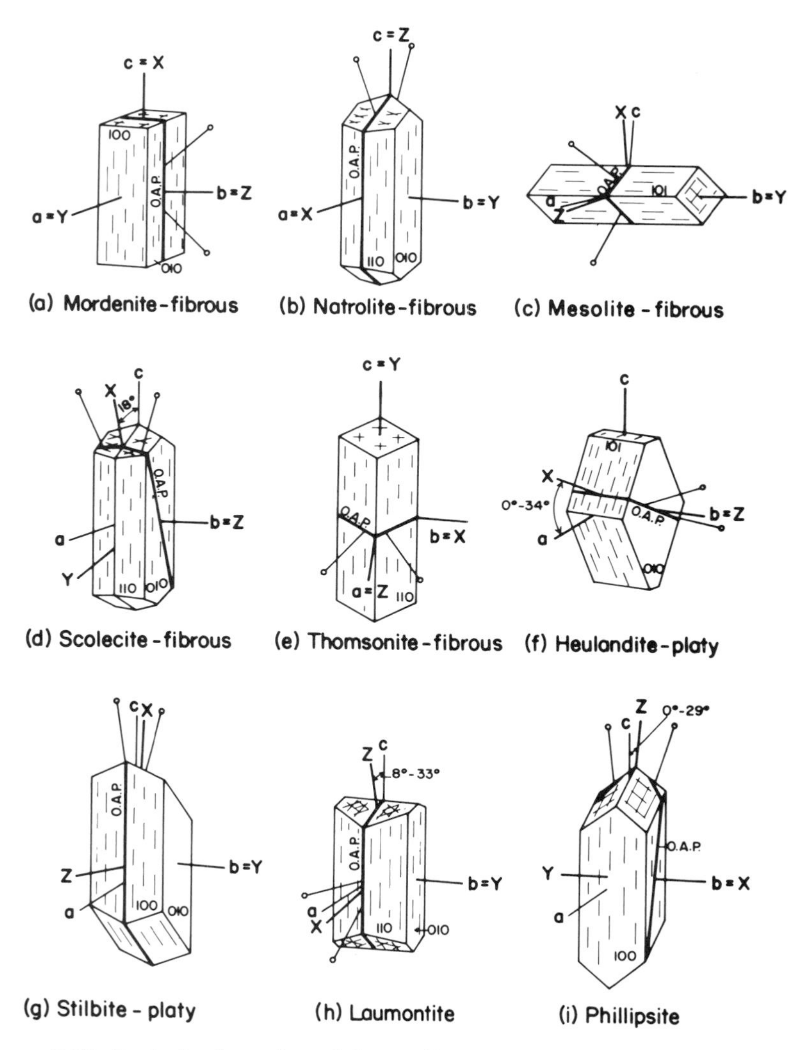

Figure 9.76 Optical orientation of the zeolites.

Phillipsite (Figure 9.76i) commonly has interpenetrant twinning. It has inclined extinction with $Z \wedge c = 10°–29°$.

Occurrence. There are three principal modes of occurrence:

(1) *Primary igneous.* Analcime is the only zeolite to form as a primary-igneous mineral (e.g., in some basalts, teschenites, and related rocks). It forms late, and is difficult to distinguish from secondary analcime.

Table 9.7 Properties of the Zeolites

Mineral	Crystal system and cleavages	Minimum RI	Maximum RI	δ	$2V$, sign and orientation
Mordenite	Orthorhombic. Perfect {010} and {100} cleavages	1.472 (α)	1.487 (γ)	0.002–0.005	+ve or −ve, $2V_x = 76°–104°$. $X = c$, $Y = a$, $Z = b$.
Natrolite	Orthorhombic. Two distinct {110} cleavages	1.473 (α)	1.496 (γ)	ca. 0.012	+ve, $2V = 58°–64°$. $r < v$ (weak) $X = a$, $Y = b$, $Z = c$
Mesolite	Monoclinic, β = ca. 90°. Two perfect {101} cleavages	1.504 (β)	1.508 (β)	0.001	+ve, $2V = 80°$. $r < v$ (strong) $X \wedge c = 8°$, $Y = b$
Scolecite	Monoclinic, β = ca. 90°. Two distinct {110} cleavages	1.507 (α)	1.521 (γ)	0.007–0.010	−ve, $2V = 36°–56°$. $r < v$ (strong) $X \wedge c = 18°$, $Z = b$
Thomsonite	Orthorhombic. Perfect {100} and distinct {010} cleavages	1.497 (α)	1.544 (γ)	0.006–0.016	+ve, $2V = 38°–75°$. $r > v$ (distinct) $Z = a$, $Y = c$, $X = b$
Heulandite	Monoclinic, $\beta = 91°$. Perfect {010} cleavage	1.476 (α)	1.512 (γ)	0.002–0.008	+ve, $2V = 0°–74°$. $r > v$ (distinct) $X \wedge a = 0°–34°$, $Z = b$

Stilbite	Monoclinic, $\beta = 129°$. Distinct {010} cleavage	1.482 (α)	1.513 (γ)	0.008–0.014	−ve, $2V = 28°–49°$. $r < v$ (weak) $X \wedge c =$ ca. 5°, $Y = b$
Analcime	Cubic. Poor cubic cleavage	1.479 (n)	1.524 (n)	0.00 –0.002	
Wairakite	Monoclinic, pseudocubic. Intersecting twins on {110}	1.498 (α)	1.502 (γ)	0.004	+ve or −ve, $2V_z = 70°–105°$. $X =$ ca. b, $Y =$ ca. a, $Z =$ ca. c.
Chabazite, gmelinite, and levyne	Trigonal. Poor $\{10\bar{1}1\}$ cleavages (pseudocubic cleavage)	1.470	1.505	0.002–0.015 usually <0.005	−ve or +ve (often anomalously biaxial), $2V = 0°–32°$
Laumonite	Monoclinic, $\beta = 111°$. Three distinct cleavages, {110} and {010}	1.502 (α)	1.526 (γ)	0.010–0.015	−ve, $2V = 26°–47°$. $r < v$ (strong) $Y = b$, $Z \wedge c = 8°–33°$
Phillipsite	Monoclinic, $\beta =$ ca. 90°. Distinct {010} and {100} cleavages. Common interpenetrant twinning	1.483 (α)	1.514 (γ)	0.003–0.010	+ve, $2V = 60°–80°$. $r < v$ (weak) $X = b$, $Z \wedge c = 10°–29°$

Note: All zeolites are colorless in thin section. The properties of the zeolites may change if heated (during thin-section making, etc.).

(2) *Secondary in igneous rocks.* Zeolites most typically occur as secondary minerals in amygdales and fissures, chiefly in basic volcanic rocks. Zonal sequences are known, so that in the Tertiary lavas of East Iceland, there is a zeolite-free zone at the top with successive zones downwards of chabazite–thomsonite, analcime, and mesolite–scolecite (Walker, 1960). Other zonal sequences have been described. The zeolites may replace feldspars and nepheline in igneous rocks.

(3) *Diagenetic, in low-grade metamorphic rocks and hot-spring deposits.* Many of the zeolites form during diagenesis in both marine (including modern deep-ocean sediments) and non-marine sediments; these zeolites commonly replace glassy tuffaceous material. In addition, zeolites may replace tuffaceous material, detrital plagioclase, and fossil material to form zonal sequences in response to burial metamorphism (the "zeolite facies") or hot-spring activity (Coombs, 1971). In zonal sequences, analcime, mordenite, and heulandite are succeeded with increasing burial by laumontite, and with increasing temperature (in hot-spring deposits) by wairakite.

Distinguishing features. The zeolites are quite distinctive as a group. All are colorless, they have a low relief with their RI commonly being less than that of Canada balsam, and their δ is low or very low. Their typical occurrence as secondary minerals in amygdales and figures is distinctive.

Analcime may be confused with leucite or sodalite, but leucite has distinctive twinning, and euhedral sodalite crystals are six-sided. Microchemical tests (see sodalite) may be used to distinguish these minerals.

Laumontite and scolecite have similar properties to the alkali-feldspars, but the cleavages of laumontite and the fibrous nature of scolecite are distinctive.

Thomsonite is similar in properties to gypsum, but the straight extinction of thomsonite is distinctive.

Refer to Table 9.7 and Figure 9.76 for distinctions amongst the zeolites.

No. 90. SCAPOLITE (−ve)

A series between *marialite* $3Na[AlSi_3O_8] \cdot NaCl$ and *meionite* $3Ca[Al_2Si_2O_8] \cdot CaCO_3$. K may also be present. Pure end-members are not known naturally.

Tetragonal. Distinct {100} and imperfect {110} cleavages. Long prismatic crystals or anhedral, granular.

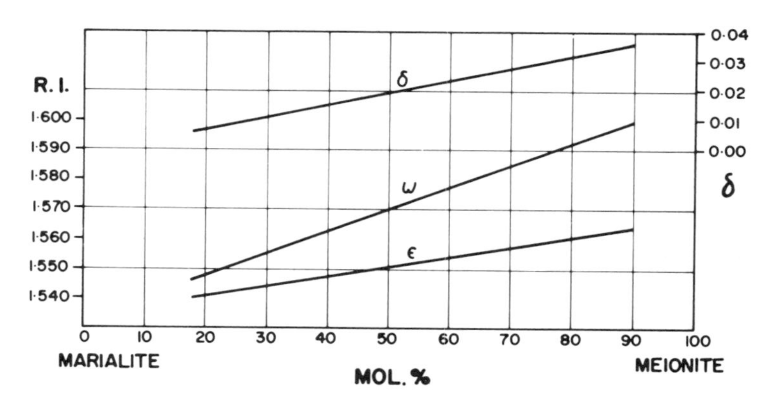

Figure 9.77 Variation of RI and birefringence in scapolite. Based on data in Deer et al. (1963).

Color in thin section. Colorless.

Optical properties. Uniaxial −ve.

$$\omega = 1.546\text{–}1.600,\ \varepsilon = 1.540\text{–}1.571.$$

$$\delta = 0.006\text{–}0.036.$$

RI and δ increase with meionite (Ca) content (Figure 9.77). Crystals are fast parallel to cleavages and crystal outlines with straight extinction.

Occurrence. Forms instead of plagioclase in a wide range of regionally metamorphosed rocks under high CO_2 pressures, or in the presence of abundant brine (NaCl). Also forms metasomatically in metamorphic rocks where NaCl is introduced, and in altered igneous rocks affected by pneumatolytic activity. Commonly found in skarns at the contact of igneous intrusions and carbonate rocks. Rarely found as phenocrysts in latite.

Distinguishing features. Low birefringent scapolite superficially resembles a number of minerals. However, it is usually distinguished by its uniaxial −ve character and straight extinction. Cancrinite has a lower RI than scapolite, and a higher δ than low RI scapolite. Nepheline and beryl differ in their habit and lack of good cleavage.

F. Volcanic Glass

Volcanic glass, of course, is not a mineral. However, it is a substance commonly encountered in microscope work, and a description of it is quite relevant to the purpose of this book.

No. 91. VOLCANIC GLASS (isotropic)

Amorphous. May be massive, vesicular, or show perlitic cracking. Some volcanic rocks are made entirely of glass, but others contain glass as a groundmass to crystals of feldspar, pyroxene, etc. Many glasses are more or less devitrified to felsite, spherulites, and/or crystallites of various kinds.

Color in thin section. Usually colorless, grey, brown or red.

Optical properties. Isotropic.
n = 1.485–1.62 (Figure 9.78).
RI generally increases with a decrease in SiO_2 content of the glass (Figure 9.78). However, many other factors (e.g., H_2O content) also control RI, and a close correlation of SiO_2 with RI is not possible. Mathews (1951) has shown that good correlations between RI and SiO_2 can be made for glasses from particular rock suites produced by artificial fusion. Curves for three suites of artificially fused glass are given in Figure 9.78.

Occurrence. In lavas and hypabyssal igneous intrusions such as dykes, occasionally making up the bulk of the rock as in obsidian, but more frequently forming the groundmass and containing abundant small crystals and phenocrysts. Also as shards, scoria, and other fragments in pyroclastic rocks. In addition to devitrification products, glass may be altered to "palagonite," a green, waxy, hydrated substance, or replaced by zeolites.

Figure 9.78 RI variation with SiO_2 content of volcanic glass for natural examples and three suites of fused glass. Based on Mathews (1951, Figure 2).

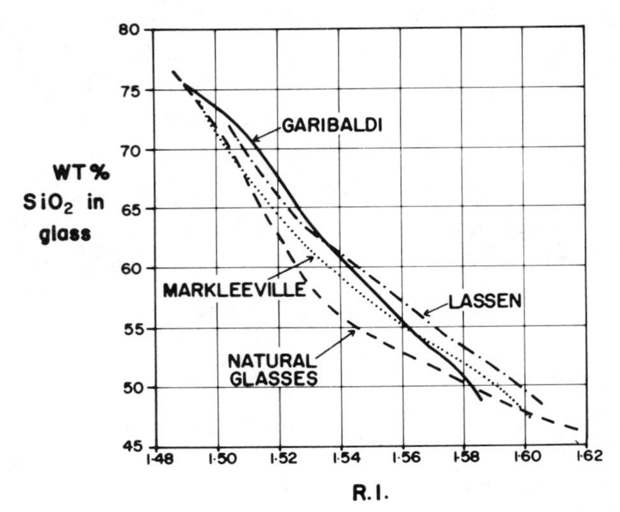

Distinguishing features. The occurrence and isotropic character are distinctive. The RI is higher than that of opal.

G. Nonsilicates

Of the nonsilicates, only the carbonates, iron oxides, the bauxite minerals, and the various salts found in evaporites occur as the prime constituents of important rock masses. Many of the other nonsilicates are important as accessory minerals in a wide range of rock types. They are described here under the headings: elements; oxides and hydroxides; sulphides; halides; sulphates; carbonates; phosphates; and tungstates. Minerals such as galena, which are important economic minerals but unimportant in petrology, are omitted.

(a) Elements

No. 92. GRAPHITE (opaque) C

Hexagonal. Perfect {0001} cleavage. Often very fine grained; crystals are hexagonal plates elongate parallel to the cleavage. Very soft, with a soapy feel.

Color in thin section. Opaque; black with a metallic lustre in reflected light.
Tabular crystals with perfect cleavage are often crinkled in a manner similar to micas.

Occurrence. In slates, schists, gneisses, marbles, and veins associated with these rocks; often finely disseminated. The grey color of marble may be due to the presence of fine-grained graphite. Also rarely found as an accessory mineral in plutonic and volcanic igneous rocks.

Distinguishing features. The opaque character, black color, platy crystals, and softness are distinctive. May be confused with magnetite or ilmenite, but both these minerals are harder; in addition, magnetite is magnetic and lacks cleavage, and ilmenite plates are not commonly crinkled or bent. Ilmenite may be altered to white (opaque) leucoxene.

(b) Oxides and Hydroxides

No. 93. PERICLASE (isotropic) MgO

Cubic. Perfect cubic cleavage. Crystals may be euhedral cubes or octahedra; also as anhedral grains.

Color in thin section. Colorless.

Optical properties. Isotropic.
n = 1.736–ca. 1.745.

Occurrence. In medium- or high-grade thermally metamorphosed marbles and dolomites, especially near igneous contacts where dolomite has dissociated. Commonly present only as relic patches in grains altered to brucite.

Distinguishing features. The high relief, isotropic character, perfect cubic cleavage, occurrence, and alteration to brucite are distinctive.

No. 94. CORUNDUM (−ve) Al_2O_3

Ruby (red) and *sapphire* (blue) are precious varieties.
Trigonal. No cleavage, but well-developed {0001} parting common. Euhedral crystals are often tapering pyramids ("barrel-shaped") with hexagonal cross sections and the {0001} form developed; also anhedral, granular. Simple or lamellar twinning with $\{10\bar{1}1\}$ composition planes common.

Color in thin section. Colorless, or pale blue, green, yellow, or pink; often color-zoned or with an irregular distribution of colors; weak to strong pleochroism with $\omega > \varepsilon$.

Color in detrital grains. As above but deeper colors, and often with marked pleochroism.

Optical properties. Uniaxial −ve; may be anomalously biaxial.

$$\omega = 1.765\text{–}1.772,\ \varepsilon = 1.759\text{–}1.763.$$

$$\delta = 0.005\text{–}0.009.$$

Fe-rich corundum with $\omega = 1.794$ and $\varepsilon = 1.785$ has been reported.

Occurrence. Found in nepheline-syenites and syenites, and associated pegmatites; also in quartz-free plagioclase-rich "dykes" cutting mafic or ultramafic rocks. Al-rich xenoliths in igneous rocks may contain corundum, often associated with spinel. Corundum also occurs relatively rarely in a wide variety of metamorphic rocks, especially Al-rich pelites, some carbonates and rocks desilicated near igneous intrusions. It is the prime constituent of emery (metamorphosed bauxite). Locally, it may be important as a detrital mineral. Corundum alters most commonly to micas, diaspore, gibbsite, or other Al-rich minerals.

Distinguishing features. The high relief combined with a low δ and uniaxial −ve character is distinctive. Corundum is also distinguished by its hardness in hand specimen.

No. 95. CASSITERITE (+ve) SnO_2

Tetragonal. Poor {100} and {110} cleavages. Euhedral crystals may be short prisms with pyramidal terminations, elongate prisms or acicular; also subhedral or anhedral. Twins common on {011} (geniculate or "kneelike" twinning).

Color in thin section. Colorless, yellow, brown or red; pleochroism weak or strong with $\varepsilon > \omega$.

Optical properties. Uniaxial +ve.

$$\omega = 1.990\text{–}2.010,\ \varepsilon = 2.093\text{–}2.100.$$

$$\delta = 0.090\text{–}0.103.$$

Crystals have extinction parallel to prism faces but oblique to twin planes; the length-slow character is difficult to determine because of the very high δ.

Occurrence. Most commonly found together with W-, Li-, B- and F-rich minerals in granite pegmatites, and associated greisen and veins; also as an accessory mineral in some granites. Locally it may be important as a detrital mineral.

Distinguishing features. The extreme relief, very high δ, and uniaxial +ve character are distinctive. May be confused with rutile and occasionally zircon, but the δ of rutile is higher and that of zircon lower; these distinctions may be difficult to make in thick detrital grains which display indeterminable very high-order interference colors.

No. 96. RUTILE (+ve) TiO_2

Tetragonal. Distinct {110} and {100} cleavages. Euhedral crystals are short to elongate or acicular prisms; less commonly anhedral, granular. Twins common on {011} (geniculate or "knee-like" twinning).

Color in thin section. Usually red, red-brown, or yellowish; often weak but sometimes strong pleochroism with $\varepsilon > \omega$.

Color in detrital grains. Usually deep colors as above, often almost opaque.

Optical properties. Uniaxial +ve; sometimes anomalously biaxial.

$$\omega = 2.605\text{–}2.616,\ \varepsilon = 2.899\text{–}2.903.$$

$$\delta = 0.286\text{–}0.294.$$

Crystals are characterized by indeterminable very high-order interference colors which may be confused with first-order white. Insertion of the sensitive-tint plate will not effect any observable change in the very high-order colors. Prismatic crystals have straight extinction.

Occurrence. A widely distributed accessory mineral, especially in metamorphic rocks and less commonly in igneous rocks. Usually as small prismatic crystals. Rutile needles are commonly embedded in other minerals forming regular patterns parallel to crystal faces, especially in amphibole and chlorite resulting from the alteration of pyroxene. Also found in veins, and as a common detrital mineral. May be altered to white (opaque) leucoxene.

Distinguishing features. The extreme relief, very high δ, crystal habit, and color are distinctive. May be confused with the less common minerals cassiterite, brookite and anatase; the δ of these three minerals is lower although this may be difficult to determine in thick detrital grains. Anatase is −ve, and brookite has indistinct extinction due to its high dispersion.

No. 97. ANATASE (−ve) TiO_2

Tetragonal. Perfect {001} and {111} cleavages. Pyramidal crystals resembling octahedra.

Color in thin section. Colorless, yellow, brown, blue or black; usually weak pleochroism, $\omega < \varepsilon$ or $\omega > \varepsilon$.

Color in detrital grains. Colorless or deep colors as above.

Optical properties. Uniaxial −ve.

$$\omega = 2.561, \varepsilon = 2.488.$$

$$\delta = 0.073.$$

Occurrence. Principally as a detrital or diagenetic mineral, sometimes prominent in heavy-mineral suites. Also in veins and altered igneous and metamorphic rocks. May be altered to white (opaque) leucoxene.

Distinguishing features. The extreme relief, color, very high δ, shape and cleavages are characteristic. Distinguished from rutile and brookite by its −ve character and lower δ. In addition, brookite differs in its indistinct extinction.

No. 98. BROOKITE (+ve) TiO_2

Orthorhombic. No good cleavage. Crystals are often tabular parallel to {100} and slightly elongate parallel to *c*.

Color in thin section. Pale to dark yellow or brown; pleochroism absent or weak.

Color in detrital grains. Deeper colors as above.

Optical properties. Biaxial +ve. $2V_z = 0°–30°$. Crossed axial plane dispersion.

$$\alpha = 2.583, \beta = 2.584 - 2.586, \gamma = 2.700–2.741.$$

$$\delta = 0.117–0.158. X = c \text{ or } a, Y = a \text{ or } c, Z = b.$$

Brookite has a very strong dispersion which causes incomplete extinction.

Occurrence. Principally as a detrital or diagenetic mineral, sometimes prominent in heavy-mineral suites. Also in veins and altered igneous and metamorphic rocks. May be altered to white (opaque) leucoxene.

Distinguishing features. The extreme relief, color, very high δ and incomplete extinction are distinctive. May be confused with rutile and anatase, but both these minerals usually extinguish completely, rutile has a higher δ, and anatase is −ve.

No. 99. PEROVSKITE (+ve, pseudo-isotropic) $CaTiO_3$

There may be substitution of Nb for Ti, and Na, Fe^{2+}, or the rare earths for Ca.
Monoclinic? (pseudocubic). Pseudocubic cleavage, not usually observed in small crystals. Crystals may be cubic or octahedral in habit. Complex multiple twinning on {111} common.

Color in thin section. Pale to very dark brown, rarely grey or green (Nb-rich varieties); pleochroism absent or very weak with $Z > X$.

Optical properties. isotropic, or biaxial with $2V$ = ca. 90°.

$$n = 2.30–2.38.$$

$$\delta = 0.00–0.002.$$

Occurrence. An accessory mineral in some feldspathoidal or melilite-bearing igneous rocks, and in some carbonatites. Also found in thermally metamorphosed carbonate rocks.

Distinguishing features. The extreme relief and very low or zero δ are distinctive. Garnet has a lower relief and a different crystal habit. Very dark, almost opaque varieties of perovskite may be confused with the other opaque minerals.

No. 100. SPINEL GROUP (isotropic or opaque)

The spinel group can be divided into the following three series:

No. 100A. Spinel series $(Mg,Fe^{2+},Zn,Mn)Al_2O_4$
No. 100B. Chromite series $(Fe^{2+},Mg)Cr_2O_4$
No. 100C. Magnetite series $(Fe^{2+},Mg,Zn,Mn,Ni)Fe^{3+}_2O_4$

Cubic. No cleavage. Euhedral crystals are octahedra; also anhedral. Twinning with {111} twin planes—not usually visible in thin section.

Spinel series

Includes *spinel* $MgAl_2O_4$, *pleonaste* $(Mg,Fe^{2+})Al_2O_4$, *hercynite* $Fe^{2+}Al_2O_4$, and *gahnite* $ZnAl_2O_4$; *picotite* is Cr-rich hercynite. Pure end members are rare.

Color in thin section. Colorless or a wide variety of colors, but especially blue and brown (spinel), green or blue-green (pleonaste), dark green to almost black and opaque (hercynite) and dark blue-green (gahnite).

Optical properties. Isotropic.
n varies in the range ca. 1.715 (spinel)–1.80 (gahnite)–1.835 (hercynite)–1.98 (picotite).
Diamond- or square-shaped sections common.

Occurrence. Spinel and pleonaste are most commonly found in high-grade metamorphosed carbonate rocks often associated with chondrodite, phlogopite, or forsterite, and in Al-rich, Si-poor schists, and xenoliths. Hercynite is found in more Fe-rich aluminous schists, and may occur with quartz in Si-rich granulites. Hercynite and pleonaste are found rarely as accessories in basic and ultramafic igneous rocks. Picotite occurs in ultramafic rocks, and gahnite in granite pegmatites and Zn ores. Members of the spinel series are found intergrown with hornblende in corona structures between olivine and plagioclase in gabbroic rocks.

Distinguishing features. The habit, high relief, isotropic nature and colors are distinctive. Periclase differs in its perfect cleavage, and garnet has a different habit and is more commonly pink than spinel.

Chromite series

Chromite (sensu stricto) is $Fe^{2+}Cr_2O_4$, but most natural crystals contain substantial Mg replacing Fe^{2+}.

Color in thin section. Usually dark brown or opaque; opaque grains are usually translucent at their edges.

Optical properties. Isotropic.
n = ca. 2.00–2.16.
Diamond- or square-shaped sections common.

Occurrence. In ultramafic rocks such as dunite, peridotite, and serpentinite, sometimes concentrated in layers. Locally it may be important in detrital heavy-mineral suites.

Distinguishing features. The deeply colored almost opaque nature, crystal habit and isotropism are distinctive. Deeply colored melanite-garnet has a different habit and occurrence.

Magnetite series

Magnetite ($Fe^{2+}Fe_2^{3+}O_4$) and varieties with some replacement of Fe^{2+} by Mg are the common members of this series. Some Ti may be present.

Color in thin section. Opaque; black with a metallic lustre in reflected light.
Diamond- or square-shaped sections; also commonly granular.

Occurrence. Very abundant as an accessory mineral or as a principal constituent in a wide variety of igneous and metamorphic rocks, especially basic igneous rocks. Also a common detrital mineral, sometimes concentrated by stream or tidal action to form magnetite sands.

Distinguishing features. The opacity, black color and metallic lustre in reflected light, strongly magnetic character, and crystal habit (if developed) are distinctive. May be confused with graphite or ilmenite, but neither of these minerals is strongly magnetic, and both commonly crystallize as hexagonal plates. In addition, graphite is very soft, and ilmenite may be altered to white (opaque) leucoxene.

No. 101. ILMENITE (opaque) $FeTiO_3$

Trigonal. No good cleavage. Euhedral crystals are thin hexagonal plates or tablets; often skeletal crystals; also anhedral, granular.

Color in thin section. Opaque; black with a metallic lustre in reflected light. Euhedral crystals appear as elongate plates in thin section. Ilmenite is commonly altered to opaque *leucoxene* (TiO_2) which is white in reflected light (often looks like cotton-wool).

Occurrence. A common accessory mineral in a wide range of igneous and metamorphic rocks, especially mafic and ultramafic types. Often found with magnetite. A common detrital mineral, sometimes concentrated by stream or tidal action to form ilmenite sands.

Distinguishing features. The opacity, black color and metallic lustre in reflected light and crystal habit (if developed) are distinctive. May be confused with graphite and magnetite, but graphite is very soft and often crinkled, and magnetite has a different habit and is strongly magnetic. The alteration of ilmenite to leucoxene is distinctive.

No. 102. HEMATITE (−ve, often opaque) Fe_2O_3

Trigonal. No good cleavage. Flaky with {0001} well-developed or rhombohedral crystals; commonly anhedral, massive, earthy.

Color in thin section. Occasionally red, translucent, especially at the edges of grains; commonly opaque and black with a metallic lustre in reflected light.

Optical properties. Opaque or uniaxial −ve.

$$\omega = 3.15\text{–}3.22, \ \varepsilon = 2.87\text{–}2.94.$$

$$\delta = 0.28.$$

Occurrence. Rare as a primary mineral in igneous rocks, but common as an alteration product and fumarole deposit. Very common in some metamorphosed Fe-rich rocks, especially the Precambrian banded iron ores in which it occurs as bands alternating with quartz. Hematite is a common red coloration and cement in sediments, and may be present in soils. Also found in veins.

Distinguishing features. The extreme relief, very high δ and red color of translucent hematite are distinctive. Opaque hematite may be difficult to distinguish from magnetite, but hematite has a distinctive red streak (the streak of magnetite is black); magnetite is also strongly magnetic. Goethite, lepidocrocite, and limonite have yellow or brown streaks, goethite is often fibrous, and limonite is isotropic.

No. 103A. GOETHITE (−ve) $FeO \cdot OH$

No. 103B. LEPIDOCROCITE (−ve) $FeO \cdot OH$

Limonite is amorphous or cryptocrystalline goethite or lepidocrocite with absorbed water.

Orthorhombic. Perfect {010} cleavage. Crystals of goethite are often fibrous parallel to *c*, whereas lepidocrocite is often platy parallel to {010}; commonly massive, earthy, stalactitic, etc.

Color in thin section. Yellow or brown; the pleochroism of goethite is variable; lepidocrocite is strongly pleochroic with $Z > Y > X$.

Optical properties. *Goethite:* biaxial −ve. $2V_x = 0°\text{–}27°$. $r > v$ (extreme).

$$\alpha = 2.217\text{–}2.275, \ \beta = 2.346\text{–}2.409, \ \gamma = 2.356\text{–}2.415.$$

$$\delta = 0.139\text{–}0.140. \ X = b, \ Y = c \text{ or } a, \ Z = a \text{ or } c.$$

Lepidocrocite: biaxial −ve. $2V_x = 83°$. Weak dispersion.

$$\alpha = 1.94,\ \beta = 2.20,\ \gamma = 2.51.$$

$$\delta = 0.57.\ X = b,\ Y = c,\ Z = a.$$

Limonite is isotropic with n = ca. 2.0.

Occurrence. Both goethite and lepidocrocite form as common alteration products of Fe-bearing minerals under oxidizing conditions. Found in soils, bogs, laterites, and sedimentary iron ores. Yellow ochre contains goethite, brown ochre contains lepidocrocite.

Distinguishing features. The extreme relief, yellow-brown color, and occurrence are distinctive. Goethite and lepidocrocite may be difficult to distinguish, although their birefringences differ. Limonite is distinguished by its isotropic character. Hematite may be most easily distinguished from all three by its red streak.

No. 104. BRUCITE (+ve) $Mg(OH)_2$

Fe and Mn may substitute for some of the Mg.
Trigonal. Perfect {0001} cleavage. Crystals are very soft micalike {0001} plates, or fibers elongate parallel to ω.

Color in thin section. Colorless.

Optical properties. Uniaxial +ve; fibers may be anomalously biaxial.

$$\omega = 1.559\text{–}1.59,\ \varepsilon = 1.579\text{–}1.60.$$

$$\delta = 0.010\text{–}0.021.$$

Anomalous red-brown interference colors may replace first-order yellow-orange. Plates and fibers are always length fast.

Occurrence. Most commonly found in thermally metamorphosed carbonate rocks, often as an alteration product of periclase. Also in veins cutting serpentinite and chlorite-rich rocks.

Distinguishing features. The soft, platy, or fibrous crystals, low to moderate relief and δ, +ve character, and occurrence are distinctive. May resemble some micas, talc, chlorite, or serpentine, but the length-fast character of brucite differs from all these except +ve chlorite which is usually green and has a lower δ.

No. 105. GIBBSITE (+ve) $Al(OH)_3$

Also known as *hydrargillite*.
Monoclinic, $\beta = 94°$. Perfect {001} cleavage. Hexagonal plates parallel to {001}; most commonly subhedral lamellae and in concretionary ag-

gregates; usually very fine grained. Multiple twinning with {001} composition planes common.

Color in thin section. Colorless or pale brown.

Optical properties. Biaxial +ve. $2V_z$ usually <20°, changes on heating. $r<$ or $>v$ (strong).

$$\alpha = \beta = 1.565\text{–}1.571,\ \gamma = 1.580\text{–}1.595.$$

$$\delta = 0.014\text{–}0.030.\ \text{At room temperatures } X = b,\ Y \wedge a = 25°,\ Z \wedge c = 21°.$$

Orientation diagrams. Crystals are usually too small for an exact determination of optical properties.
(001) section (Figure 9.79a): off-centered acute bisectrix figure; δ' very low.
(010) section (Figure 9.79b): obtuse bisectrix (flash) figure; δ' (moderate) usually ca. 0.02; inclined extinction with the angle between the fast direction and the good cleavage = 25°.
(100) section (Figure 9.79c): off-centered flash figure; δ' (moderate) usually ca. 0.02; straight extinction and fast parallel to the cleavage.

Figure 9.79 Orientation diagrams for gibbsite.

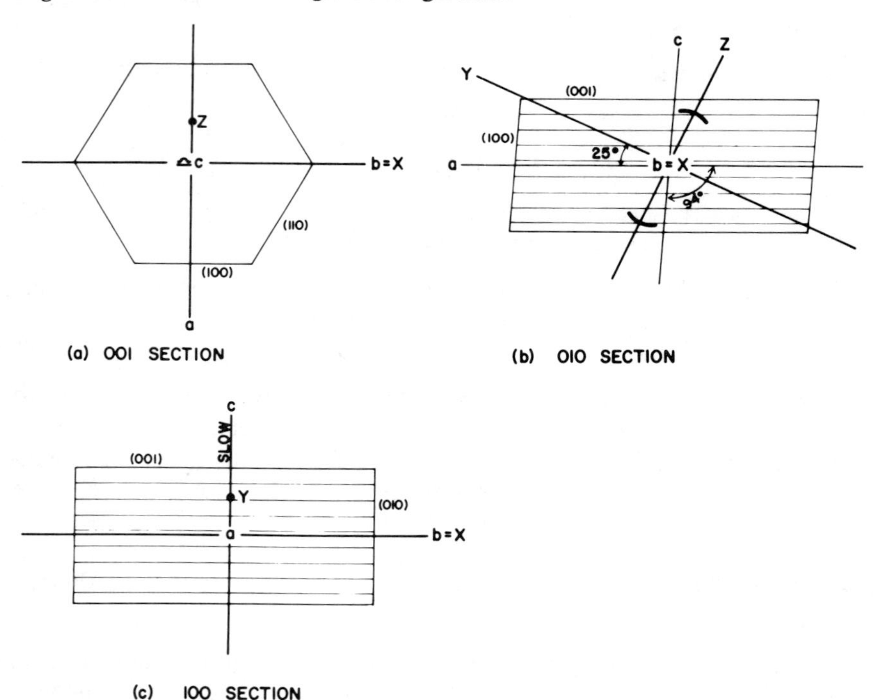

Occurrence. One of the prime constituents of bauxite, usually with diaspore and boehmite, and in laterites and clays formed under similar conditions; also in emery deposits. Gibbsite occurs in veins and cavities in some Al-rich igneous rocks, and as an alteration product of corundum.

Distinguishing features. Superficially resembles brucite, the micas and clay minerals. The inclined extinction in (010) sections and the occurrence of gibbsite distinguish it from brucite, and the length-fast character distinguishes it from micas and the clay minerals. Gibbsite differs from associated diaspore and boehmite in its lower RI.

No. 106. DIASPORE (+ve) AlO(OH)

Orthorhombic. Perfect {010} cleavage, and other imperfect prismatic cleavages. Crystals are tabular parallel to {010} and/or elongate parallel to *c;* often in massive fine-grained aggregates.

Color in thin section. Usually colorless, sometimes pink or brown pleochroic with $Z > Y > X$.

Optical properties. Biaxial +ve. $2V_z = 84°–96°$. $r < v$ (weak).

$$\alpha = 1.685–1.706, \beta = 1.705(?)–1.725, \gamma = 1.730–1.752.$$

$$\delta = 0.040–0.050. X = c, Y = b, Z = a.$$

Crystals are usually length fast with the OAP parallel to the length and good cleavage traces.

Occurrence. One of the prime constituents of bauxite, usually with gibbsite and boehmite, and in laterites and clays formed under similar conditions. Diaspore also occurs in emery deposits, in altered Al-rich rocks, and as an alteration product of Al-rich minerals such as corundum, kyanite, and andalusite.

Distinguishing features. The high relief and δ, the OAP parallel to the good cleavage, and occurrence are distinctive. May be confused with sillimanite or micas, but diaspore differs in its length-fast character, and it has a higher RI than the micas. Diaspore differs from associated gibbsite and boehmite in its higher RI and δ.

No. 107. BOEHMITE (+ve) AlO(OH)

Orthorhombic. Perfect {010} and good {100} and {001} cleavages. Euhedral crystals dominated by {111} and {010} faces, but usually extremely fine-grained.

Color in thin section. Colorless.

Optical properties. Biaxial +ve. $2V_z = 74° - 86°$.

$$\alpha = 1.640\text{–}1.649,\ \gamma = 1.660\text{–}1.668.$$

$$\delta = 0.015\text{–}0.020.\ X = c,\ Y = b,\ Z = a.$$

Crystals have straight or symmetrical extinction.

Occurrence. One of the prime constituents of bauxite, usually with gibbsite and diaspore, and in laterites and clays formed under similar conditions. Also found as an alteration product of feldspar and nepheline in syenite, etc.

Distinguishing features. The occurrence is distinctive. The fine-grained nature of boehmite does not normally allow determination of its optical properties. It differs from associated gibbsite in its higher RI, and from diaspore in its lower RI and δ.

(c) Sulphides

No. 108A. PYRITE (opaque) FeS_2

No. 108B. MARCASITE (opaque) FeS_2

Pyrite is cubic; marcasite is orthorhombic. Neither has a good cleavage. Pyrite crystallizes as cubes or pyritohedra (Figure 1.5), marcasite as {010} tablets, fibrous aggregates and concretions; pyrite may also be massive, granular.

Color in thin section. Opaque; brassy yellow with a metallic lustre in reflected light.

Occurrence. Pyrite has a wide range of occurrences, as a primary and secondary mineral in many igneous and metamorphic rocks, in skarns, in fumaroles, and as a diagenetic mineral in sediments, especially muds laid down in shallow water and reducing conditions. Pyrite is also common in veins and fissures. Marcasite is restricted to sediments and veins. Pyrite and marcasite readily alter to iron oxides, particularly limonite, sometimes accompanied by gypsum or other sulphates.

Distinguishing features. The opacity and brassy-yellow color in reflected light are distinctive. Pyrite and marcasite may be confused with pyrrhotite and chalcopyrite. The euhedral cubes and pyritohedra of pyrite are distinctive, but massive material is less easy to identify. Pyrrhotite is magnetic and has a bronze color in reflected light and chalcopyrite is softer and golden in color. Marcasite is usually distinguished from pyrite by its generally fibrous nature and more restricted occurrence.

No. 109. PYRRHOTITE (opaque) $Fe_{1-x}S(x = 0\text{–}0.2)$

Monoclinic (pseudohexagonal), β = ca. 90°. No good cleavage, but frequently has a {0001} parting. Usually massive, granular, or lamellar; euhedral crystals are hexagonal tablets.

Color in thin section. Opaque; bronze color with a metallic lustre in reflected light.

Occurrence. As a primary mineral in some basic igneous rocks. More rarely found in thermally metamorphosed carbonate rocks, and some schists. Also forms as a diagenetic mineral in muds laid down under reducing conditions, and in veins. Pyrrhotite, like pyrite, alters to limonite.

Distinguishing features. Pyrrhotite resembles pyrite, but pyrite is brassy yellow in reflected light, and euhedral crystals differ in habit. Pyrrhotite, unlike pyrite, is magnetic, and will react to a hand-magnet. Chalcopyrite differs in its golden-yellow color.

No. 110. CHALCOPYRITE (opaque) $CuFeS_2$

Tetragonal. No good cleavage. Usually massive; euhedral crystals are tetrahedral in shape.

Color in thin section. Opaque; golden-yellow with a metallic lustre in reflected light.

Occurrence. As a primary mineral in some basic igneous rocks, often associated with pyrite and pyrrhotite, and as a diagenetic or secondary mineral in sediments. Also in veins and a wide variety of metasomatic deposits.

Distinguishing features. The opacity and golden-yellow color in reflected light are distinctive. It resembles pyrite and pyrrhotite, but pyrite is harder and has a brassy color, and pyrrhotite is magnetic and has a bronze color.

(d) Halides

No. 111. FLUORITE (isotropic) CaF_2

Cubic. Perfect {111} cleavages. Euhedral crystals are cubes; often anhedral, interstitial.

Color in thin section. Colorless or commonly blue or purple; often color zones or with an irregular distribution of colors.

Optical properties. Isotropic.

n = 1.433–ca. 1.44, increases with rare-earth content.
The {111} cleavages are usually visible in thin section as two sets intersecting at ca. 70°. Crushed cleavage-bound fragments are triangular in outline.

Occurrence. Found as an interstitial accessory mineral (possibly of hydrothermal origin) in igneous rocks, especially acid and alkaline types; also in pegmatites and fumarole deposits. Fluorite is common in some hydrothermal veins, and also occurs as a detrital mineral and more rarely as a cement in sediments.

Distinguishing features. The low RI, isotropism, perfect cleavages and purple color (if present) are distinctive. Opal lacks cleavage and color.

No. 112. HALITE (isotropic) NaCl

Common rock-salt.
See Figure 5.8.
Cubic. Perfect cubic cleavage. Euhedral crystals are cubes; commonly massive, anhedral, granular. Soluble in water and has a salty taste.

Color in thin section. Colorless.

Optical properties. Isotropic.

n = 1.544.

Occurrence. An evaporite from seawater or salt lakes. May be found as scattered euhedra in some sediments, but most commonly occurs in massive beds, often intruded as massive salt domes or walls.

Distinguishing features. The isotropism, very low relief, occurrence and solubility in water are distinctive. For the properties of associated salts see Table 9.8.

Nos. 113–118. OTHER WATER-SOLUBLE SALTS FOUND IN EVAPORITES

Halite is by far the most abundant of the water-soluble salts found in evaporites. The properties of halite and the other water-soluble salts most commonly encountered are summarized in Table 9.8, the data coming mainly from Borchert and Muir (1964) to which reference should be made for details of other evaporite minerals. Because of their solubility in water, they require special methods of thin-section preparation, and a suitable procedure has been described by Bennett (1958).

Table 9.8 Properties of Water-Soluble Salts Found in Evaporites

Mineral	Crystal system and cleavage	Taste	RI and δ	$2V$ and orientation
No. 112 Halite $NaCl$	Cubic. Perfect cubic cleavage	salty	$n = 1.544$	isotropic
No. 113 Sylvite KCl	Cubic. Perfect cubic cleavage	salty, bitter	$n = 1.490$	isotropic
No. 114 Carnallite $KCl \cdot MgCl_2 \cdot 6H_2O$	Orthorhombic. No cleavage	bitter	$\alpha = 1.466$, $\beta = 1.475$ $\gamma = 1.494$, $\delta = 0.028$	$2V_z = 66°$ $r < v$ (weak). $X = c$, $Y = b$, $Z = a$
No. 115 Langbeinite $K_2SO_4 \cdot 2MgSO_4$	Cubic. No cleavage	tasteless	$n = 1.534$	isotropic
No. 116 Kainite $4KCl \cdot 4MgSO_4 \cdot 11H_2O$	Monoclinic. Distinct {100} and {110} cleavages	salty, slightly bitter	$\alpha = 1.494$, $\beta = 1.505$ $\gamma = 1.516$, $\delta = 0.022$	$2V_x = 85°$ $r > v$ (weak). $Y = b$, $Z \wedge c = 13°$
No. 117 Kieserite $MgSO_4 \cdot H_2O$	Monoclinic. Distinct {111} and {110} cleavages	tasteless	$\alpha = 1.520$, $\beta = 1.533$ $\gamma = 1.584$, $\delta = 0.064$	$2V_z = 55°$ $r > v$ (distinct). $Y = b$, $X \wedge c = 41°$
No. 118 Polyhalite $K_2SO_4 \cdot MgSO_4 \cdot 2CaSO_4 \cdot 2H_2O$	Triclinic. Distinct {100} cleavage	tasteless	$\alpha = 1.547$, $\beta = 1.560$ $\gamma = 1.567$, $\delta = 0.020$	$2V_x = 64°$ $r < v$ (weak). orientation uncertain

Note: Gypsum, anhydrite, and the carbonates are common in evaporites.

(e) Sulphates

The properties of a number of water-soluble sulphates found in evaporites are given in Table 9.8.

No. 119A. BARYTE (+ve) $BaSO_4$

No. 119B. CELESTINE (+ve) $SrSO_4$

Orthorhombic. Perfect {001}, good {210} and distinct {010} cleavages. Euhedral crystals are {001} tablets which may be elongate parallel to *a* or *b*; often anhedral, granular. Baryte may have lamellar twinning on {110}.

Color in thin section. Colorless.

Optical properties. *Baryte*: biaxial +ve. $2V_z = 37°$. $r < v$ (weak).

$$\alpha = 1.636,\ \beta = 1.637,\ \gamma = 1.648.$$

$$\delta = 0.012.\ X = c,\ Y = b,\ Z = a.$$

Celestine: biaxial +ve. $2V_z = 51°$. $r < v$ (distinct).

$$\delta = 1.622,\ \beta = 1.624,\ \gamma = 1.631.$$

$$\delta = 0.009.\ X = c,\ Y = b,\ Z = a.$$

Orientation diagrams. *(100) section* (Figure 9.80a): acute bisectrix figure; δ' (very low) < 0.002; two cleavages at 90° to each other and straight extinction; extinction may be indistinct due to proximity of optic axes; OAP across the best cleavage.
(001) section (Figure 9.80b): obtuse bisectrix figure; δ' (low) $= 0.011$ (baryte) or 0.007 (celestine); two good cleavages at 102° to each other with symmetrical extinction and the OAP parallel to a third cleavage.
(010) section (Figure 9.80c): flash figure; δ (low) $= 0.012$ (baryte) or 0.009 (celestine); straight extinction with the slow direction parallel to the well-developed {001} cleavage; an oblique cross-cleavage may be visible.

Occurrence. Baryte is most commonly found in hydrothermal veins, often with Pb and Zn minerals; also as concretions in sediments, especially limestones, and more rarely as a detrital mineral or as a cement in sandstones. Celestine is found in evaporites, and in veins and fissures in carbonate sediments; also in some metalliferous veins.

Distinguishing features. The cleavages, moderate relief, low δ and occurrence are distinctive. Baryte is not easy to distinguish from celestine; an accurate measurement of RI or $2V$ is necessary.

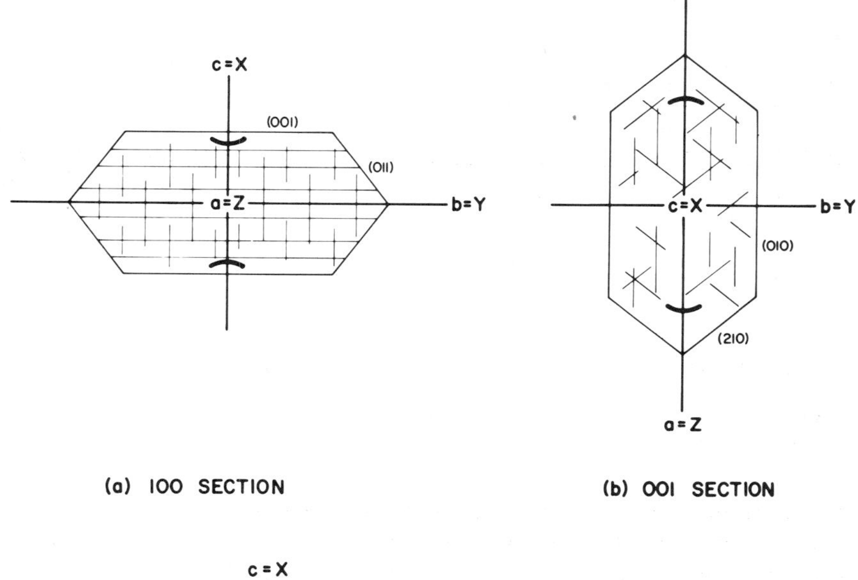

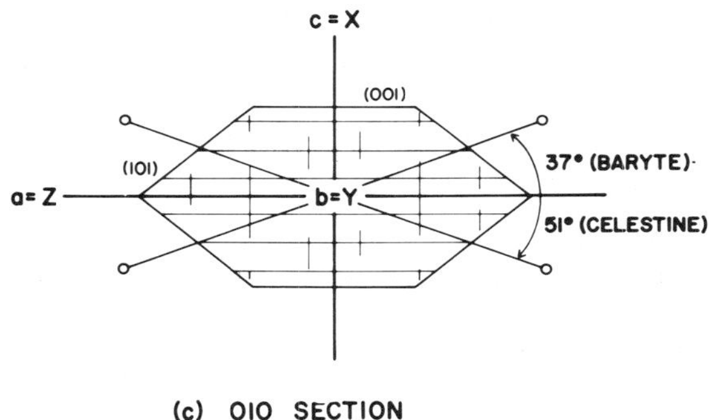

Figure 9.80 Orientation diagrams for baryte and celestine.

No. 120. ANHYDRITE (+ve) $CaSO_4$

Orthorhombic. Perfect {001} and good {010} and {100} cleavages which produce cube-shaped cleavage fragments. Usually massive anhedral, granular; sometimes fibrous. Simple or lamellar twinning on {010}; if two sets of twins are developed they intersect at ca. 96°.

Color in thin section. Colorless.

Optical properties. Biaxial +ve. $2V_z = 42°–44°$. $r < v$ (distinct).

$$\alpha = 1.569–1.574,\ \beta = 1.574–1.579,\ \gamma = 1.609–1.618.$$

$$\delta = 0.040–0.047.\ X = c,\ Y = b,\ Z = a.$$

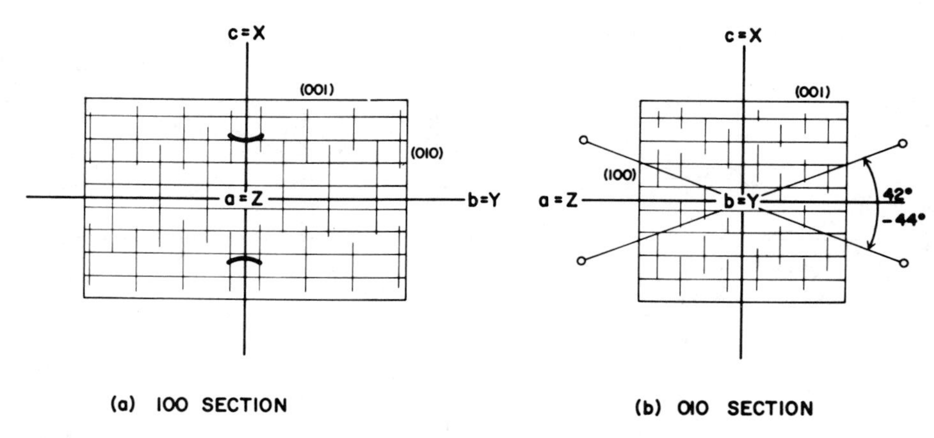

Figure 9.81 Orientation diagrams for anhydrite.

Orientation diagrams. *(100) section* (Figure 9.81a): acute bisectrix figure; δ′ (low) = 0.005; two good cleavages at 90° and straight extinction. *(010) section* (Figure 9.81b): flash figure; δ (high) = 0.040–0.047; two cleavages at 90°, straight extinction with the slow direction parallel to the best cleavage.
The (001) section displays two cleavages at 90°, provides an obtuse bisectrix figure and has a high δ′ (ca. 0.035).

Occurrence. One of the prime constituents of evaporite deposits, often with gypsum and halite. It may hydrate to gypsum or form by the dehydration of gypsum. It also occurs in sediments and veins as a result of the weathering of sulphides. Also found occasionally in amygdales, and in fumarole deposits.

Distinguishing features. The occurrence, pseudocubic cleavages, and high δ are distinctive. Gypsum, baryte and celestine all have a much lower δ than anhydrite.

No. 121. GYPSUM (+ve) $CaSO_4 \cdot 2H_2O$

See Figure 5.6.
Monoclinic, β = 114°. Perfect {010} and distinct {100} and {011} cleavages. Euhedral crystals are {010} tablets with well-developed {011} and {$\bar{1}$11} forms; often massive, subhedral to anhedral, granular; also fibrous parallel to *c*. Twinning on {100}; lamellar twinning may be produced during thin-section making.

Color in thin section. Colorless.

Optical properties: biaxial +ve. $2V_z = 58°$. $r > v$ (strong).

$$\alpha = 1.519–1.521,\ \beta = 1.522–1.526,\ \gamma = 1.529–1.531.$$

$$\delta = 0.010.\ X \wedge c = 38°,\ Y = b,\ Z \wedge a = 14°.$$

The $2V$ decreases with increase in temperature so that at 91°C, $2V_z = 0°$.
Gypsum may partially dehydrate during thin-section making to form the fibrous hemihydrate $CaSO_4 \cdot \frac{1}{2}H_2O$ (*plaster of paris*). The hemihydrate has $\alpha = 1.559$ and $\gamma = 1.583$, i.e., an RI and δ intermediate between those of gypsum and anhydrite; it readily reabsorbs water to form gypsum.

Orientation diagrams. *(010) section and cleavage fragments* (Figure 9.82a): flash figure; δ (low) = 0.010; two distinct cleavages at 114° to each other; inclined extinction with $Z \wedge$ {100} cleavage traces = 52°. *Sections parallel to b* (Figure 9.82b): OAP is parallel to the perfect {010} cleavage; acute or obtuse bisectrix or optic-axis figure, centered or off-centered; may be two oblique cross cleavages {011} or a single cross-cleavage {100} at 90° to {010}; δ' is low or very low.

Occurrence. One of the prime constituents of evaporite deposits, often with anhydrite and halite. It may dehydrate to anhydrite or form by hydration of anhydrite. Also found in muddy and calcareous sediments, veins, fumarole deposits, and may be produced by the weathering of sulphides such as pyrite.

Figure 9.82 Orientation diagrams for gypsum.

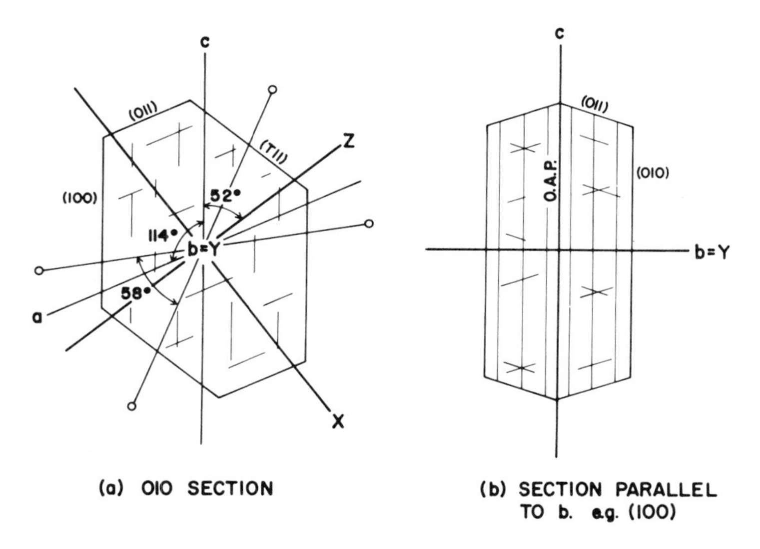

Distinguishing features. The low relief and δ, perfect cleavage, and occurrence are distinctive. The optical properties are similar to those of some zeolites, e.g., thomsonite, but their orientations and parageneses differ.

No. 122. ALUNITE (+ve) $KAl_3(SO_4)_2(OH)_6$

Trigonal. Good {0001} cleavage. Euhedral crystals may be pseudocubic rhombohedra or {0001} plates; often anhedral aggregates.

Color in thin section. Colorless.

Optical properties. Uniaxial +ve.

$$\omega = 1.572, \varepsilon = 1.592.$$

$$\delta = 0.020.$$

Platelike crystals and cleavage traces are length fast with straight extinction.

Occurrence. Found in hydrothermally altered acid volcanic rocks; also in veins.

Distinguishing features. The moderate relief and δ, uniaxial +ve character, and cleavage are distinctive, except for brucite which is very similar. However, the occurrences of brucite and alunite are different. Muscovite superficially resembles alunite but is length slow.

(f) Carbonates

No. 123. ARAGONITE (−ve) $CaCO_3$

May contain Sr.

Orthorhombic. Imperfect {010} cleavage. Euhedral crystals are acicular with steep pyramidal terminations, or {010} tablets with six-sided cross sections; also subhedral in columnar or fibrous aggregates, or anhedral. Lamellar twinning with {110} twin planes; also pseudohexagonal cyclic twinning on {110}. Effervesces like calcite in dilute HCl.

Color in thin section. Colorless.

Optical properties. Biaxial −ve. $2V_x = 18°$. $r < v$ (weak).

$$\alpha = 1.530\text{–}1.531, \beta = 1.680\text{–}1.682, \gamma = 1.685\text{–}1.686.$$

$$\delta = 0.155. X = c, Y = a, Z = b.$$

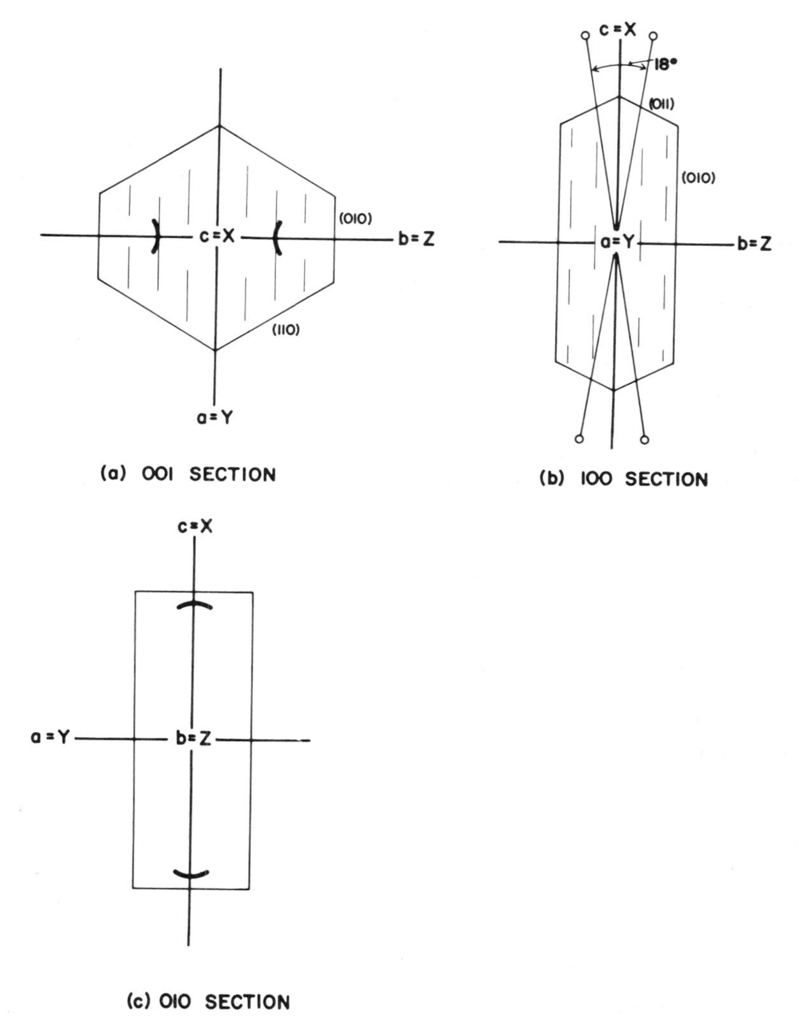

Figure 9.83 Orientation diagrams for aragonite.

Orientation diagrams. *(001) section* (Figure 9.83a): acute bisectrix figure; δ′ (low or very low) <0.005; poor single cleavage; extinction may be indistinct due to proximity of optic axes; intersecting sets of twins on {110} may be visible.

(100) section (Figure 9.83b): flash figure; δ (very high) = 0.155; single poor cleavage, straight extinction and length fast; twin planes are very oblique to this section.

(010) section (Figure 9.83c): obtuse bisectrix (flash) figure; δ' (very high) = 0.150; no cleavage, straight extinction and length fast in prismatic crystals; {110} twins parallel to *c* may be visible.

Occurrence. Aragonite is metastable at ordinary temperatures and pressures. Nevertheless, many calcareous shells are built of aragonite, sometimes together with calcite as in some bivalves, and aragonite is an important constituent of modern shallow-water calcareous sediments. It readily changes to calcite so that it is not found in most ancient limestones. Aragonite also occurs in vesicles and cavities in volcanic rocks, especially basic volcanics, and as a high-pressure modification of calcite in regionally metamorphosed rocks of the lawsonite–glaucophane schist facies.

Distinguishing features. The very high δ and effervescence in dilute acid may make distinction between aragonite and calcite difficult, especially in modern fine-grained carbonate sediments. Aragonite is distinguished by its single poor cleavage, small $2V$, and orientation of twin sets, but in fine-grained material, stain tests (referred to under calcite) may be necessary.

No. 124. THE TRIGONAL (RHOMBOHEDRAL) CARBONATE GROUP

No. 124A. Calcite (−ve) $CaCO_3$
No. 124B. Dolomite (−ve) $CaMg(CO_3)_2$
No. 124C. Ankerite (−ve) $Ca(Mg,Fe^{2+},Mn)(CO_3)_2$
No. 124D. Magnesite (−ve) $MgCO_3$
No. 124E. Siderite (−ve) $FeCO_3$
No. 124F. Rhodochrosite (−ve) $MnCO_3$

See Plates 39 and 40, Figures 5.5 and 9.85.

Chemical composition. The following solid-solution series are possible at low temperatures:

(1) Up to approximately 50% Mn may substitute for Ca to form Mn-rich calcite, but there is an immiscibility gap between this and rhodochrosite.
(2) Only small amounts of Mg may substitute for Ca in calcite. High-magnesium calcites in modern sediments commonly contain up to about 10 mol% $MgCO_3$, but these are unstable, and calcite in ancient limestones is usually close to pure $CaCO_3$.
(3) Dolomite is a true double salt with equal proportions of Ca and Mg. There is no solid solution series between dolomite and either calcite

or magnesite. There is complete solid solution with ankerite in which there is substantial replacement of Mg by Fe^{2+} and Mn.

(4) There are solid solution series between magnesite and siderite and between siderite and rhodochrosite.

Crystallography. Trigonal. Perfect $\{10\bar{1}1\}$ cleavages (rhombohedral), *Caicite* may be euhedral, commonly as elongate prisms with scalenohedral or rhombohedral terminations, or as rhombohedra; also anhedral, massive, stalactitic, etc. *Dolomite, ankerite,* and *siderite* may form euhedral rhombohedra; also massive, anhedral. *Magnesite* and *rhodochrosite* are usually massive, granular. Growth twins, especially on $\{0001\}$ and $\{10\bar{1}1\}$ are common in calcite, dolomite, and ankerite. Lamellar twins produced by deformation are very common on $\{01\bar{1}2\}$ in calcite (Figure 9.85); lamellar glide-twins on $\{02\bar{2}1\}$ are produced in dolomite only at temperatures > ca. 300° C. Twinning is rare in rhodochrosite and siderite, and absent in magnesite. A short discussion of glide mechanisms in calcite and dolomite is given after the section on distinguishing features.

Color in thin section. Usually colorless; siderite may be pale yellow-brown and rhodochrosite pale pink.

Optical properties. Uniaxial −ve. May be anomalously biaxial as a result of strain or twinning (Turner, 1975 a and b).
RI and δ values for pure end-members are:

Calcite: $\omega = 1.658$, $\varepsilon = 1.486$, $\delta = 0.172$.

Dolomite: $\omega = 1.679$, $\varepsilon = 1.500$, $\delta = 0.179$.

Magnesite: $\omega = 1.700$, $\varepsilon = 1.509$, $\delta = 0.191$.

Siderite: $\omega = 1.875$, $\varepsilon = 1.635$, $\delta = 0.242$.

Rhodochrosite: $\omega = 1.816$, $\varepsilon = 1.597$, $\delta = 0.219$.

The variation in the value of ω with chemistry is given in Figure 9.84 (ω can be measured in any grain). Because crushed material always lies on the perfect $\{10\bar{1}1\}$ cleavages, ε is difficult to measure. However, the value of ε can be determined from ε' measured on cleavage fragments with the use of a chart provided by Loupekine (1947).

Occurrence. *Calcite* is found as a primary or secondary mineral in a very wide range of rock types. It is the prime constituent of most limestones as a primary precipitate, as a mineral in shell fragments, and as a diagenetic mineral. Metastable aragonite and high-Mg calcite are common in modern shallow-sea limestones but eventually change to calcite, the Mg being flushed out or sometimes forming dolomite. Cal-

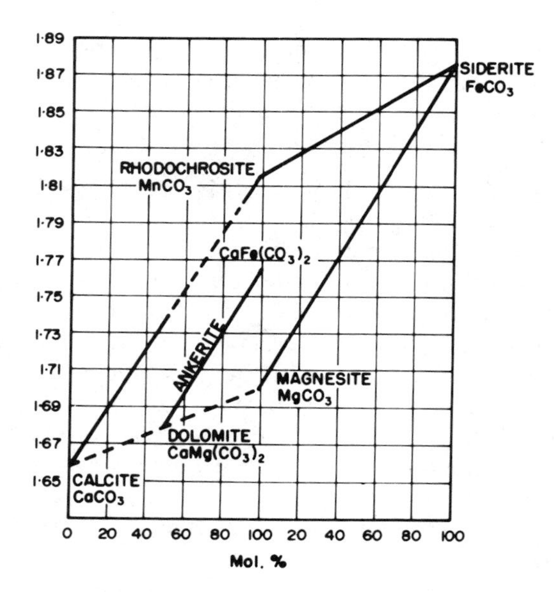

Figure 9.84 Variation of ω with composition in the trigonal carbonates. Based on Kennedy (1947, Figure 8).

cite occurs as ooliths in some limestones. Calcite is a common cement in other sediments, and is precipitated from groundwaters to form travertine deposits and stalactites, etc. It is found in many metamorphic rocks of low to high grade, particularly metamorphosed limestones (marbles). In some circumstances, CO_2 may be lost during metamorphism, and the Ca combines with silica to form minerals such as wollastonite, diopside, tremolite, grossular, vesuvianite, etc. In igneous rocks, calcite is usually secondary, occurring in vesicles, fissures, and replacing a wide range of minerals, especially plagioclase; it also occurs as a primary magmatic mineral in carbonatites and associated rocks. Calcite is common in hydrothermal veins.

Dolomite is most commonly found in sediments, often with calcite; some is primary, but much is secondary. Secondary dolomite is often prominent as euhedral rhombohedra. Low-grade metamorphism of dolomite sediments produces dolomitic marble, but at high grades the dolomite may dissociate to calcite and periclase (often altered to brucite). Dolomite may react with silica during metamorphism to form minerals such as talc and tremolite. Dolomite also occurs as a primary mineral in some carbonatites, and as a hydrothermal mineral in veins.

Ankerite is most common as a secondary mineral in Fe-rich sediments and coal, especially in veins. It is also found as a primary magmatic mineral in carbonatites, and in metamorphosed Fe-rich sediments.

Magnesite forms as an alteration product of Mg-rich igneous and metamorphic rocks. Talc-magnesite schists are produced by the low-temperature hydrothermal metamorphism of serpentinite. Magnesite also occurs in some evaporites.

Siderite is a secondary mineral in some Fe-rich sediments; it also occurs in hydrothermal veins, and in metamorphosed Fe-rich sediments. It alters to limonite.

Rhodochrosite is rare and found in association with other Mn-rich minerals in ore deposits, veins and sediments. It alters to black opaque pyrolusite (MnO_2).

Distinguishing features. The very high δ and the perfect rhombohedral cleavages are distinctive of the group. The "twinkling" of calcite, dolomite, ankerite, and magnesite is characteristic, and serves to distinguish them from other minerals with a very high δ such as titanite (sphene). Distinction among the trigonal carbonates is less easy. Measurement of ω (Figure 9.84) is not always sufficient to identify a particular species. Dolomite, ankerite, and siderite are characterized by their common euhedral (rhombohedral) habit. The alteration of siderite to limonite is distinctive. In deformed rocks, calcite is characterized by abundant twin lamellae (Figure 9.85); dolomite does not develop glide

Figure 9.85 Typical deformation twinning in calcite, Arthur Marble, New Zealand (view measures 3.1 × 2.0 mm). Crossed-polarized light.

twins below 300° C, and magnesite always lacks glide twinning. The twin lamellae of calcite are $\{01\bar{1}2\}$ and those of dolomite $\{02\bar{2}1\}$; in sections displaying two intersecting sets of twins, ω (the direction of higher relief) lies in the acute angle between the lamellae in calcite but in the obtuse angle in dolomite. Calcite effervesces readily in cold dilute HCl whereas dolomite does not.

Stain tests. In order to make modal and textural analyses of carbonate sediments, it may be necessary to distinguish the various carbonate minerals using stain tests. Numerous methods have been devised, and they are discussed fully in Wolf et al. (1967) and Lewis (1983). Some experience is necessary in their use since the effects of staining vary with differing grain size and porosity.

Glide mechanisms in calcite and dolomite. Calcite deforms very readily by translation and twin gliding. Extensive laboratory investigations (summarized in Turner and Weiss, 1963) have shown that the principal glide mechanisms are: twin gliding on $\{01\bar{1}2\}$ effective under all conditions of temperature and pressure investigated; translation on $\{10\bar{1}1\}$ especially at temperatures up to 400° C; translation on $\{02\bar{2}1\}$ especially at temperatures >500° C. Twinning generally occurs in preference to translation if the sense of shear is suitable. Gliding causes rotation, bending and flattening of grains; complex patterns of glide twins are commonly observed in thin sections (Figure 9.85). The strain caused by gliding may assist nucleation of new grains which grow at the expense of earlier-strained crystals (process of recrystallization). A preferred orientation of *c*-axes develops during deformation and recrystallization.

Dolomite deforms less easily, but suffers translation gliding on $\{0001\}$ at low temperatures if suitably oriented in the stress field. Dolomite is unusual in that it becomes stronger with increasing temperature; twin gliding on $\{02\bar{2}1\}$ does not occur below 300° C and becomes more important than translation at high temperatures.

(g) Phosphates

No. 125. APATITE (−ve) $Ca_5(PO_4)_3(OH,F,Cl)$

See Plates 11 and 12.

C replaces some of the *P* in the related minerals *dahllite* (carbonate-apatite with $F < 1\%$) and *francolite* (carbonate-apatite with $F > 1\%$). *Collophane* is a term used for cryptocrystalline dahllite or francolite. Hexagonal. No good cleavage. Euhedral crystals of apatite are short, long or acicular prisms with a hexagonal cross section; also subhedral or anhedral grains. Dahllite is the inorganic phase in bone, and dahllite and francolite form concretions and fibrous radiating or spherulitic aggregates.

Color in thin section. Usually colorless, but some inclusion-rich apatite may appear colored and pleochroic; dahllite, francolite and collophane are commonly yellow-brown or grey.

Optical properties. *Apatite:* uniaxial −ve; occasionally biaxial with a small $2V$.

$$\omega = 1.632\text{–}1.668,\ \varepsilon = 1.628\text{–}1.665.$$

$$\delta = 0.002\text{–}0.008.$$

The RI of chlorapatite > hydroxyapatite > fluorapatite.
Related minerals: the RI of dahllite varies in the range 1.52–1.61 (usually 1.55–1.59), and that of francolite is <1.63. The δ of dahllite and francolite varies from very low to as high as 0.016, and they may be biaxial. Collophane is isotropic.

Occurrence. Apatite is a very common accessory mineral in almost all types of igneous and metamorphic rocks, normally as very small grains or prisms; it may be abundant in some alkaline types and carbonatites. It is found as a detrital mineral in many sediments. The inorganic phase in bone material is dahllite, and dahllite, francolite, or collophane are prime constituents in rock phosphate, secondary phosphatic material in sedimentary rocks, and coprolites.

Distinguishing features. The moderate relief, low or very low δ, −ve character, and lack of cleavage or color in apatite are distinctive. The brown color, occurrence, and amorphous, fibrous, or skeletal character of collophane, francolite, and dahllite are distinctive.

No. 126. MONAZITE (+ve) $(Ce,La)PO_4$

Th, Nd, and other rare elements present.
Monoclinic, $\beta = 104°$. Distinct {100} and {001} cleavages. Usually small crystals elongate parallel to *b* or {100} tablets. Twinning with {100} twin planes common.

Color in thin section. Colorless or pale yellow.

Color in detrital grains. Yellow, brown, or red; pleochroism weak with $Y > X = Z$.

Optical properties. Biaxial +ve. $2V_z = 3°\text{–}19°$. $r <$ or $> v$ (weak).

$$\alpha = 1.770\text{–}1.800,\ \beta = 1.777\text{–}1.801,\ \gamma = 1.825\text{–}1.850.$$

$$\delta = 0.045\text{–}0.075.\ X = b,\ Y \wedge a = 7°\text{–}12°,\ Z \wedge c = 2°\text{–}7°.$$

{100} tablets have slightly inclined or parallel extinction. Crystals elongate parallel to *b* are length fast with straight extinction. {001} cleavage fragments provide an almost centered acute bisectrix figure.

Occurrence. A rare accessory mineral in acid and alkaline igneous rocks and gneisses. May cause pleochroic haloes in biotite. It is resistant to weathering and may be concentrated as a detrital mineral in sufficient quantities to be economically important as a source of Ce and other rare elements.

Distinguishing features. The very high relief, high δ, and color are characteristic. Detrital grains are usually well rounded or egg-shaped. It may be confused with zircon, titanite (sphene), and xenotime, but zircon has a higher RI and is uniaxial, and both titanite (sphene) and xenotime have a higher δ.

No. 127. XENOTIME (+ve) YPO_4

Other rare-earths are usually present.
Tetragonal. Distinct {110} cleavages. Crystals are prismatic with pyramidal terminations, and resemble zircon.

Color in thin section. Colorless, yellow or brown.

Color in detrital grains. Generally yellow or brown; weak pleochroism.

Optical properties. Uniaxial +ve.

$$\omega = 1.720\text{–}1.724,\ \varepsilon = 1.810\text{–}1.828.$$

$$\delta = 0.086\text{–}0.107.$$

Prismatic crystals are length slow with straight extinction.

Occurrence. A rare accessory mineral in acid and alkaline igneous rocks and gneisses, often associated with zircon. Xenotime may cause pleochroic haloes in biotite. Commonly found as detrital grains in heavy-mineral suites.

Distinguishing features. Often mistaken for zircon, but xenotime is distinguished by its lower RI and higher δ.

(h) Tungstates

No. 128. SCHEELITE (+ve) $CaWO_4$

Molybdenum may substitute for tungsten.
Tetragonal. Distinct {101} cleavages. Crystals are dominated by pyramidal faces; commonly massive. Twinning on {110} common.

Color in thin section. Colorless.

Color in detrital grains. Usually colorless; sometimes yellow-brown.

Optical properties. Uniaxial +ve.

$$\omega = 1.920,\ \varepsilon = 1.937.$$

$$\delta = 0.017.$$

Fluoresces blue in ultraviolet light.

Occurrence. In granite pegmatites, and in veins in schists and contact metamorphic rocks. Found as detrital grains in heavy mineral suites.

Distinguishing features. The very high relief, moderate birefringence, and uniaxial +ve character are distinctive.

References

Ashworth, J.R. 1972. Myrmekites of exsolution and replacement origins. *Geol. Mag.* 109: 45–62.

Bailey, E.H. and Stevens, R.E. 1960. Selective staining of plagioclase and K-feldspar on rock slabs and thin sections. *Geol. Soc. Am. Bull.* 71: 2047.

Barker, D.S. 1970. Compositions of granophyre, myrmekite, and graphic granite. *Geol. Soc. Am. Bull.* 81: 3339–3350.

Barth, T.F.W. 1969. *Feldspars.* Wiley-Interscience, New York, N.Y. 261 pp.

Bennett, R.L. 1958. Evaporite sections. *J. Inst. Sci. Technol.* 4: 358.

Bloss, F.D. 1981. *The Spindle Stage: Principles and Practice.* Cambridge University Press, Cambridge, England. 340 pp.

Bloss, F.D. and Light, J.F. 1973. The detent spindle stage. *Am. J. Sci.* 273-A: 536–538.

Borchert, H. and Muir, R.O. 1964. *Salt Deposits.* D. van Nostrand, London. 338 pp.

Borg, I.Y. 1967. Optical properties and cell parameters in the glaucophane-riebeckite series. *Contrib. Mineral. Petrol.* 15: 67–92.

Bottinga, Y., Kudo, A. and Weill, D. 1966. Some observations on oscillatory zoning and crystallization of magmatic plagioclase. *Am. Mineral.* 51: 792–806.

Bowen, N.L. 1913. The melting phenomena of the plagioclase feldspars. *Am. J. Sci.* 35: 577–599.

Bowen, N.L. and Tuttle, O.F. 1950. The system $NaAlSi_3O_8$–$KAlSi_3O_8$–H_2O. *J. Geol.,* 58: 489–511.

Boyd, F.R. and Schairer, J.F. 1964. The system $MgSiO_3$–$CaMgSi_2O_6$. *J. Petrol.* 5: 275–309.

Brindley, G.W. and Brown, G. 1980. *Crystal Structures of Clay Minerals and Their X-Ray Identification*. Mineralogical Society Monograph No. 5. London. 495 pp.

Brown, G.M. and Vincent, E.A. 1963. Pyroxenes from the late stages of fractionation of the Skaergaard Intrusion, East Greenland. *J. Petrol.* 4: 175–197.

Burri, C., Parker, R.L. and Wenk, E. 1967. *Die optische Orientierung der Plagioklase*. Birkhäuser Verlag, Basel. 334 pp.

Carver, R.E. 1971. *Procedures in Sedimentary Petrology*. Wiley-Interscience, New York, N.Y. 653 pp.

Coombs, D.S. 1971. Present status of the zeolite facies. In: *Molecular Sieve Zeolites, I. Advances in Chemistry Series No. 101*. American Chemical Society, Washington, D.C. pp. 317–327.

Deer, W.A., Howie, R.A. and Zussman, J. 1962. *Rock-Forming Minerals, 1, 3 and 5*. Longmans, London. Vol. 1: 333 pp., Vol. 3: 270 pp., Vol. 5: 371 pp.

Deer, W.A., Howie, R.A. and Zussman, J. 1963. *Rock-Forming Minerals, 2 and 4*. Longmans, London. Vol. 2: 379 pp., Vol. 4: 435 pp.

Deer, W.A., Howie, R.A. and Zussman, J. 1966. *An Introduction to the Rock-Forming Minerals*. Longmans, London. 528 pp.

Deer, W.A., Howie, R.A. and Zussman, J. 1978. *Rock-Forming Minerals: 2A Single-Chain Silicates*. Longman, London. 2nd ed. 668 pp.

Deer, W.A., Howie, R.A. and Zussman, J. 1982. *Rock-Forming Minerals: 1A Orthosilicates*. Longman, London. 2nd ed. 919 pp.

Devismes, P. 1978. Atlas photographie des mineraux d'alluvions. *Mémoire du Bureau de recherches géologique et minières*. 95: 203 pp.

Emmons, R.C. 1943. The universal-stage. *Geol. Soc. Am. Mem.* 8: 205 pp.

Ewart, A. 1976. Pyroxene and magnetite phenocrysts from the Taupo quaternary rhyolitic pumice deposits, New Zealand. *Mineral. Mag.* 36: 180–194.

Flinn, D. 1973. Two flow-charts of orthoscopic U-stage techniques. *Mineral. Mag.* 39: 368–370.

Foord, E.E. and Mills, B.A. 1978. Biaxiality in isometric and dimetric crystals. *Am. Mineral.* 63: 316–325.

Fraser, W.E. and Downie, G. 1964. The spectrochemical determination of feldspars within the field microcline-albite-labradorite. *Mineral. Mag.* 33: 790–798.

Frondel, C. 1962. *Dana's System of Mineralogy, III. Silica Minerals*. John Wiley and Sons, New York, N.Y. 334 pp.

Galopin, R. and Henry, N.F.M. 1972. *Microscopic Study of Opaque Minerals*. Heffer and Sons, Cambridge, England. 322 pp.

Garaycochea, I. and Wittke, O. 1964. Determination of the optic angle 2V from the extinction curve of a single crystal mounted on a spindle stage. *Acta Cryst.* 17: 183–189.

Gregnanin, A. and Viterbo, C. 1965. Metodo di colorazione per identificare la cordierite in sezione sottile. *Rend. Soc. Mineral. Ital.* 21: 111–120.

Griffen, D.T. and Ribbe, P.H. 1973. The crystal chemistry of staurolite. *Am. J. Sci.* 273A: 479–495.

Grim, R.E. 1968. *Clay Mineralogy*. McGraw-Hill, New York, N.Y. 2nd. ed. 596 pp.

Harrington, V.F. and Buerger, M.J. 1931. Immersion liquids of low refraction. *Am. Mineral.* 16: 45–54.

Hartshorne, N.H. and Stuart, A. 1970. *Crystals and the Polarising Microscope.* Arnold, London. 4th ed. 614 pp.

Hess, H.H. 1949. Chemical composition and optical properties of common clinopyroxenes, I. *Am. Mineral.* 34: 621–666.

Holdaway, M.J. 1971. Stability of andalusite and the aluminium silicate phase diagram. *Am. J. Sci.* 271: 97–131.

Houghton, H.F. 1980. Refined techniques for staining plagioclase and alkali-feldspars in thin section. *J. Sedim. Petr.* 50: 629–631.

Hurlbut, C.S. and Klein, C. 1977. *Manual of Mineralogy (after Dana).* Wiley-Interscience, New York, N.Y. 19th ed. 532 pp.

Hutchison, C.S. 1974. *Laboratory Handbook of Petrographic Techniques.* Wiley-Interscience, New York, N.Y. 527 pp.

Jaffe, H.W., Robinson, P., Tracey, R.J. and Ross, M. 1975. Orientation of pigeonite exsolution lamellae in metamorphic augite: correlation with composition and calculated optimal phase boundaries. *Am. Mineral.* 60: 9–28.

Jones, J.B., Sanders, J.V. and Segnit, E.R. 1964. Structure of opal. *Nature* 204: 990–991.

Kamb, W.B. 1958. Isogyres in interference figures. *Am. Mineral.* 43: 1029–1067.

Kennedy, G.C. 1947. Charts for correlation of optical properties with chemical composition of some common rock-forming minerals. *Am. Mineral.* 32: 561–573.

Kuno, H. 1955. Ion substitution in the diopside–ferropigeonite series of clinopyroxenes. *Am. Mineral.* 40: 70–93.

Laduron, D.M. 1971. A staining method for distinguishing paragonite from muscovite in thin section. *Am. Mineral.* 56: 1117–1119.

Lancelot, Y. 1973. Chert and silica diagenesis in sediments from the Central Pacific. In: E.L. Winterer, J.L. Ewing et al., *Initial reports of the Deep Sea Drilling Project, XVII.* U.S. Government Printing Office, Washington, D.C. pp. 377–405.

Laniz, R.P., Stevens, R.E. and Norman, M.B. 1964. Staining of plagioclase feldspar and other minerals with F.D. and C. Red. No. 2. *U.S. Geol. Survey Prof. Paper.* 501-B: 152–153.

Larsen, E.S. and Berman, H. 1934. The microscopic determination of the nonopaque minerals. *U.S. Geol. Survey Bull.* 848: 266 pp.

Laskowski, T.E., Scotford, D.M. and Laskowski, D.E. 1979. Measurement of refractive index in thin section using dispersion staining and oil immersion techniques. *Am. Mineral.* 64: 440–445.

Leake, B.E. 1968. Optical properties and composition in the orthopyroxene series. *Mineral. Mag.* 36: 745–747.

Leake, B.E. 1978. Nomenclature of amphiboles. *Mineral. Mag.* 42: 533–563.

Lewis, D.W. 1983. *Practical Sedimentology.* Hutchison Ross, Stroudsburg, Pa. 288 pp.

Lindholm, R.C. and Dean, D.A. 1973. Ultra-thin thin sections in carbonate petrology. *J. Sed. Pet.* 43: 295–297.

Lister, B. 1978. Preparation of polished sections. *Rept. Inst. Geol. Sci.* 78: 20 pp.

Lister, G.S., Paterson, M.S. and Hobbs, B.E. 1978. The simulation of fabric development in plastic deformation and its application to quartzite: the model. *Tectonophysics*. 45: 107–158.

Loomis, T.P. 1982. Numerical simulations of crystallisation processes of plagioclase in complex melts: the origin of major and oscillatory zoning in plagioclase. *Contrib. Mineral. Petrol.* 81: 219–229.

Louisnathan, S.J., Bloss, F.D. and Korda, E.J. 1978. Measurement of refractive indices and their dispersion. *Am. Mineral.* 63: 394–400.

Loupekine, I.S. 1947. Graphical derivation of refractive index ε for the trigonal carbonates. *Am. Mineral.* 32: 502–507.

Lumpkin, G.R. and Ribbe, P.H. 1979. Chemistry and physical properties of axinites. *Am. Mineral.* 64: 635–645.

Lyons, P.C. 1971. Staining of feldspars on rock-slab surfaces for modal analysis. *Mineral. Mag.* 38: 518–519.

MacKenzie, W.S. and Smith, J.V. 1956. The alkali-feldspars, III. An optical and X-ray study of the high-temperature feldspars. *Am. Mineral.* 41: 405–427.

Mathews, W.H. 1951. A useful method for determining approximate composition of fine-grained igneous rocks. *Am. Mineral.* 36: 92–101.

Meyrowitz, R. 1955. A compilation and classification of immersion media of high index of refraction. *Am. Mineral.* 40: 398–409.

Milner, H.B. 1962. *Sedimentary Petrography II: Principles and Applications.* Allen & Unwin, London. 4th ed. 715 pp.

Morse, S.A. 1970. Alkali feldspars with water at 5 kb pressure. *J. Petrol.* 11: 221–251.

Muir, I.D. 1951. The clinopyroxenes of the Skaergaard intrusion, eastern Greenland. *Mineral. Mag.* 29: 690–714.

Müller, G. 1967. *Methods in Sedimentary Petrology.* Schwiezerbart'sche Verlagsbuchhandlung, Stuttgart (transl. by Schmincke, H.-U). 283 pp.

Myer, G.H. 1966. New data on zoisite and epidote. *Am. J. Sci.* 264: 364–385.

Nakamura, Y. and Kushiro, I. 1970. Equilibrium relations of hypersthene, pigeonite and augite in crystallising magmas: microprobe study of a pigeonite andesite from Weiselberg, Germany. *Am. Mineral.* 55: 1999–2015.

Nicolas, A. and Poirier, J.P. 1976. *Crystalline plasticity and solid state flow in metamorphic rocks.* Wiley-Interscience, New York, N.Y. 444 pp.

Oehler, J.H. 1973. Tridymite-like crystals in cristobalitic "cherts". *Nature Phys. Sci.* 241: 64–65.

Phillips, F.C. 1971. *An Introduction to Crystallography.* Oliver and Boyd, Edinburgh, 4th ed. 351 pp.

Phillips, W.R. 1964. A numerical system of classification for chlorites and septechlorites. *Mineral. Mag.* 33: 1114–1124.

Phillips, W.R. and Griffen, D.T. 1981. *Optical Mineralogy—the Nonopaque Minerals.* W.H. Freeman, San Francisco, Calif. 677 pp.

Poldervaart, A. 1950. Correlation of physical properties and chemical composition in the plagioclase, olivine, and orthopyroxene series. *Am. Mineral.* 35: 1067–1079.

Poldervaart, A. and Hess, H.H. 1951. Pyroxenes in the crystallisation of basaltic magna. *J. Geol.* 59: 472–489.

Reed, F.S. and Mergner, J.L. 1953. Preparation of rock thin sections. *Am. Mineral.* 38: 1184–1203.

Robinson, P., Jaffe, H.W., Ross, M. and Klein, C. 1971. Orientation of exsolution lamellae in clinopyroxenes and clinoamphiboles: consideration of optimal phase boundaries. *Am. Mineral.* 56: 909–939.

Saggerson, E.P. and Turner, L.M. 1982. General comments on the identification of chlorites in thin sections. *Mineral. Mag.* 46: 469–473.

Selkregg, K.R. and Bloss, F.D. 1980. Cordierites: compositional controls of Δ, cell parameters, and optical properties. *Am. Mineral.* 65: 522–533.

Shand, S.J. 1939. On the staining of feldspathoids, and on zonal structure in nepheline. *Am. Mineral.* 24: 508–513.

Shelley, D. 1973. Myrmekites from the Haast Schists, New Zealand. *Am. Mineral.* 58: 332–338.

Shelley, D. 1979. Plagioclase preferred orientation, Foreshore Group metasediments, Bluff, New Zealand. *Tectonophysics* 58: 279–290.

Shelley, D. 1980. Quartz [0001]-axes preferred orientation, Bluff, New Zealand: origin elucidated by grain-size measurements. *Tectonophysics* 62: 321–337.

Shurcliff, W.A. 1962. *Polarised Light.* Harvard and Oxford University Presses. 207 pp.

Slemmons, D.B. 1962. Determination of volcanic and plutonic plagioclases using a three or four-axis universal stage. *Geol. Soc. Am. Spec. Paper* 69: 64 pp.

Smith, J.R. 1958. The optical properties of heated plagioclases. *Am. Mineral.* 43: 1179–1194.

Sriramadas, A. 1957. Diagrams for the correlation of unit cell edges and refractive indices with the chemical composition of garnets. *Am. Mineral.* 42: 294–298.

Starkey, J. 1967. On the relationship of periclase and albite twinning to the composition and structural state of plagioclase feldspars. *Schweiz. Mineral. Petrogr. Mitt.* 47: 257–268.

Taggart, J.E., Jr. 1977. Polishing techniques for geologic samples. *Am. Mineral.* 62: 824–827.

Tobi, A.C. 1956. A chart for measurement of optic axial angles. *Am. Mineral.* 41: 516–519.

Tobi, A.C. and Kroll, H. 1975. Optical determination of the An-content of plagioclases twinned by Carlsbad-law: a revised chart. *Am. J. Sci.* 275: 731–736.

Tröger, W.E. 1979. *Optical Determination of Rock-Forming Minerals.* E. Schweizerbart'sche Verlagsbuchhandlung, Stuttgart. 4th ed. (English edition by Bambauer, H.U., Taborszky, F. and Trochim, H.D.) 188 pp.

Tullis, J., Christie, J.M. and Griggs, D.T. 1973. Microstructures and preferred orientations of experimentally deformed quartzites. *Geol. Soc. Am. Bull.* 84: 297–314.

Tunell, G. 1952. The angle between the a-axis and the trace of the rhombic section on the {010} pinacoid in the plagioclases. *Am. J. Sci.* Bowen Volume. pp. 547–551.

Turner, F.J. 1975(a). Biaxial calcite: occurrence, optics, and associated minor strain phenomena. *Contrib. Mineral. Petrol.* 50: 247–255.

Turner, F.J. 1975(b). Biaxiality in relation to visible twinning in experimentally deformed calcite. *Contrib. Mineral. Petrol.* 53: 241–252.

Turner, F.J. and Verhoogen, J. 1960. *Igneous and Metamorphic Petrology*. McGraw-Hill, New York, N.Y., 2nd ed., 694 pp.

Turner, F.J. and Weiss, L.E. 1963. *Structural Analysis of Metamorphic Tectonites*. McGraw-Hill, New York, N.Y. 545 pp.

Tuttle, O.F. 1952. Optical studies on alkali-feldspars. *Am. J. Sci.*, Bowen Volume. pp. 553–567.

Tuttle, O.F. and Bowen, N.L. 1958. Origin of granite in the light of experimental studies in the system $NaAlSi_3O_8$—$KAlSi_3O_8$—SiO_2—H_2O. *Geol. Soc. Am. Mem.* 74: 153 pp.

Vance, J.A. 1961. Polysynthetic twinning in plagioclase. *Am. Mineral.* 46: 1097–1119.

Vance, J.A. 1969. On synneusis. *Contrib. Mineral. Petrol.* 24: 7–29.

Walker, G.P.L. 1960. Zeolite zones and dike distribution in relation to the structure of the basalts of eastern Iceland. *J. Geol.* 68: 515–528.

Weaver, C.F. and McVay, T.N. 1960. Immersion oils with indices of refraction from 1.292 to 1.411. *Am. Mineral.* 45: 469–470.

Whittaker, E.J.W. 1981. *Crystallography: an Introduction for Earth Science (and Other Solid State) Students*. Pergamon, Oxford, 254 pp.

Wicks, F.J. and Whittaker, E.J.W. 1977. Serpentine textures and serpentinization. *Can. Mineral.* 15: 459–488.

Widmark, T. 1979. Staining of albite. *Geol. Fören. Stockholm Förh.* 101: 357–358.

Wilcox, R.E. 1959. Use of the spindle stage for determination of principal indices of refraction of crystal fragments. *Am. Mineral.* 44: 1272–1293.

Williams, H., Turner, F.J. and Gilbert, C.M. 1982. *Petrography*. W.H. Freeman, San Francisco, Calif. 2nd ed. 626 pp.

Wilson, M.D. and Sedeora, S.S. 1979. An improved thin section stain for potash feldspar. *J. Sedim. Petr.* 49: 637–638.

Winchell, A.N. 1937. *Elements of Optical Mineralogy. I. Principles and Methods*. John Wiley and Sons, New York, N.Y. 5th ed. 263 pp.

Winchell, A.N. 1939. *Elements of Optical Mineralogy, III. Determinative Tables*. John Wiley and Sons, New York, N.Y. 2nd ed. 231 pp.

Winchell, A.N. 1951. *Elements of Optical Mineralogy, II. Description of Minerals*. John Wiley and Sons, New York, N.Y. 4th ed. (in collaboration with H. Winchell). 551 pp.

Winchell, H. 1958. The composition and physical properties of garnet. *Am. Mineral.* 43: 595–600.

Winchell, H. 1965. *Optical Properties of Minerals: A Determinative Table*. Academic Press, New York, N.Y. 91 pp.

Wolf, K.H., Easton, A.J. and Warne, S. 1967. Techniques of examining and analysing carbonate skeletons, minerals, and rocks. In: G.V. Chilingar, H.J. Bissell and R.W. Fairbridge (Editors), *Carbonate Rocks. Developments in Sedimentology, 9B*. Elsevier, Amsterdam. pp. 253–342.

Wright, H.G. 1964. The use of epoxy resins in the preparation of petrographic thin sections. *Mineral. Mag.* 33: 931–933.

Zeck, H.P. 1972. Irrational composition planes in cordierite sector trillings. *Nature Phys. Sci.* 238: 47–48.

Index

Note: The main page reference for each mineral is printed in bold face, and the consecutive numbering system is given in parentheses.

B

C

V

W

X

Y

Z